FINITELY ADDITIVE MEASURES AND RELAXATIONS OF EXTREMAL PROBLEMS

MONOGRAPHS IN CONTEMPORARY MATHEMATICS

Formerly CONTEMPORARY SOVIET MATHEMATICS

Series Editor: Revaz Gamkrelidze, Steklov Institute, Moscow, Russia

---

ARITHMETIC OF ALGEBRAIC CURVES
Serguei A. Stepanov

ASYMPTOTIC METHODS IN SINGULARLY PERTURBED SYSTEMS
E. F. Mishchenko, Yu. S. Kolesov, A. Yu. Kolesov, and N. Kh. Rozov

ASYMPTOTICS OF OPERATOR AND PSEUDO-DIFFERENTIAL EQUATIONS
V. P. Maslov and V. E. Nazaikinskii

COHOMOLOGY OF INFINITE-DIMENSIONAL LIE ALGEBRAS
D. B. Fuks

DIFFERENTIAL GEOMETRY AND TOPOLOGY
A. T. Fomenko

FINITELY ADDITIVE MEASURES AND RELAXATIONS OF EXTREMAL PROBLEMS
A. G. Chentsov

HOMOLOGY OF ANALYTIC SHEAVES AND DUALITY THEOREMS
V. D. Golovin

LINEAR DIFFERENTIAL EQUATIONS OF PRINCIPAL TYPE
Yu. V. Egorov

THE OBLIQUE DERIVATIVE PROBLEM OF POTENTIAL THEORY
A. I. Yanushauskas

OPTIMAL CONTROL
V. M. Alekseev, V. M. Tikhomirov, and S. V. Fomin

THEORY OF OPERATORS
V. A. Sadovnichiĭ

THEORY OF SOLITONS: The Inverse Scattering Method
S. Novikov, S. V. Manakov, L. P. Pitaevskii, and V. E. Zakharov

TOPICS IN MODERN MATHEMATICS: Petrovskii Seminar No. 5
Edited by O. A. Oleinik

FINITELY ADDITIVE MEASURES AND RELAXATIONS OF EXTREMAL PROBLEMS

A. G. Chentsov

Mathematics and Mechanics Institute
Urals Division of the Russian Academy of Sciences
Ekaterinburg, Russia

Translated from Russian by
A. V. Lutsuk

CONSULTANTS BUREAU • NEW YORK, LONDON, AND MOSCOW

Library of Congress Cataloging-in-Publication Data

On file

This translation is published under an agreement with the
Russian Authors' Society (RAO)

ISBN 0-306-11038-5

©1996 Consultants Bureau, New York
A Division of Plenum Publishing Corporation
233 Spring Street, New York, N.Y. 10013

10 9 8 7 6 5 4 3 2 1

Preface

This monograph is devoted to the extension of extremal problems in the class of finitely additive measures [9]. The need for extensions of this kind is due to the absence of existence theorems for optimal solutions and also to the necessity of "regularizing" many settings that are of interest in practice. In this book, we shall focus our attention on the latter case, which is very essential for applications. We shall speak about perturbations of the conditions under which a "real" problem is considered; they virtually cannot be avoided in concrete situations; the only thing one can hope for is the "smallness" of these perturbations. But even in the world of "small" perturbations serious troubles may be encountered. In the sequel, we shall confine ourselves to perturbation of the system of constraints of a corresponding extremal problem. As is well known, any problem of this kind is characterized by the following two important components: the criterion $\mathcal{K}$ and the constraints $\mathcal{O}$, which determine the domain where one can select a solution with the aim of minimizing $\mathcal{K}$ (we restrict ourselves to problems of this kind). The value of the problem, $\mathrm{val}\,(\mathcal{K}, \mathcal{O})$ (finite or infinite), characterizes the potentially attainable quality. If, instead of $\mathcal{O}$, one uses "close" (to $\mathcal{O}$) constraints $\mathcal{O}'$, then the "new" value, $\mathrm{val}\,(\mathcal{K}, \mathcal{O}')$, can no longer be close to $\mathrm{val}\,(\mathcal{K}, \mathcal{O})$. If we speak about weakening of the constraints $\mathcal{O}$, then $\mathcal{O}'$ means $\mathcal{O} \subset \mathcal{O}'$, i.e., is a less strict condition, and then the "new" value may differ from the "old" one only by being smaller. If this difference is significant even for close $\mathcal{O}'$, then one should question the value itself, which could be used as a measure of optimality in the extremal problem subject to perturbations. This "value" can be interpreted as, for instance, a (generalized) limit of $\mathrm{val}\,(\mathcal{K}, \mathcal{O}')$ under the conditions of "convergence" of $\mathcal{O}'$ to $\mathcal{O}$. Here it is no longer sufficient to speak about any concrete perturbation $\mathcal{O}'$ of "severe" constraints $\mathcal{O}$; rather one should speak about the asymptotic behavior of these perturbed constraints, i.e., about a series of constraints $\mathcal{O}'$, and it is natural to interpret the limit value obtained as an asymptotic value. The coincidence of the latter with val $(\mathcal{K}, \mathcal{Q})$ has the meaning of computational stability, and we subject the series of constraints of the form $\mathcal{O}'$ to the condition of "convergence" to $\mathcal{O}$. But such a stability is often absent, and in that case it is the asymptotic value that plays a more important role than the conventional one. We treat it as a regularization of the conventional one, thus interpreting the term as a refinement or correction of the usual sense of its meaning. The indicated asymptotic value (the regularization of the conventional one) can, however, be considered again as a conventional value, but this time for a problem reduced in some sense. The latter has some nice properties which can be absent in the original setting; for example, the minimum may not be attainable in the original problem whereas in the "new one" it is attainable, as a rule. The point is that the transition to the new problem is realized by improving the solution space. This improvement has the meaning of compactification or extension; such procedures have much in common with extensions of topological spaces [19, 54]. It turns out that the achievement of topological perfection of the space is directly connected with a realization of the regularization mentioned above. It is of interest that extensions (compactifications) allow us to consider and to compare different regularizations in order to establish their nonsensitivity to partial perturbations, structural stability, etc. Furthermore, these are the properties that we shall examine in detail in order to obtain an easily verifiable "external product."

There are many publications dealing with extensions of the space of solutions and with their application to extremal problems (see, for example, [4, 6, 55] for problems of optimal control, [17, 53, 55] for calculus of variations, [7, 10] for mathematical programming). We mention here only those papers which are the closest to our work. It should be noted that in the extension of constructions a major role has traditionally been played by measures (see, e.g., [20, 21, 32, 46, 55]) which were usually assumed to be countably additive and regular [9, Chap. IV]. This is due, in particular, to the familiar theorem of Riesz (see [9, Chap. IV; 4, Chap. I] and others), and to the procedures of convexification (local or global) of a problem whose role is also substantial [53]; here one also has to bear in mind some natural analogies with extensions of games in the class of mixed strategies [26, 27]. However, in a number of cases it becomes necessary to use, for similar purposes, finitely additive measures [9, Chaps. III, IV]. Let us consider this question more thoroughly because these extensions will be the main subject of our discussion.

The extension of constructions in the class of finitely additive measures is a nontraditional application of (finitely additive) measure theory; we observe, among other applications, the theory of stochastic processes, game theory, and utility theory. The use of these measures as a material for compactifications is connected both with various discontinuous dependences in the constraints of the respective extremal problems (see integral representations [9, Chap. IV]), and with some important topological properties. This, on one hand, enables one to realize, on the basis of finitely additive measures, some universal extension of constructions [34] with the use of the Stone representation, and, on the other hand, to deal with some rather concrete and specific procedures for the respective classes of problems. In this book, we deal mainly with the second approach. The author, however, did not try to embrace all familiar extension of constructions in the class of finitely additive measures that provide us with particularly useful representations for the respective extremal problems. In fact, we consider one scheme of imbedding the "conventional" solutions, identified with step-functions on the main space, into a space of finitely additive measures of bounded variation; within the framework of this scheme the imbedding is realized by integrating a step-function with respect to a given finitely additive measure (the problem is to construct an indefinite integral). In this respect, one can observe a certain conceptual analogy with extensions in the class of distributions (see, for example, [13]); however, there are some obvious and significant distinctions. But even this approach to the problem of "finitely additive" extension is not treated here in full generality; a number of results were omitted due to space limitations (but we give the corresponding references). We focus our attention on conceptual questions; much attention is devoted to obtaining qualitative conclusions for natural applications to the problems of control of linear systems. The extension of constructions that we consider (see Chap. 4 and further) was advanced in [42] and then subsequently used for investigation of various classes of extremal problems (see, for example, [40, 41, 43–45] and others). Some methods put forward in [40–43] were successfully applied later to a different extension scheme [45, 46] that exploits the idea of closure in the space of measures of the corresponding set of finite weighted sums of points of the main space (see [47], where a fairly detailed discussion of the approach is given).

The author expresses his profound gratitude to his teacher, N. N. Krasovskii, for his constant interest and support, his valuable advice, and his detailed discussion of the results covered in this monograph.

My colleagues at the Department of Control Systems greatly helped in preparing the manuscript. Special thanks are due to L. N. Bestuzheva, Yu. I. Berdyshev, S. I. Morina, and V. E. Pak.

Contents

Contents

Chapter 1

Relaxations of Extremal Problems (cogent discussion and examples)

1.1. Introduction

The main purpose of this chapter is to prepare the reader to use, in the sequel, some rather abstract constructions in order to make correct extensions of some initial problems, and also (and to an even greater extent) to consider some motivating examples and use them to obtain some inferences. The point is that the problem of "filter" optimization that will be considered later in the general case, is, at first glance, absolutely unnatural. Nevertheless, it is important for us for two reasons; first, many concrete settings concerning the issues of optimization by perturbing a complex of constraints can be reduced to this general problem; second, it is an abstract problem and the constructions used in its investigation provide a general technique by which concrete problems can be reduced to general problems, and in the sequel we refer to this technique as "filter" optimization procedures. Consequently, the indicated technique must be understood not only at a formal but also at a substantive level in order to use it as an instrument for investigating concrete extremal problems. With this aim in view, we shall first of all discuss a number of examples dealing with optimization under perturbed constraints; recall that in this book perturbations have the character of weakening of the constraints of a problem being solved.

1

1.2. Relaxations of Integral Constraints

Consider the simplest scalar control system

$$\dot{x}(t) = f(t), \qquad x(0) = 0, \tag{1.2.1}$$

functioning on the interval $[0,1]$ by virtue of a program $f = (f(t) \geqslant 0, 0 \leqslant t < 1)$, assumed to be piecewise constant (p.c.) and right continuous (r.c.). In this section, we denote by F the set of all these (p.c. and r.c.) nonnegative control function-programs f. We shall consider, in addition, some additional restraints on the choice of f which we shall regard as the simplest and defined by inequalities. We shall agree to use the inequalities with one sign, namely, as inequalities of the form $p(f) \leqslant q$. In this example, we shall discuss the situation where the supplementary conditions on the choice of $f \in F$ have the form

$$\int_0^1 f(t)\, dt \leqslant 1, \qquad \int_0^1 s(t) f(t)\, dt \leqslant a. \tag{1.2.2}$$

The function s in $(1.2.2)$ will be assumed to be glued from constrictions to intervals of the form $[\alpha, \beta]$, $0 \leqslant \alpha < \beta \leqslant 1$, from real functions continuous on $[0,1]$, i.e., piecewise continuous (pw.c.) and r.c.; constraints $(1.2.2)$ specify the possibilities of realizing in $(1.2.1)$ the motion $x_f(\cdot) = (x_f(t), 0 \leqslant t \leqslant 1)$;

$$x_f(\theta) = \int_0^\theta f(\tau)\, d\tau \qquad (0 \leqslant \theta \leqslant 1). \tag{1.2.3}$$

Namely, from all possible f-solutions, $f \in F$, $(1.2.2)$ leaves only those which correspond to programs f admissible in the sense of $(1.2.2)$. Let us first consider the following version of the pair (s, a) in $(1.2.2)$; the function s is defined on $[0,1]$ by the condition $s(t) = -t$, $a = -1$. Then $(1.2.2)$ degenerates into the inconsistent condition

$$\int_0^1 f(t)\, dt \leqslant 1 \leqslant \int_0^1 t f(t)\, dt. \tag{1.2.4}$$

Therefore, irrespective of the choice of a criterion, the value of our extremal problem should be identified with ∞. Nevertheless, we shall introduce a criterion in the form of a relation associating the program $f \in F$ with the quantity

$$-x_f(1) = -\int_0^1 f(t)\, dt.$$

Consider three versions of perturbation (1.2.4) by the value $\varepsilon > 0$:

$$\int_0^1 f(t)\, dt \leqslant 1 + \varepsilon, \qquad 1 \leqslant \int_0^1 t f(t)\, dt, \tag{1.2.5}$$

$$\int_0^1 f(t)\, dt \leqslant 1, \qquad 1 - \varepsilon \leqslant \int_0^1 t f(t)\, dt, \tag{1.2.6}$$

$$\int_0^1 f(t)\, dt \leqslant 1 + \varepsilon, \qquad 1 - \varepsilon \leqslant \int_0^1 t f(t)\, dt. \tag{1.2.7}$$

Each one of the three systems of constraints (1.2.5)–(1.2.7) is consistent, and the control programs f, satisfying the three relaxed conditions, are constructed by the same scheme: to construct a program f admissible in the sense of (1.2.5)–(1.2.7) for $\varepsilon > 0$, one should set it equal to zero outside of the interval $[1 - \delta, 1)$, $0 < \delta < 1$, and on the interval itself set $f(t)$ constant and maximal within the corresponding constraint imposed on the total impulse. Of course, the number δ, $\delta > 0$, should be selected sufficiently small.

Thus, we have a common asymptotics of ε-admissible control programs (some nonessential differences in the conditions of normalization that manifest themselves in the restriction on the total impulse are ignored for simplicity). This circumstance can also be interpreted as the possibility of a common regularization since the values of the criterion "along" the asymptotics mentioned above converge as $\varepsilon \downarrow 0$ to the same value -1.

So, an inconsistent and, at first glance, poorly arranged problem with constraint (1.2.4) possesses in reality good asymptotic properties because ε-perturbations (1.2.5)–(1.2.7) are equivalent in the limit with respect to the result. This circumstance is in reality rather rare good luck. To illustrate this inference, let us consider another specification of (s, a), namely, define the function s by the condition $s(t) = t$ for $0 \leqslant t < 1$, and set the number a equal to 0. Thus, the new system of constraints on the choice of $f \in F$ has the form

$$\int_0^1 f(t)\,dt \leqslant 1, \qquad \int_0^1 t f(t)\,dt \leqslant 0. \tag{1.2.8}$$

The quality functional remains the same: it is the value of the phase coordinate at the last moment, taken with the opposite sign. Of course, (1.2.8) is a consistent system; its only admissible element is a function which is identically zero so that the conventional value is finite and becomes 0. Now, following (1.2.5)–(1.2.7), we introduce three families of ε-relaxations ($\varepsilon > 0$):

$$\int_0^1 f(t)\,dt \leqslant 1 + \varepsilon, \qquad \int_0^1 t f(t)\,dt \leqslant 0, \tag{1.2.9}$$

$$\int_0^1 f(t)\,dt \leqslant 1, \qquad \int_0^1 t f(t)\,dt \leqslant \varepsilon, \tag{1.2.10}$$

$$\int_0^1 f(t)\,dt \leqslant 1 + \varepsilon, \qquad \int_0^1 t f(t)\,dt \leqslant \varepsilon. \tag{1.2.11}$$

In the case of (1.2.9), we unfortunately have the former one-element admissible set; the result obtained under this condition coincides with the conventional value and is independent of ε, $\varepsilon > 0$. However, condition (1.2.10) gives much greater freedom, allowing one to form admissible control programs f in the form of sufficiently "narrow" rectangular impulses of unit area near the point $t = 0$. Consequently, under constraints (1.2.10) we can attain the value -1 on the range of our quality functional. In other words, the value of the minimization problem of the terminal criterion mentioned above is -1 since a better result cannot be obtained under the restriction on the total impulse corresponding to inequality (1.2.10).

Condition (1.2.11) behaves practically like (1.2.10) but admits of a greater freedom of action at the expense of weakening the resource constraint. The value of the ε-problem corresponding to (1.2.11) is equal to $-(1 + \varepsilon)$. We have again a factual equivalence of (1.2.10) and (1.2.11) with respect to the result, so that we may say that the asymptotic value is equal to -1, which is a common value for these two versions of constructing relaxations. However, the relaxations constructed on the basis of (1.2.9) have an asymptotic behavior that is quite different and (in the sense of a result) substantially worse. At the same time, to the problems perturbed in accordance with (1.2.9) one can certainly assign an asymptotic value corresponding to the limiting process as $\varepsilon \downarrow 0$; it coincides with the conventional value, i.e., with the value for an unperturbed problem, and can hardly be regarded as a regularization of the latter. We see that perturbations (1.2.10) have regularizing properties (for the time being we do not endow this statement with an exact meaning) and perturbations (1.2.9) do not. This shows the essential distinction between the second example and the first. The meaning of the "correct" optimization in the second example consists in approaching the δ-impulse at the initial moment of time by selecting the control program in the form of a rather narrow rectangular (realized) impulse of an area close to unity. Note, by the way, that a similar mode arises in a number of mechanical systems when they are controlled under the condition of bounded total impulse. A similar situation arises when a single particle is controlled with the aim of approaching the origin at the last moment.

These examples show that relaxations can modify the structure of a problem crucially; in particular, inconsistent problems can be transformed into consistent ones. One can also see from these examples that the solution space F is rather imperfect; it lacks delta-functions with which one should operate in order to satisfy the constraints and minimize the criterion. It may so happen, however, that even delta-functions will not suffice. We shall show this by giving a very simple example. For this purpose we set $a = 0$ in (1.2.2) and define the function s by the conditions

$$
s(t) = \begin{cases} \dfrac{1}{2} - t, & 0 \leqslant t < \dfrac{1}{2}, \\[2mm] 1, & \dfrac{1}{2} \leqslant t < 1. \end{cases}
$$

The quality criterion remains the same. Then the conventional value for this terminal problem with constraints (1.2.2) on the choice of $f \in F$ is 0, and we do not consider it in detail. The asymptotic value corresponding to an infinite series of problems with constraints on the choice of $f \in F$ of the form

$$
\int_0^1 f(t)\, dt \leqslant 1, \qquad \int_0^1 s(t) f(t)\, dt \leqslant \varepsilon, \qquad \varepsilon > 0, \tag{1.2.12}
$$

is equal to -1. To verify this, it suffices to consider the sequence $(f_1, f_2, \ldots)$

in F for which

$$f_k(t) = \begin{cases} 0, & 0 \leqslant t < \dfrac{1}{2} - \dfrac{1}{2k}, \\[2mm] 2k, & \dfrac{1}{2} - \dfrac{1}{2k} \leqslant t < \dfrac{1}{2}, \\[2mm] 0, & \dfrac{1}{2} \leqslant t < 1 \end{cases}$$

for $k = 1, 2, \ldots$; the fact is that for $\varepsilon > 0$ and for almost all k the program f_k is admissible in problem (1.2.12). At first glance, the sequence (f_k) is closable by the delta-function supported at the point $t = 1/2$. If we want to make our argument more accurate, we must introduce Dirac's measure $\mu = \delta_t|_{t=1/2}$ supported at the point $t = 1/2$; the latter must be interpreted as the functional

$$g \mapsto g\left(\frac{1}{2}\right) \colon C([0,1]) \to \mathbf{R}, \tag{1.2.13}$$

where $C([0,1])$ is the set of all continuous real-valued functions over $[0,1]$. In turn, every control program f_k can be identified with the functional

$$g \mapsto 2k \int\limits_{1/2(1-1/k)}^{1/2} g(t)\, dt \colon C([0,1]) \to \mathbf{R}. \tag{1.2.14}$$

Then the sequence (f_k) can be regarded as convergent to μ since

$$2k \int\limits_{1/2(1-1/k)}^{1/2} g(t)\, dt \to g\left(\frac{1}{2}\right) \tag{1.2.15}$$

for $g \in C([0,1])$; we speak about the convergence of the sequence defined by (1.2.14) to (1.2.13). Of course, functional (1.2.13) can be extended (continued) to a larger set of functions; in any case, one can define its action on the function s, the result of which will be the number

$$s\left(\frac{1}{2}\right) = 1.$$

This is natural for Dirac's measure μ, which closes the sequence (f_k) in the sense of convergence (1.2.15). Alas! This number (and this is the μ-integral of s) violates the condition of nonnegativeness to which the sequence (f_k) tended. In the exact formalization, to which we shall not resort now, it is easy to check that Dirac's measure μ cannot be admissible in the sense of satisfying constraints of the type of inequalities which are limiting with respect to (1.2.12) as $\varepsilon \downarrow 0$. Thus, an attempt at closing in the sense of convergence (1.2.15), where g is a continuous function, does not make up for the absent element for which one could reject in (1.2.12) the parameter $\varepsilon > 0$. Nevertheless, (f_k) has another limit, which is different from the Dirac limit. This limit is defined on a larger, compared to $C([0,1])$, set of functions but coincides with the values of functional (1.2.13) on continuous functions; the respective precise definitions require the use of finitely additive (f.a.) measures, which we shall consider later on.

1.3. Constraints Imposed on the Vector Integrand of a Control

The examples of the preceding section were of a purely illustrative nature. We shall now consider, at a substantive level, a class of control problems having practical significance. However, the last statement is certainly relative since we shall deal with model situations reflecting some peculiarities of real processes. Later we shall complicate the set-up to some extent in order to bring it closer to more specific control problems. Let us consider the linear controllable system

$$\dot{x}(t) = A(t)x(t) + f(t)b(t), \qquad x(t_0) = x_0, \qquad (1.3.1)$$

operating on a given time interval $I_0 \triangleq [t_0, \theta_0]$ $(t_0 < \theta_0)$ in a k-dimensional phase space under the action of a (scalar) control program $f = (f(t) \geqslant 0,$ $t_0 \leqslant t < \theta_0)$, whose choice is in our hands but which must satisfy some natural conditions, some of which have technical character ($f \in F$, where F is the set of all nonnegative real p.c. and r.c. functions f which are defined on the interval $t_0 \leqslant t < \theta_0$ and take values on the number line). In (1.3.1), x_0 is a given k-dimensional vector of the initial conditions (i.c.), $A(\cdot)$ is a matrix-valued mapping onto I_0 ($A(t)$ is a $(k \times k)$-matrix for $t \in I_0$) assumed, for simplicity, to be componentwise continuous, $b = b(\cdot)$ is a k-measure-valued vector function on

$$I \triangleq [t_0, \theta_0) = I_0 \setminus \{\theta_0\}.$$

We assume that every component $b_i(\cdot)$ of the vector function $b(\cdot)$ is obtained by a finite gluing, on intervals of the form $[\alpha, \beta)$, $t_0 \leqslant \alpha < \beta \leqslant \theta_0$, of constrictions of functions continuous on I_0 and taking values in $\mathbf{R}$; in other words, the vector function $b = b(\cdot)$ is pw.c. and r.c. We denote by $x_f(\cdot) = (x_f(t), t_0 \leqslant t \leqslant \theta_0)$ the f-solution of (1.3.1) corresponding to the specified i.c., $x_f(t)$ is a k-dimensional vector for all $t \in I_0$. The vector function $x_f(\cdot)$ can be determined from Cauchy's relation

$$x_f(t) = \Phi(t, t_0)x_0 + \int_{t_0}^{t} f(\tau)\Phi(t, \tau)b(\tau)\, d\tau, \qquad (1.3.2)$$

$t \in I_0$, where $\Phi(\cdot, \cdot)$ is a fundamental matrix function of solutions of the homogeneous system

$$\dot{x} = A(t)x.$$

It stands to reason that we must assume that $f \in F$ in (1.3.2) and perform the integration componentwise. As in Sec. 1.2, we determine the criterion as being dependent on the final state of the system, so that the control quality $f \in F$ is estimated by the relation

$$f \mapsto g_0(x_f(\theta_0))\colon\ F \to \mathbf{R},$$

where g_0 is a continuous function of k variables. The criterion of this type is widespread, but, from the standpoint of subsequent arguments, it also admits essential generalizations from which we shall now refrain. We define here the constraints on the choice of $f \in F$ in the form

$$\left(\int_{t_0}^{\theta_0} s_1(t)f(t)\, dt, \ldots, \int_{t_0}^{\theta_0} s_n(t)f(t)\, dt \right) \in Y, \qquad (1.3.3)$$

where the vector function $s(\cdot) = (s_1(\cdot), \ldots, s_n(\cdot))$ has the same character as $b(\cdot)$; it is defined on I. The set Y lying in an n-dimensional space is assumed to be closed in the natural sense of coordinatewise convergence. Furthermore, we assume that the Y-constraint (1.3.3) includes the (resource) constraint on the total impulse of f: a certain function s_j, where $j \in \overline{1, n}$, is identically 1, and in this case the set Y satisfies the condition $y_j \leqslant c$ for $y \in Y$ ($c > 0$ is a given constant). An important special case of condition (1.3.3) has the form

$$\left(\int_{\alpha_1}^{\beta_1} f(t)\, dt, \ldots, \int_{\alpha_n}^{\beta_n} f(t)\, dt \right) \in Y. \qquad (1.3.4)$$

Here α_i and β_i satisfy the conditions $t_0 \leqslant \alpha_i < \beta_i \leqslant \theta_0$, and with this $\alpha_j = t_0$ and $\beta_j = \theta_0$ for some $j \in \overline{1, n}$ (for Y the former assumption remains valid). Here is one specific version of condition (1.3.4). We assume $n \geqslant 2$.

Let $\alpha_n = t_0$ and $\beta_n = \theta_0 = \beta_{n-1}$, so that we identify the subscript j, for definiteness, with the number n, $n \geqslant 2$. This means that we introduce the resource constraint by means of the function s_n, $s_n(t) \equiv 1$; substantially, this means that the inequality

$$\int\limits_{t_0}^{\theta_0} f(t)\, dt \leqslant c \qquad (1.3.5)$$

holds. Along with (1.3.5), by means of the condition

$$\left(\int\limits_{\alpha_1}^{\beta_1} f(t)\, dt, \ldots, \int\limits_{\alpha_{n-1}}^{\beta_{n-1}} f(t)\, dt \right) \in Y_1, \qquad (1.3.6)$$

we introduce a supplementary constraint, assuming (in the case of $n > 2$) that $\alpha_2 = \beta_1, \ldots, \alpha_{n-1} = \beta_{n-2}$. In other words, we associate (1.3.6) with the given partition of I into $n - 1$ mutually disjoint intervals $[\alpha_i, \beta_i)$. We define the subset Y_1 in the arithmetic space of dimension $n - 1$ so that we can formalize the following substantive condition as (1.3.6).

Condition of alternation of impulses and pauses. Let $0 < a < b$. We say that the control $f \in F$ "operates" on the clock period $[\alpha_r, \beta_r)$, $r \in \overline{1, n - 1}$, in the impulse regime if the integral of

$$(f|[\alpha_r, \beta_r)) = (f(t), \alpha_r \leqslant t < \beta_r)$$

(i.e., the integral of the function f on the interval $[\alpha_r, \beta_r)$) is larger than or equal to b, and we say that in this period f behaves like a pause if that integral does not exceed a. The supplementary condition will be as follows: those and only those programs $f \in F$ are admissible which satisfy (1.3.5) and operate in every period either in the impulse or in the pause regime, and any two impulses are separated in time by at least one pause.

The meaning of this condition is quite clear: the case at hand is the requirement of alternation of the periods of intensive work of the system and of unloading. We now have to reduce our substantive condition to representation (1.3.6), letting, for brevity, $m \triangleq n - 1$. Namely, we obviously ought to introduce a set Y_1, for which the condition of alternation of impulses and pauses is equivalent to inclusion (1.3.6), and, moreover, to make sure that the set Y_1 is closed, which is very important in the subsequent constructions.

Let Y_1 be the set of all m-dimensional vectors

$$y = (y_1, \ldots, y_m) \qquad (1.3.7)$$

with nonnegative components, for each of which: (1) $\forall\, i \in \overline{1,m}$ either $y_i \leqslant a$ or $b \leqslant y_i$, (2) $\forall\, k \in \overline{1,m}$, $r \in \overline{1,m}$, $k < r$, the following implication holds:

$$((b \leqslant y_k)\,\&\,(b \leqslant y_r)) \Rightarrow (\exists\, s \in \overline{k,r}\colon y_s \leqslant a). \tag{1.3.8}$$

We have thus introduced Y_1 making a strict use of conditions (1) and (2). Let us show that Y_1 is closed as a subset of an m-dimensional space (the equivalence of (1.3.6) with the set Y_1 defined by means of conditions (1), (2) mentioned above and the substantive condition of alternation of impulses and pauses is obvious). Thus, let $(y^{(l)};\, l = 1, 2, \ldots)$ be a sequence of elements of Y_1 converging to some vector (1.3.7). If we fix the subscript $i \in \overline{1,m}$, then $\forall\, l = 1, 2, \ldots$ we have

$$(y_i^{(l)} \leqslant a) \vee (b \leqslant y_i^{(l)}).$$

But then, since $b - a > 0$ and

$$y_i^{(l)} \to y_i \quad \text{as} \quad l \to \infty,$$

we have $y_i \notin (a, b)$. So, the "limit" vector y satisfies the condition

$$(y_i \leqslant a) \vee (b \leqslant y_i)$$

for any number $i \in \overline{1,m}$. Let now $k \in \overline{1,m}$ and $r \in \overline{1,m}$, $k < r$, satisfy the condition of (1.3.8):

$$(b < y_k)\,\&\,(b \leqslant y_r).$$

Then, for almost all natural l, we have

$$(b \leqslant y_k^{(l)})\,\&\,(b \leqslant y_r^{(l)}).$$

Indeed, if there existed, for instance, arbitrarily large numbers l, for which one had $y_k^{(l)} < b$, then, by the definition of the set Y_1, we would have $y_k^{(l)} \leqslant a$ for these numbers; the convergence of $(y_k^{(l)};\, l = 1, 2, \ldots)$ to y_k is in this case impossible because $a < b$. This contradiction shows that $b \leqslant y_k^{(l)}$ for almost all l. The following assertion can be proved similarly: $b \leqslant y_r^{(l)}$ for almost all l. But in this case, since we have $y^{(l)} \in Y_1$ for all l, we obtain, according to (1.3.8),

$$\{s \in \overline{k,r}\mid y_s^{(l)} \leqslant a\} \neq \varnothing \tag{1.3.9}$$

for almost all l. Since $(y_s^{(l)},\, l = 1, 2, \ldots)$ converges to y_s for $s \in \overline{k,r}$, condition (1.3.9) excludes the condition

$$b \leqslant y_s \qquad (s \in \overline{k,r}), \tag{1.3.10}$$

because otherwise (i.e., under condition (1.3.10)) one could find, for arbitrarily large l, numbers $s = s_l$ for which

$$0 < b - a \leqslant y_s - y_s^{(l)},$$

which contradicts the convergence stated above. The negation of (1.3.10) means the validity (for the vector y) of property (2) used in the definition of Y_1. Thus (the nonnegativity of $y_1, \ldots, y_m$ is obvious) $y \in Y_1$, and the fact that Y_1 is closed is established.

We have thus completed the reduction of the alternation of impulses and pauses condition to the form (1.3.5) and (1.3.6). The problems of optimization of the (terminal) criterion (1.3.3) under the conditions of the constraint stated above have, as can be seen from the statement of the substantive constraint, practical significance because this condition concerns the possibilities of control with unloadings between peak levels. One of the qualitative questions connected with this problem is the following: how will the optimum (value) of the problem of minimizing functional (1.3.3) change if we replace in (1.3.5) the resource parameter c by $c + \varepsilon$ and the set Y_1 in (1.3.6) by an ε-neighborhood? Will the value change substantially for small ε, or will the change of the value be just as small? In the second case, we shall speak about computational stability.

We shall confine ourselves to a more substantive case of a consistent system (1.3.5), (1.3.6). It turns out that in this case we have the property of computational stability in the sense mentioned above: by weakening constraints (1.3.5) and (1.3.6) by a small value $\varepsilon > 0$, we change the value just a little. This testifies to a certain structural stability of the problem with respect to "constraints," and this can give, in particular, some possibilities in the sense of structural stability with respect to numerical parameters a, b, and c that are specified more or less subjectively when the system is being constructed. It should be pointed out that the inference concerning computational stability needs, of course, a proper justification which will be given further for situations that are substantially more general than those of (1.3.5), (1.3.6), and even (1.3.3). Here, we only point out that this justification will require the replacement of physically realizable programs f by some generalized objects, namely, measures, where the measures are finitely additive.

Note one more point related to a terminal problem under constraints of the form

$$\int_{t_0}^{\theta_0} f(t)\,dt \leqslant c, \qquad \left(\int_{t_0}^{\theta_0} s_1(t)f(t)\,dt,\ldots,\int_{t_0}^{\theta_0} s_m(t)f(t)\,dt\right) \in Y_1. \qquad (1.3.11)$$

Here, we preserve, for the vector function $(s_1(\cdot),\ldots,s_m(\cdot))$, the same assumptions as in (1.3.3) and assume that the number m coincides with $n-1$. For complete reduction of (1.3.11) to the form (1.3.3), we must define s_n as a unit function and represent Y as $Y_1 \times (-\infty, c]$; of course, here we mean a representation in the form of a product, up to one-to-one correspondence: Y must be composed of all the vectors $(y_1,\ldots,y_m,y_n)$ for which $(y_1,\ldots,y_m) \in Y_1$ and $y_n \leqslant c$. Every problem (1.3.11) is characterized by a finite or infinite value. Consider an increase of the perturbations of the resource parameter c. In other words, the question is of replacing (1.3.11) by the problem

$$\begin{cases} g_0(x_f(\theta_0)) \to \inf, \qquad \displaystyle\int_{t_0}^{\theta_0} f(t)\,dt \leqslant c+\varepsilon, \\[2em] \left(\displaystyle\int_{t_0}^{\theta_0} s_1(t)f(t)\,dt,\ldots,\int_{t_0}^{\theta_0} s_m(t)f(t)\,dt\right) \in Y_1, \end{cases} \qquad (1.3.12)$$

where $\varepsilon \geqslant 0$. Comparing (1.2.4) and (1.2.5), one can easily derive the inference about the absence of stability of the result in problems (1.3.12) under the perturbation

$$c \to c+\varepsilon, \qquad \varepsilon > 0. \qquad (1.3.13)$$

Thus, (1.3.13) is a disastrous way of perturbing for the conventional setting. Let us make an attempt to somehow regularize the value function, providing this term with a meaning which is certainly different from that used in the theory of ill-posed problems. We wish to find a "true" quality which must be (in our perception) stable with respect to perturbations (1.3.13). To do this, we associate each number δ, $\delta > 0$, with a closed δ-neighborhood Y_1^δ of the set Y_1. Now, $\forall \varepsilon \in [0,\infty)$, $\delta \in (0,\infty)$, we introduce the problem

$$\begin{cases} g_0(x_f(\theta_0)) \to \inf, \qquad \displaystyle\int_{t_0}^{\theta_0} f(t)\,dt \leqslant c+\varepsilon, \\[2em] \left(\displaystyle\int_{t_0}^{\theta_0} s_1(t)f(t)\,dt,\ldots,\int_{t_0}^{\theta_0} s_m(t)f(t)\,dt\right) \in Y_1^\delta; \end{cases} \qquad (1.3.14)$$

the conventional value of (1.3.14) does not exceed the similar value of (1.3.12). Let us agree to express the value of (1.3.14) in terms of $\mathrm{val}\,(\varepsilon, \delta)$, $\varepsilon \geqslant 0$, $\delta > 0$, assuming, for simplicity, that every problem (1.3.14) is consistent. The parameters ε and δ in (1.3.14) are disparate: the first one is connected with a real perturbation (1.3.13) and the second one was introduced artificially and can be assumed to be arbitrarily small. The parameter δ will be "correcting" in the subsequent reasoning. Then, we can use, as a kind of substitute for the conventional value (optimum) of problem (1.3.12), the number $\sup(\{\mathrm{val}\,(\varepsilon, \delta): \delta > 0\})$, thus obtaining a new dependence on ε, $\varepsilon \geqslant 0$. This new function proves to be continuous at the point $\varepsilon = 0$. This also follows from general aspects as a consequence of using natural procedures of compactification of the solution space. The constructed regularizing function shows that the "true" value of problem (1.3.12) is continuous at the point $\varepsilon = 0$ (moreover, it depends on ε, $\varepsilon \geqslant 0$, monotonically and satisfies some other properties which are natural for a value function). Indeed, in concrete problems, small violations of a constraint connected with the set Y_1 are practically unavoidable; the new "value function" takes them into account at the approximation level. In this section, we gave some concrete examples and some assertions for them following from the general facts established below; we considered the simplest cases, in which, however, the corresponding arguments are not obvious. They require justification; the effects arising in these examples are connected with the noncompactness of the control space, and the regularizations of the respective dependences are based on the use of compactifications. In the next section, we shall consider a more complicated example, which at the same time admits natural analogies with the constructions carried out in this section. However, the necessary justifications will be omitted (they will be given later in a more general form); we confine ourselves to the setting part and the discussion of the results, adding only some fragments of the arguments that are essential for some separate elements of the problem setting.

1.4. Asymptotic Effects in a Cone Optimization Problem (an Example)

In this section, we continue to consider control problems; however, the effects considered further on are much more complicated owing to the specific character of the problem under study, for which we shall also formulate the main assertions in the form of corollaries of general propositions. At the same time, we shall also complicate the control system (1.3.1) by neglecting the assumption that the control parameter $f(t)$ is one-dimensional; the situation where the control f is a vector function is more frequent. From the standpoint of quality index, the problem under consideration is a generalization of the case of multicriteria optimization and has the meaning of cone optimization. As before, we confine ourselves to a substantial discussion without claiming to be exact. Thus, consider the nonstationary linear system

$$\dot{x}(t) = A(t)x(t) + B(t)f(t), \qquad x(t_0) = x_0. \tag{1.4.1}$$

We assume that the phase space, the control interval I_0, and the matrix-valued mapping $A(\cdot)$ are in conformity with the conditions imposed on system (1.3.1). As for $B(\cdot)$, we shall assume in (1.4.1) that it is the matrix-valued mapping onto I from Sec. 1.3, all of whose components have the same character as those of the vector function $b(\cdot)$ in (1.3.1), i.e., they are pw.c. and r.c. functions on I, obtained by gluing together a finite number of constrictions of continuous functions to I_0. Here $B(t)$ is a $(k \times r)$-matrix for $t \in I$, where k corresponds to Sec. 1.3 and r is an integer that determines the dimensionality of the control vector. We assume that $f = (f_1, \ldots, f_r)$, which is a control vector function on I, has nonnegative components $f_1, \ldots, f_r$ and satisfies some constraints, which we shall divide conventionally into two groups:

(1) the resource constraint

$$\sum_{i=1}^{r} \int_{t_0}^{\theta_0} f_i(t)\, dt \leqslant c, \tag{1.4.2}$$

where $c > 0$ is a given, as in Sec. 1.3, number (in this section we assume that $f_1, \ldots, f_r$ are p.c. and r.c. functions, and therefore the integration in (1.4.2) is understood in its simplest form,

(2) the vector integrand constraint

$$\int_{t_0}^{\theta_0} S(t)f(t)\, dt \in \mathcal{Y}, \tag{1.4.3}$$

where $S(\cdot)$ is a matrix-valued function on I of the same character as $B(\cdot)$ in (1.4.1), $S(t)$ is an $(m \times r)$-matrix for $t \in I$ (m is an integer), $\mathcal{Y}$ is a closed nonempty set of an m-dimensional space. In the sequel, we shall call (1.4.2) and (1.4.3) the c- and the $\mathcal{Y}$-constraints respectively.

Consider a nonempty set Λ of the k-dimensional phase space of system (1.4.1) and associate with every vector $x = (x_1, \ldots, x_k)$ the criterion value as a distance $\rho(x, \Lambda)$, providing, for definiteness, the k-dimensional space with a natural Euclidean metric. Let, moreover, in this section Q be a nonempty subset I_0 from Sec. 1.3, i.e., $\varnothing \neq Q \subset [t_0, \theta_0]$. Suppose that in this section $\mathbf{F}$ denotes the set of all p.c. and r.c. componentwise nonnegative vector functions f with r-dimensional values. Then every control program $f \in \mathbf{F}$ generates a motion $x_f(\cdot) = (x_f(t), t_0 \leqslant t \leqslant \theta_0)$ defined on I_0 by a Cauchy's formula similar to (1.3.2), so that

$$x_f(t) = \Phi(t, t_0)x_0 + \int_{t_0}^{t} \Phi(t, \tau)B(\tau)f(\tau)\,d\tau.$$

Let us associate every control program $f \in \mathbf{F}$ with an estimate v_f defined as the mapping

$$q \mapsto \rho(x_f(q), \Lambda)\colon Q \to \mathbf{R}.$$

Using the symbol $\leq$ to denote the pointwise ordering of the set $\mathbf{R}^Q$ of all (real-valued) functions over Q, we shall consider the control f_1 to be not worse than f_2 if $v_{f_1} \leq v_{f_2}$. Now it is natural to raise the question of $\leq$-minimization of v_f on the set $\mathbf{F}_\partial$ of all controls $f \in \mathbf{F}$ respecting the c- and $\mathcal{Y}$-constraints. This minimization should, however, be interpreted as a process of seeking $\leq$-minimal estimates and not the $\leq$-least estimates because the latter exist very seldom in problems of this kind. But even with the $\leq$-minimal estimates not everything is satisfactory in our case since the class of controls that we use is not compact in the necessary sense (we do not specify what sense we mean; it will be clear in the sequel). However, one uses a simple technique to get rid of this problem, which is similar, in a sense, to the situation where the minimum is not attained in a problem of scalar optimization and one has to speak about the greatest lower bound of values of the quality criterion. For our purposes, together with the "true" estimates v_f, $f \in \mathbf{F}_\partial$, we introduce the "estimates" $v\colon Q \to \mathbf{R}$ which are arbitrarily close to the true ones; the closeness should be understood here in the sense of the topology $\mathcal{Q}$ of the pointwise convergence in $\mathbf{R}^Q$. In other words, one has to "replace" the set $\{v_f\colon f \in \mathbf{F}_\partial\}$ by its $\mathcal{Q}$-closure, which we shall denote by $\mathcal{V}_1$ for brevity. It turns out that $\mathcal{V}_1$ can be regarded as a continuous image of a closed subset of a compact set (lying in the space of the so-called generalized controls), which is nonempty if $\mathbf{F}_\partial \neq \varnothing$; moreover, the closed set mentioned above can be regarded as a "closure" of $\mathbf{F}_\partial$ when the latter is imbedded into the indicated compact set. When $\mathcal{V}_1$ is represented as a continuous image of a closed set, one has to take into account the familiar property of closeness of a continuous operator that acts from a compact space into a Hausdorff one and also the obvious consequence of this property to preserve this closure by the image operation. We also take into account the continuity of the dependence $x \to \rho(x, \Lambda)$ within the limits of the k-dimensional space. Now (the corresponding precise assertions are given in the next chapter), we can verify, with due account of a certain coordination of $\leqq$ and $\mathcal{Q}$, that if $\mathbf{F}_\partial \neq \varnothing$, then the set V_1 of all $\leqq$-minimal elements of $\mathcal{V}_1$ is necessarily nonempty. Stretching the point, we can now regard V_1 as an ordinary "value" of an extremal problem with the constraints $f \in \mathbf{F}_\partial$, a sort of optimum for this problem. Note that when introducing the set $\mathcal{V}_1$, we actually replaced the conventional solution $f \in \mathbf{F}_\partial$ by a net of solutions, i.e., by a generalized sequence in $\mathbf{F}_\partial$. It makes sense, however, to admit now for consideration nets (generalized sequences) in $\mathbf{F}$, which no longer respect the constraints absolutely but "approach" them from outside. In other words, as in Sec. 1.3, we admit the possibility of replacing the set $\mathcal{Y}$ by its closed ε-neighborhood $\mathcal{Y}_\varepsilon$ in the sense of the sup-norm of the m-dimensional space, $\varepsilon > 0$. The selection of this norm is, of course, not necessary, but is convenient in some special cases when one considers constraints of the inequality type. Now we shall also admit for consideration "estimates" $\tilde{v}$ for every one of which one can indicate a generalized sequence (f_α) in $\mathbf{F}$ such that (v_{f_α}) converges to v in the sense of $\mathcal{Q}$, and in this case we have

$$\int_{t_0}^{\theta_0} S(t) f_\alpha(t) \, dt \in \mathcal{Y}_\varepsilon$$

from some moment if only $\varepsilon > 0$; the c-constraint is still assumed to be unshakable. The "estimates" $\tilde{v}$ attainable in the sense indicated above now constitute a certain new set $\mathcal{V}_2$ corresponding only to the asymptotic satisfaction of the $\mathcal{Y}$-constraint. We denote the set of all $\leqq$-minimal elements of $\mathcal{V}_2$ by V_2, thus obtaining some new "value," which we now have some reason to consider to be asymptotic.

Finally, let us also introduce the set $\mathcal{V}_3$ of all "estimates" $\tilde{v}$ realizable in the class of generalized limits (v_{f_α}) along the "solutions" (f_α), which are now allowed to violate not only $\mathcal{Y}$- but also the c-constraint. Here we require, of course (for $\varepsilon > 0$) the ε-respect of these two constraints from some moment, i.e., we consider approximate net-solutions "retracting" into the constraints (and, in particular, sequence-solutions). For $\mathcal{V}_3$, we introduce the set of all minimal (in the sense of $\leq$) elements, which we shall denote by V_3. Then from general assertions, presented in the subsequent chapters, we can make the following inferences of a qualitative character:

(1) the equality $V_2 = V_3$ always holds true, which testifies to some structural stability of the problem with respect to the resource parameter,

(2) if the mapping $S(\cdot)$ has p.c. and r.c. components, then $V_1 = V_2 = V_3$ (computational stability).

Note that actually the condition on $S(\cdot)$ formulated in (2) is not exotic. It is fulfilled, for instance, in the following case of practical interest that generalizes the situation with the "impulse–pause" constraints from Sec. 1.3. Indeed, the control problem considered here is, in particular, multichannelled: the functions $(f_1, \ldots, f_r)$ are essentially controls in isolated channels and each one of them can be subjected to the condition of "impulse" and "pause" alternation from Sec. 1.3 (of course, the threshold levels and the systems of time periods in the channels may differ). Then, in every channel one can introduce its own constraint of the form (1.3.6) with an appropriate choice of the set Y_1 just as was done in Sec. 1.3 (see (1.3.7), (1.3.8)). The set $\mathcal{Y}$ can actually be obtained as a product of such copies of $Y_1 = Y_{1i}$, $i = 1, \ldots, r$. We shall now omit the relevant rather obvious constructions, observing only that the resulting system of constraints can be easily represented in the form (1.4.2), (1.4.3) and the components of the matrix-valued function $S(\cdot)$ will be p.c. and r.c., or, to be more precise, these components will be defined by the characteristic functions of the time periods. As a result, we obtain that the main condition imposed on $S(\cdot)$ in (2) is realized in this specific example (the resulting problem turns out to be, generally speaking, nonconvex).

Chapter 2

Compactificators in Extremal Problems

2.1. Introduction

The substantive examples considered in the preceding chapter showed
a number of "unpleasant" effects typical of concrete settings of extremal
problems and due essentially to the "bad" nature ("bad" properties) of the
solution space of the initial problem. These circumstances concerned vari
ous kinds of instability, but one could also observe, in a number of cases,
the absence of existence theorems, which is of no little importance from the
standpoint of the methods of solution and optimality conditions. At the
same time, there are spaces that are quite perfect in this sense, namely,
compact spaces [54] for which troubles of the indicated kind are absent. In
this connection, it is natural to embed the initial problem with a "bad" solution space into an appropriately generalized problem on a compact space
in order to carry over some of the useful properties of the "good" problem
to the initial setting. The essence of such an embedding must consist at
least in realizing the initial space as an everywhere dense subset in the space
of the indicated generalized problem. Furthermore, the criterion must be
transformed into a continuous goal dependence (criterion) of the generalized problem. Then we shall clearly not lose the specificity of the initial
problem, but realize the limit effects of this problem as points of a compact
set. The techniques of the above embedding (compactification) can differ in
principle; there is no uniqueness property for compactification. Thus, the
simplest version of embedding the set of all rational points from $[0, 1]$ into

a compact space is a canonical embedding into $[0, 1]$, i.e., the embedding which establishes a correspondence between every rational point $r \in [0, 1]$ and the point r itself. However, one could suggest more complicated methods, using, for example, the embedding with the aid of Dirac's measures into the space of f.a. two-valued $(0, 1)$-measures. Nevertheless, the nonuniqueness of compactification of a solution space is not an obstacle; moreover, a researcher may select for his purposes a suitable version of compactification. Thus, in the nonlinear optimal control problems with geometric constraints on the choice of "instantaneous" controls, wide use is made of compactification in the corresponding class of Radon measures, or, what is equivalent, in the class of $*$-weakly measurable functions of time that assume values in the form of normalized Borel measures on the set defining the geometric constraints mentioned above. The class of these generalized objects has turned out to be not only natural but also convenient from the standpoint of existence theorems and the corresponding extension of the theory of Pontryagin's maximum principle. Actually, such a compactification was applied by Krasovskii to formalize the stability property in nonlinear differential games [20]. This compactification plays an important role in game problems of programmed control that make up auxiliary constructions for nonlinear differential games (see, e.g., [20, 21, 32]). In this book, we are chiefly concerned with the extension of constructions in the class of f.a. measures that can be used both in the "universal" extension schemes (see in this connection the compactification in the class of two-valued f.a. $(0, 1)$-measures; we should also mention here a link with the Stone representation [5, 31] and with the constructions of nonstandard analysis [11]), and in the constructions specialized for a certain class of problems. Nevertheless, it appears reasonable to construct the corresponding extension scheme in axiomatic form, without specifying the form of generalized objects, and for this we need a short summary of topological concepts [2, 3].

2.2. Some Notions from Topology

The topology on the arbitrary set T is a nonempty family τ of the subsets of T that contains T and possesses the property $G_1 \cap G_2 \in \tau$ for $G_1 \in \tau$ and $G_2 \in \tau$, and the property

$$\bigcup_{G \in \mathcal{G}} G \in \tau, \tag{2.2.1}$$

if $\mathcal{G}$ is a subfamily of τ. If in (2.2.1) one takes for $\mathcal{G}$ the empty subfamily of τ, then one finds that every topology contains $\varnothing$. The pair (T,τ), where T is a set and τ is a topology of T, is called a topological space; in this case the sets $G \in \tau$ are open and the sets belonging to the family

$$\mathcal{F}_\tau \triangleq \{T \setminus G : G \in \tau\} \tag{2.2.2}$$

are closed in (T,τ). The families τ and $\mathcal{F}_\tau$ form some natural duality corresponding to the given topological space (T,τ); if H is a subset of T, then $\mathrm{cl}\,(H,\tau) \in \mathcal{F}_\tau$ is, by definition, the closure of H in (T,τ), i.e., the smallest, with respect to inclusion, set from $\{F \in \mathcal{F}_\tau|\ H \subset F\}$. If (T,τ) and (S,θ) are two topological spaces, $T \neq \varnothing$, $S \neq \varnothing$, then the function $g\colon T \to S$ is continuous if $g^{-1}(G) \in \tau$ for $G \in \theta$. We denote by $\tau_{\mathbf{R}}$ (or τ_∂) the natural (resp. discrete) topology of the number line $\mathbf{R}$; in the first case we mean the $|\cdot|$-topology and in the second the family of all subsets of $\mathbf{R}$.

We denote by $\mathbf{C}(T,\tau)$, where (T,τ) is a nonempty $(T \neq \varnothing)$ topological space, the set of all continuous, with respect to (T,τ) and $(\mathbf{R},\tau_{\mathbf{R}})$, functionals on T. These functionals will play a major role in the extension constructions for extremal problems.

The notion of a neighborhood is one of the most important in general topology; we shall follow here [19, p. 62], namely, if (T,τ) is a topological space, then the set H, $H \subset T$, is a neighborhood, or more precisely, a τ-neighborhood of the point $\theta \in T$ (of a set Θ, $\Theta \subset T$) if we have $\theta \in G$ (inclusion $\Theta \subset G$) for some set $G \in \tau$, $G \subset H$. Observe that the set G, $G \subset T$, is τ-open if and only if it is a τ-neighborhood of any of its points.

We use the standard [54, p. 196] notion of compactness of a topological space; the term "bicompactness" is not used. Here (and henceforth) we shall need the concept of a subspace, which we shall now recall, introducing the appropriate notation. Namely, if (T,τ) is a topological space and S is a subset of T, then the family

$$\tau|_S \triangleq \{S \cap G : G \in \tau\} \tag{2.2.3}$$

is a topology of S and is usually called the topology induced by (T,τ); the pair $(S,\tau|_S)$ is called a subspace of (T,τ). If, under the conditions defining (2.2.3), $(S,\tau|_S)$ is a compact space, then S is said to be a compact set of (T,τ). This definition is also standard and, therefore, is not discussed in detail. In every compact topological space (T,τ), the family $\mathcal{F}_\tau$ consists only of compact (in (T,τ)) subsets of T. The converse is true for the so-called separated (Hausdorff) topological spaces, i.e., for spaces (T,τ), for which $\forall\, t_1 \in T,\ t_2 \in T,\ t_1 \neq t_2$, one can indicate a τ-neighborhood H_1 of the point t_1 and a τ-neighborhood H_2 of the point t_2 which satisfy the condition $H_1 \cap H_2 = \varnothing$. We call a compact Hausdorff space a compactum [54, p. 208].

We shall recall here the following important property of continuous mappings: if (T,τ) and (S,θ) are nonempty topological spaces and $f\colon T \to S$ a continuous function, then the f-image of the τ-compact subset of T is a θ-compact subset of S. This property is connected with the familiar property of continuous functionals on compact spaces, namely, for every nonempty compact space (T,τ) we have $\forall\, g \in \mathbf{C}(T,\tau)$:

$$\{t \in T \mid \forall\, \tilde{t} \in T\colon g(t) \leqslant g(\tilde{t})\} \neq \varnothing. \tag{2.2.4}$$

A very important (although obvious) consequence follows from (2.2.4). It uses the fact that for every nonempty topological space (T,τ) and arbitrary nonempty set S, $S \subset T$, one has

$$(g|S) \triangleq (g(s))_{s \in S} \in \mathbf{C}(S, \tau|_S), \tag{2.2.5}$$

if $g \in \mathbf{C}(T,\tau)$. On the left-hand side of (2.2.5) we introduced the traditional notation for the constriction of a mapping onto a nonempty subset of its domain of definition; this notation is used repeatedly in the sequel. Combining (2.2.4) and (2.2.5), we clearly obtain the following inference: if (T,τ) is a nonempty topological space, $g \in \mathbf{C}(T,\tau)$, and K is a nonempty τ-compact subset of T, then

$$\{k_0 \in K \mid \forall\, k \in K\colon g(k_0) \leqslant g(k)\} \neq \varnothing. \tag{2.2.6}$$

Relation (2.2.6) will be directly used in the extension constructions.

We shall need to consider nonempty families of subsets of a set given a priori. From all families of this kind we shall need families satisfying the following semimultiplicativity property (recall that a multiplication for the set spaces is usually identified with an intersection), namely, if X is a set and $\mathcal{X}$ is a nonempty family of subsets of X, then we say that $\mathcal{X}$ is semimultiplicative if $\forall\, A \in \mathcal{X},\ B \in \mathcal{X}\ \exists\, C \in \mathcal{X}$:

$$C \subset A \cap B \tag{2.2.7}$$

(this property is naturally connected with the important notion of direction (in $\mathcal{X}$) [19, p. 95], but we shall not discuss it now). If X is a nonempty set, then by a filter base on X we mean any (nonempty) semimultiplicative family of subsets of X not containing $\varnothing$. This definition is consistent with

[3, p. 81]. By induction it immediately follows from the definition that every filter base $\mathcal{X}$ of the (nonempty) set X is a centered system of subsets of the latter: for any choice of $n \in \mathcal{N} \triangleq \{1; 2; \ldots\}$ (this notation for the set of positive integers will be used in the sequel) and a "collection"

$$(X_i)_{i \in \overline{1,n}} \colon \overline{1, n} \to \mathcal{X}$$

necessarily $\exists X^* \in \mathcal{X} \colon X^* \subset X_j$ for $j \in \overline{1, n}$. If $\mathcal{X}$ is a filter base on the nonempty set X; (Y, τ) a nonempty topological space; $r \colon X \to Y$, then the family

$$\{\operatorname{cl}(r^1(H), \tau) \colon H \in \mathcal{X}\},$$

where, by definition, $r^1(S)$ is the r-image of every set S, $S \subset X$, is a filter base of the set Y. The combination of the two properties mentioned above leads, in the case of mappings assuming values in a compact space, to the following assertion: if, under the conditions of the preceding inference, (Y, τ) is a compact space, then

$$\bigcap_{A \in \mathcal{X}} \operatorname{cl}(r^1(A), \tau) \neq \varnothing; \tag{2.2.8}$$

the set on the left-hand side of (2.2.8) is an element of $\mathcal{F}_\tau$, and, in particular, it is compact in (Y, τ). The last assertion is based on the familiar property of centered families of sets $\mathcal{H}$, $\mathcal{H} \subset \mathcal{F}_\tau$, in the compact space (Y, τ). Moreover, the following statement, related to (2.2.7) and (2.2.8), is valid. If X and Y are nonempty sets, τ is a compact topology of Y (i.e., a topology of Y for which (Y, τ) is a compact space), $r \colon X \to Y$, and $\mathcal{X}$, $\mathcal{X} \neq \varnothing$, is a semimultiplicative family of subsets of X, then (2.2.8) holds if and only if $\mathcal{X}$ is a filter base of X. Here we have taken account of the fact that the image and the closure of an empty set are empty sets. The mapping (operator) r in these simple assertions can be quite arbitrary. It is now important to observe a natural concord of (2.2.6) and the inferences connected with (2.2.8). In fact, under the conditions defining (2.2.8), for $g \in \mathbf{C}(Y, \tau)$ and

$$\bigcap_{A \in \mathcal{X}} \operatorname{cl}(r^1(A), \tau) \tag{2.2.9}$$

we have property (2.2.6). As a matter of fact, precisely the specification of K in (2.2.6) by means of (2.2.9) defines in the subsequent extension constructions the transformation of constraints of an asymptotic character (the nature of whose appearance is, in particular, illustrated in Chap. 1) into constraints that are "standard" for the theory of extremal problems. Generally, in the constraints defining (2.2.6), to denote the optimum (value) of a "standard" minimization problem

$$g(k) \to \min, \qquad k \in K, \qquad\qquad (2.2.10)$$

we use the natural symbol

$$\min_{k \in K} g(k) \quad \text{and not} \quad \inf_{k \in K} g(k),$$

leaving the latter for use in situations where relation (2.2.6) can be violated. Problems of the type (2.2.10), where g is a continuous functional on a topological space containing K as a nonempty compact set, constitute the main element of extensions.

In the subsequent constructions, we shall have to use nets [19, p. 96] and not just sequences of elements of a set. As a rule, this circumstance is connected with the cardinality of the respective filter base defining the constraints of asymptotic character. Here we have to introduce the notion of direction [19, p. 95] which requires us, in turn, to recall some familiar simple concepts concerning preorders; we shall do so briefly, referring the interested reader to the special literature (see [35, 58, and others]). Given a set T, a binary relation in T (on T) is any subset of the product set $T \times T$; if ρ is a binary relation in T, $a \in T$ and $b \in T$, then we assume, as usual,

$$(a\rho b) \Leftrightarrow ((a, b) \in \rho).$$

The binary relation ρ on the set T is called a preorder if it is: (1) transitive, $\forall\, a \in T,\, b \in T,\, c \in T$ we have

$$((a\rho b)\&(b\rho c)) \Rightarrow (a\rho c),$$

this definition of transitivity is quite traditional [19, p. 95]; (2) reflexive, i.e., $t\rho t$ for $t \in T$. The preorder ρ in the set T is said to be a (partial) order if ρ is antisymmetric, i.e., $\forall\, a \in T,\, b \in T$

$$((a\rho b)\&(b\rho a)) \Rightarrow (a = b).$$

In the sequel, we shall most often use the natural order $\leqslant$ in $\mathbf{R}$ and also the pointwise order in the space of functionals with a common domain. If H is a nonempty set,

$$u\colon H \to \mathbf{R}, \qquad v\colon H \to \mathbf{R},$$

then, by definition, we identify the inequality $u \leq v$ with the property $u(h) \leqslant v(h)$ for $h \in H$. Generally, for the preordered set (T, ρ), in the case of $a\rho b$, where $a \in T$ and $b \in T$, we say that b follows a in the sense of ρ (a is smaller than or equal to b with respect to ρ); if $\widetilde{T}$ is a subset of T, $x \in T$, then x is said to be a majorant (minorant) of $\widetilde{T}$ with respect to ρ if one has $t\rho x$ for $t \in \widetilde{T}$ (one has $x\rho t$ for $t \in \widetilde{T}$). The nonempty preordered set (T, ρ) is said to be directed if $\forall\, a \in T$, $b \in T$, the set $\{a, b\}$ has (in T) a majorant with respect to ρ (i.e., a ρ-majorant); in this case the preorder ρ is called a direction in T (on T). If (D, ρ) is a directed set, H is an arbitrary nonempty set,

$$h \colon D \to H, \qquad\qquad (2.2.11)$$

then the triple (D, ρ, h) is called a net in H. This notation is a little redundant and we use it for purely methodical reasons (in the literature, it may occur that a net is denoted by a pair (ρ, h) since D can be uniquely determined from h as the domain of the mapping mentioned above; sometimes we can confine ourselves to writing the function h itself if it is clear on what directed set it is defined). Nets play for us a special role, particularly sequences with values in topological spaces. If (D, ρ) is a directed set (recall that this presupposes that D is nonempty), (H, τ) is a topological space, h is the operator (2.2.11), and $x \in H$, then the net (D, ρ, h) is said to be convergent to x in (H, τ) if, for any choice of a τ-neighborhood X of x, one can find $d \in D$ such that $\forall\, \delta \in D$

$$(d\rho\delta) \Rightarrow (h(\delta) \in X). \qquad\qquad (2.2.12)$$

In this case, it is said about (2.2.12) that the net (D, ρ, h) lies in X from some moment (f.s.m.). We shall also use this terminology.

Henceforth, an important role will be played by the Birkhoff theorem [19, p. 97], allowing one to identify the closure of the arbitrary set A in the corresponding topological space with the set of various generalized limits of nets in A. Moreover, we shall widely use the representation of the continuity property in terms of convergence of nets, namely, if (T, τ) and (S, θ) are nonempty topological spaces, $g \colon T \to S$, then the operator g is continuous with respect to the topologies τ and θ if and only if for any choice of the net (D, ρ, h) in T and the point $t \in T$ the τ-convergence of (D, ρ, h) to t (convergence in (T, τ)) implies the θ-convergence of the net $(D, \rho, g \circ h)$ to $g(t)$. In the sequel, we shall use this representation without any additional explanation; its obvious justification can be found, e.g., in [19, 54].

If (D, ρ) is a directed set and M is a subset of D, then M is said to be cofinal (with (D, ρ)) if

$$\forall\, x \in D \quad \exists\, y \in M: x\rho y.$$

It is easy to see that if M is cofinal, then the pair (M, ρ_M), where $\rho_M \triangleq \rho \cap (M \times M)$, is a directed set. Other applications of cofinality will be considered somewhat later, when we introduce the notion of a subnet. The notion of a directed product is also rather important for the sequel. Namely, if (D_1, ρ_1) and (D_2, ρ_2) are directed sets, then the (nonempty) set $D_1 \times D_2$ can also be supplied with a direction $\rho_1 \odot \rho_2$, for which $\forall\, x_1 \in D_1,\ x_2 \in D_2,\ y_1 \in D_1,\ y_2 \in D_2$, by definition,

$$((x_1, x_2)\rho_1 \odot \rho_2(y_1, y_2)) \Leftrightarrow ((x_1\rho_1 y_1)\&(x_2\rho_2 y_2)). \tag{2.2.13}$$

We shall need definition (2.2.13) to investigate generic compactificators. Given directed sets (D_1, ρ_1), (D_2, ρ_2), and $\varphi\colon D_1 \to D_2$, the operator φ is said to be isotonic if $\forall\, x \in D_1,\ y \in D_1$

$$(x\rho_1 y) \Rightarrow (\varphi(x)\rho_2\varphi(y)). \tag{2.2.14}$$

We call φ isotonic-cofinal if φ is an isotonic operator and the φ-image of D_1 is a set cofinal to (D_2, ρ_2). The last term is not traditional but is convenient for our purposes from the standpoint of one important characteristic of the compactness property. Namely, if (K, τ), $K \neq \varnothing$, is a compact space and (D, ρ, h) is a net in K, then one can find a point $x \in K$, a directed set (Δ, r), $\Delta \neq \varnothing$, and an isotonic-cofinal (with respect to (Δ, r) and (D, ρ)) operator

$$\varphi\colon \Delta \to D,$$

for which the net $(\Delta, r, h \circ \varphi)$ converges to x in (K, τ). In the sequel, we shall call this procedure thinning out an arbitrary net in a compact space to a convergent subnet (we shall not consider any more general procedure of selecting a subnet). Here we have a certain analogy with the process of thinning out a sequence to a convergent subsequence. This ends our short digression into topology; however, we shall supplement our knowledge of the subject in the sequel when needed.

2.3. Compactificator of an Extremal Problem with Constraints of Asymptotic Character

We shall consider, in abstract form, the procedure of reducing extremal problems subjected to perturbations to a certain generalized form of a "standard" extremal problem on a compact space. For this purpose, we recall some familiar notions, fixing in this section the nonempty set X as the main solution space. Henceforth, we shall denote by $\mathbf{B}(X)$ the set of all bounded functionals from $\mathbf{R}^X$; the latter symbol is used to denote the set of all functionals on X. We define in $\mathbf{R}^X$ (and, hence, in $\mathbf{B}(X)$) the linear operations and carry out the multiplication pointwise. By a linear subspace of $\mathbf{B}(X)$ we shall mean any nonempty set H, $H \subset \mathbf{B}(X)$, for which

$$(\forall\, a \in \mathbf{R},\ g \in H\colon \alpha g \in H)\&(\forall\, g \in H,\ h \in H\colon g + h \in H).$$

In this section, we shall assume, if the contrary is not specified, that the objective functionals of the extremal problems considered later are elements of $\mathbf{B}(X)$. In other words, we shall deal with minimization problems (we shall consider these settings) of bounded functionals on X, and the selection of a specific point $x \in X$ will obey some other constraints corresponding to the respective specific problem. At the same time, when perturbations are introduced into the setting of the problem, these constraints become fuzzy and acquire asymptotic character. We shall now discuss this situation at a formalized level with the aim of isolating an abstract "block" which can be "built into" various substantive problems as an essential part. This relates to the substantive constructions of Chap. 1.

Thus, we are going to replace the minimization problems

$$s(x) \to \inf, \qquad x \in X_1, \tag{2.3.1}$$

where $s \in \mathbf{B}(X)$ and $X_1 \subset X$, by some of their asymptotic analogs.

Let $\mathcal{X}$ be a filter base (we shall confine ourselves to this case) on the set X. We introduce the notion of an $\mathcal{X}$-admissible net in X, defining it as an arbitrary net $(D, \angle, d)$, $d \in X^D$, which $\forall\, A \in \mathcal{X}$ lies in A f.s.m. (see Sec. 2.2); here and henceforth, to denote directions, we shall try to use the notation that is natural for the theory of ordered spaces and to avoid, as far as possible, the letter notation.

DEFINITION 2.1. If $s \in \mathbf{B}(X)$ and $\mathcal{X}$ is a filter base on X, we shall interpret as an approximate $(s, \mathcal{X})$-solution every $\mathcal{X}$-admissible net $(D, \angle, d)$ in X for which the net $(D, \angle, s \circ d)$ converges to a number in the sense of the natural $|\cdot|$-topology $\tau_{\mathbf{R}}$ of the number line $\mathbf{R}$.

The condition of convergence in this definition is not essential. However, this agreement provides some conveniences, enabling one to associate every approximate $(s, \mathcal{X})$-solution $(D, \angle, d)$ with a natural quality index determined by means of the generalized limit of the net $(D, \angle, s \circ d)$.

DEFINITION 2.2. If $s \in \mathbf{B}(X)$ and $\mathcal{X}$ is a filter base on X, then we say that the number $c \in \mathbf{R}$ is $(s, \mathcal{X})$-attainable if there is an $\mathcal{X}$-admissible net $(D, \angle, d)$ in X such that $(D, \angle, s \circ d)$ converges to c in $(\mathbf{R}, \tau_{\mathbf{R}})$.

In other words, the $(s, \mathcal{X})$-attainable numbers are the generalized limits of nets $(D, \angle, s \circ d)$ "along" various possible approximate $(s, \mathcal{X})$-solutions. Now it is natural to identify the process of optimization with the minimization of $(s, \mathcal{X})$-attainable numbers. In fact, if $s \in \mathbf{B}(X)$ and $\mathcal{X}$ is a filter base on X, then we denote by $\Sigma_0(s, \mathcal{X})$ the (nonempty and bounded, as one can easily see) set of all $(s, \mathcal{X})$-attainable numbers:

$$(\mathcal{X} - \min)[s] \triangleq \inf(\Sigma_0(s, \mathcal{X})) \in \mathbf{R}; \qquad (2.3.2)$$

we are going to use the quantity (2.3.2) instead of the values for problems of the type (2.3.1). For a number of reasons, it is appropriate to represent (2.3.2) in terms of the solution of a "standard" extremal problem, or, more precisely, the minimum problem for a continuous functional defined on a compact space, under constraints in the form of a closed subset of this space. This representation was realized in a rather general form in [34], although, essentially, similar representations have been used for a long time in various parts of the theory of extremal problems (see, for example, the problems of optimal control [4]). We shall discuss this construction [34] very briefly for a sketchy illustration of a more general procedure presented below and concerning the cone optimization problem. Thus, in this section we speak about "replacing" the process of determining the quantity (2.3.2) by solving problems of the type (2.2.10). Following [34], we introduce this procedure as universal within the respective linear subspace $\mathbf{B}(X)$. This approach allows us to consider, together with the investigation of all kinds of instabilities related to the perturbation of the system of constraints, the question of dependence of the optimal result and the solutions providing it on the objective function (functional) of the initial problem if the perturbations of this objective function "occur" in the linear subspace $\mathbf{B}(X)$ (corresponding to the compactificator). In more complicated cases of asymptotic cone optimization, we confine ourselves to individual compactificators.

In the sequel, we use the standard notation $v \circ u$ to denote the superposition of the functions $u: A \to B$ and $v: B \to C$, where A, B, and C are nonempty sets; $v \circ u: A \to C$, and with this, $(v \circ u)(a) = v(u(a))$, $a \in A$.

Given a linear subspace H of $\mathbf{B}(X)$, we say that an H-compactificator is any triple (Y, τ, m), where (Y, τ) is a nonempty compact space and $m \in Y^X$ satisfies the following two conditions: (1) by virtue of the operator m, the image $m^1(X) \triangleq \{m(x): x \in X\}$ (here and below we use this notation for the image of an operator without any additional explanations) of the set X is an everywhere dense, in the sense of (Y, τ), subset of Y; (2) $\forall\, h \in H$ $\exists g \in \mathbf{C}(Y, \tau)$

$$h = g \circ m.$$

The functional $g = g_h$ in the last relation is uniquely determined by $h \in H$ so that every H-compactificator (Y, τ, m) generates an operator $h \mapsto g_h$ acting from H into $\mathbf{C}(Y, \tau)$ and henceforth denoted by $R_H^0[Y, \tau, m]$ (this operator is actually an isometric isomorphism of H onto a linear subspace of $\mathbf{C}(Y, \tau)$ in the sense of natural sup-norms induced from $\mathbf{B}(X)$ and $\mathbf{B}(Y)$ respectively); this defines the construction of the transformations of possible objective functionals into generalized criteria used in extension constructions. Furthermore, we define a transformation of asymptotic constraints into standard ones by the law

$$\mathcal{X} \mapsto \bigcap_{U \in \mathcal{X}} \mathrm{cl}\,(m^1(U), \tau). \qquad (2.3.3)$$

In relation (2.3.3), we are concerned with an operator mapping nonempty families of subsets $\mathcal{X}$ into τ-closed subsets of Y, where (Y, τ, m) is an H-compactificator for some linear subspace H of the "ambient" space $\mathbf{B}(X)$. It is clear that in reality we shall apply (2.3.3) to families $\mathcal{X}$ possessing some specific properties that are essential for problems of asymptotic optimization. We shall use for $\mathcal{X}$ a filter base on X; thus in this section $\mathcal{X}$ is a nonempty family of nonempty subsets of X such that $\forall\, A \in \mathcal{X}, B \in \mathcal{X}$ $\exists\, C \in \mathcal{X}: C \subset A \cap B$. These conditions on the choice of $\mathcal{X}$ are typical of a large variety of practical problems that satisfy a sort of consistency condition (it is formally expressed by the requirement $\varnothing \notin \mathcal{X}$); this property of "asymptotic" consistency allows us, however, to discuss the question of computing the value of the problem. In more general cases of cone optimization this condition will not be imposed until it becomes essential; then we shall introduce an appropriate notation. In this preliminary section we avoid detailed formalization and notation connected with set-theoretic concepts.

Thus, for every choice for $\mathcal{X}$ of a filter base on X, we get, as a result of operation (2.3.3), a τ-closed subset of Y.

Till the end of this section we fix for H a linear subspace of $\mathbf{B}(X)$; for (Y, τ, m) we fix an H-compactificator (we suppose that the triple (Y, τ, m) is somehow defined); and for $\mathcal{X}$ a filter base on X. Under these conditions, the intersection-set on the right-hand side of (2.3.3) is nonempty. Thus, we have again, for this intersection (see (2.2.8)), the nonempty set

$$\bigcap_{U \in \mathcal{X}} \mathrm{cl}\left(m^1(U), \tau\right) \in \mathcal{F}_\tau, \qquad (2.3.4)$$

which is consequently τ-compact; the set (2.3.4) plays the role of an admissible set, or a set of admissible elements for the generalized problems of minimization of the continuous functionals $R^0[Y, \tau, m](h) \triangleq R_H^0[Y, \tau, m](h)$, $h \in H$. Here, in particular, we once again emphasize the universality of the compactificator (Y, τ, m) for investigating the whole complex of problems of asymptotic optimization of the functionals $h \in H$. The following theorem plays a key role in the sequel.

THEOREM 2.3.1. *Given $h \in H$. The optimization problem for the functional h with asymptotic constraints $\mathcal{X}$ is equivalent with respect to the result to the generalized problem*

$$R^0[Y, \tau, m](h)(y) \to \min, \qquad y \in \bigcap_{U \in \mathcal{X}} \mathrm{cl}\left(m^1(U), \tau\right), \qquad (2.3.5)$$

so that

$$(\mathcal{X} - \min)[h] = \min_{y \in \bigcap_{U \in \mathcal{X}} \mathrm{cl}\left(m^1(U), \tau\right)} R^0[Y, \tau, m](h)(y).$$

The proof of this theorem is an easy combination of familiar concepts of general topology; these constructions will be considered in detail for cone asymptotic optimization problems. For the time being we confine ourselves to a formal discussion of the statement of the theorem. Under our "constraints" $\mathcal{X}$ every functional $h \in H$ generates an asymptotic optimization problem, whose meaning is well clarified by (2.3.2) and by Definition 2.2; every problem of this kind is associated with a "standard" problem (2.3.5), which always possesses an optimal solution.

REMARK. In Definition 2.1, the approximate solution is subjected, in addition to the essential condition of $\mathcal{X}$-admissibility, to the condition of convergence of the image net. However, as one can easily verify, for any choice of an $\mathcal{X}$-admissible net $(D, \angle, d)$, we have $\forall s \in \mathbf{B}(X),\ \varepsilon \in (0, \infty)$

$$(\mathcal{X} - \min)[s] - \varepsilon < (s \circ d)(\delta)$$

f.s.m. This allows us to treat (2.3.2) as an asymptotic "quality" attainable in the limit without any additional stipulations.

Theorem 2.3.1 realizes a generalized representation of the value. However, in the most general case of the problem of optimizing the values of h

"by the filter" generated by the base $\mathcal{X}$, the compactificator also provides the construction of an asymptotically optimal solution (more precisely, an approximate solution) [34, Sec. 5]. We shall not discuss it now in detail (this will be done for a more general cone optimization problem). We just observe that this optimal approximate solution can be related (for an appropriate choice of parameters) to any point of minimum of the generalized problem (2.3.5) according to a quite specific law which is not used in the main part of this monograph. At the same time, this shows that the set of solutions of problem (2.3.5) characterizes, to some extent, the "totality" of the approximate solution-nets possessing the property of asymptotic optimality. It turns out that this characterization is exhaustive "up to a subnet" [34, p. 26]. Namely, the set of all (optimal) solutions of problem (2.3.5) exactly coincides with the set of all points $y \in Y$ for each of which one can indicate an approximate $(h, \mathcal{X})$-solution $(D, \angle, d)$ which is optimal in the sense that $(D, \angle, h \circ d)$ converges to $(\mathcal{X} - \min)[h]$ in $(\mathbf{R}, \tau_{\mathbf{R}})$ and satisfies, in addition, the property of convergence of the net $(D, \angle, m \circ d)$ to y in (Y, τ). This can be proved by using the well-known procedure of thinning out an arbitrary net with values in a compact space to a convergent subnet by means of an appropriate isotonically cofinal operator (see Sec. 2.2). This construction also clarifies the inference on the exhaustive character of identifying the optimal set of problem (2.3.5) and the "totality" of all asymptotically optimal approximate $(h, \mathcal{X})$-solutions; any solution $(D, \angle, d)$ of this kind can be "thinned out" by the indicated method, without losing the property of (asymptotic) optimality, to an approximate $(h, \mathcal{X})$-solution $(T, \ll, d \circ r)$, where $(T, \ll)$ is a nonempty directed set possessing the property of convergence of the net $(T, \ll, m \circ d \circ r)$ to some point $y \in Y$. Here,

$$r \colon T \to D$$

is an operator isotonically cofinal in the sense of $(T, \ll)$ and $(D, \angle)$. The characteristic of the optimal set for problem (2.3.5) can be used to obtain some indirect characteristic of the stability of the "totality" of asymptotically optimal approximate $(h, \mathcal{X})$-solutions upon the variation of $h \in H$. For a similar dependence of the asymptotic value we can note the (computational) stability under the variation of the objective function: $\forall h_1 \in \mathbf{B}(X)$, $h_2 \in \mathbf{B}(X)$

$$|(\mathcal{X} - \min)[h_1] - (\mathcal{X} - \min)[h_2]| \leqslant \|h_1 - h_2\|_X, \qquad (2.3.6)$$

where $\|\cdot\|_X$ denotes the sup-norm in $\mathbf{B}(X)$. Relation (2.3.6) can be applied, in particular, to the case $h_1 \in H$, $h_2 \in H$ and expresses the simplest form of computational stability. For an indirect characterization of stability of the approximate solutions themselves, we note the following important property of upper semicontinuity of the (multivalued) map

$$\mathbf{M}: H \to 2^Y, \tag{2.3.7}$$

defined by the following condition: $\forall\, h \in H$

$$\mathbf{M}(h) \triangleq \left\{ y_0 \in \bigcap_{U \in \mathcal{X}} \mathrm{cl}\,(m^1(U), \tau) \,\middle|\, \forall\, y \in \bigcap_{U \in \mathcal{X}} \mathrm{cl}\,(m^1(U), \tau) : \right.$$

$$\left. R^0[Y, \tau, m](h)(y_0) \le R^0[Y, \tau, m](h)(y) \right\}. \tag{2.3.8}$$

Namely, let $\mathcal{U}_H(X)$ be the metric topology H generated by the norm induced from $(\mathbf{B}(X), \|\cdot\|_X)$. Then, as is easy to check, the following theorem holds.

THEOREM 2.3.2. *Mapping* (2.3.7), (2.3.8) *is upper semicontinuous in the following traditional sense: for any choice of the point $h^0 \in H$ and the neighborhood G^0 of the set* $\mathbf{M}(h^0)$ *in the space (Y, τ) there exists a neighborhood G_0 of h^0 in the space $(H, \mathcal{U}_H(X))$ for which*

$$\bigcup_{h \in G_0} \mathbf{M}(h) \subset G^0.$$

We omit the proof of this theorem since it reduces to the use of standard techniques of modern topology connected with "thinning out" a net to a convergent subnet.

To conclude the section, we observe that various special cases of the general setting of the problem of asymptotic optimization of the values of the functional $h \in H$ under "constraints" $\mathcal{X}$ were discussed in [34, 39, 48]. In particular, in [34, p. 21] conditions are given under which (at the expense of introducing a certain agreement between $\mathcal{X}$ and (Y, τ, m)) it becomes possible to simplify conceptually the general procedure of constructing asymptotically optimal approximate solutions. In the subsequent chapters we shall advance in this direction even further, namely, for the class of extremal problems under consideration the corresponding optimal approximate net-solution will be constructed explicitly. The problem of sufficiency of the class of sequential approximate solutions was considered in [49]. For compactificators that are not using the f.a. measures, see [4, 55]; there are some other papers, too (see, for example, [20, 21, 32], where some elements of compactification were used in game control problems). Finally, the question of equivalence of exact and approximate solutions was considered in various modifications in connection with different classes of extremal problems (see, for example, [7]). In particular, in the typical problems of convex programming with a finite number of constraints of inequality type, a sufficient condition for this equivalence is the well-known Slater condition (and other regularity conditions), which results in the absence of duality break. In Chap. 9, we shall consider, for nonconvex problems, some other sufficient conditions for this equivalence (note here [34], where one more general type of conditions of this kind is suggested).

2.4. Some Notions of Set Theory. Products of Topological Spaces

The requirements of modern extremal problems often force us to introduce settings that are characterized, to some extent, by being multicriteria. We came across such a situation in Sec. 1.4, where we had to minimize an infinite-dimensional, generally speaking, vector of deviations with respect to a given set in the space of phase states. In the special case of the finite set Q of Sec. 1.4, we have a multicriteria problem (in other terms, a vector optimization problem). Starting with Sec. 2.5, we shall consider questions concerning the "regularization" of these problems; in this case, the object of perturbation will be, as before, the constraints of the problem, and these perturbations will again have the meaning of weakening of its conditions. The main attention is henceforth paid to the cone asymptotic optimization problem; we have already noted that the control problem of Sec. 1.4 provides an example of such a problem. However, for a number of reasons, it is appropriate to widen the respective questions by using some fragments of the theory of extremal problems with values in preordered spaces (see, in particular, [8]). This is because of some conceptual factors; in the course of exposition, we shall clarify some statements of the preceding section as well. At the same time, this exposition will concern some rather abstract constructions, and this will require the introduction of some notation connected with set-theoretic concepts. In the sequel, we shall not abuse the aforementioned general notation and agreements; they are mainly used in this chapter.

Given a set S, we denote by $\mathcal{P}(S)$ (by 2^S) the family of all (all nonempty) subsets of S and by $\mathrm{Fin}\,(S)$ the family of all finite subsets from 2^S. Given sets A and B, we denote by B^A, as before, the set of all operators acting from the set A into the set B; if $g \in B^A$ and $C \in \mathcal{P}(A)$, then $(g|C) \in B^C$ is, by definition, the constriction of g to C: $(g|C)(x) \triangleq g(x)$ for $x \in C$. Note that we regard the function $f \in B^A$, where A and B are sets, as a binary relation $f \in \mathcal{P}(A \times B)$ such that $\forall\, a \in A$ the set $\{b \in B|\ (a,b) \in f\}$ is one-element (one-point). In this case, we can also use the traditional notation $f\colon A \to B$. If we accept the choice axiom: A and B are nonempty sets, and

$$\widetilde{F}\colon A \to 2^B,$$

then the product set

$$\prod_{a \in A} \widetilde{F}(a) \triangleq \{f \in B^A |\ \forall\, \alpha \in A\colon f(\alpha) \in \widetilde{F}(\alpha)\}$$

is nonempty. The use of the quantifiers $(\exists, \forall)$, the connectives $(\vee, \&, \Rightarrow, \Leftrightarrow)$, and the special symbols def (by definition) and $\triangleq$ (equal by definition) is not specified since it is quite traditional. In logically homogeneous propositions with a repetition of one and the same quantifier, we shall write it only once, replacing the other entries by commas. This does not lead to ambiguity but shortens the principal assertions, so that, for instance, expressions of the form

$$\exists\, a \in A,\ b \in B(a),\ldots \quad \text{instead of} \quad \exists\, a \in A\ \exists\, b \in B(a)\ldots \qquad (2.4.1)$$

are used in the sequel without additional explanations. Here, however, we think it necessary to express a reservation: in some cases we have to speak about choosing an arbitrary set possessing a certain property, and the collection of objects possessing this property cannot always be regarded as a set (for instance, the expression "the set of all sets" is contradictory from the standpoint of logic). Therefore, the use of quantifiers should be stated explicitly compared to (2.4.1); similar conventions can be found, for instance, in [23]. We shall agree, first of all, that the expression $S[A \neq \varnothing]$ (any other letter may be used for A) stands for the statement: A is a nonempty set. Then the expressions

$$\underset{B}{\forall}\ S[B \neq \varnothing], \qquad \underset{B}{\exists}\ S[B \neq \varnothing] \qquad (2.4.2)$$

will be used to denote the statements: "for any nonempty set B," "there exists a nonempty set B," respectively (any other letter may stand for B). Finally, a need may arise for combining expressions of the form (2.4.1), (2.4.2); for example, in the sequel we may choose expressions of the form

$$\underset{A}{\exists}\ S[A \neq \varnothing]\ \exists\, a \in A,\ b \in B(a),\ldots, \qquad (2.4.3)$$

where we repeat the relevant quantifier ($\exists$ in this case) when joining the constituent expressions (2.4.1) and (2.4.2) despite the fact that no other quantifiers appear in (2.4.3). To illustrate (2.4.3), we give an expression characterizing the Birkhoff theorem.

Given a nonempty set T (i.e., $\mathrm{S}[T \neq \varnothing]$), we denote by $(\mathrm{DIR})\,[T]$ the set of all directions on T; moreover, if (Θ, θ) is a topological space ($\Theta \neq \varnothing$), $\ll\, \in\, (\mathrm{DIR})\,[T]$, $f \in \Theta^T$, $y \in \Theta$, then the (somewhat redundant from the standpoint of symbolics) expression

$$(T, \ll, f) \xrightarrow{\theta} y$$

will denote the following statement: the net $(T, \ll, f)$ is convergent to y in topology θ. Now, the Birkhoff theorem (see Sec. 2.2) can be briefly written as follows: if (Θ, θ) is a topological space and $A \in \mathcal{P}(\Theta)$, then

$$\mathrm{cl}\,(A, \theta) = \{x \in \Theta|\ \underset{B}{\exists}\ \mathrm{S}[B \neq \varnothing]\,\exists\ \ll\, \in\, (\mathrm{DIR})\,[B],\ f \in A^B \colon (B, \ll, f) \xrightarrow{\theta} x\}.$$
$$(2.4.4)$$

Below we give one more representation which is important for the extension constructions. It is similar to (2.4.4); we again speak about the representation of a set of limit elements along approximate net-solutions admissible from the standpoint of various asymptotic constraints. Now, however, we shall give some more notation and natural agreements that are important for the sequel.

Everywhere in the sequel, we denote by $\mathcal{N}$ the set of all positive integers $\mathcal{N} \triangleq \{1, 2, \ldots\}$. Given $k \in \mathcal{N}$, we denote by $\overline{1, k}$ the set of all $i \in \mathcal{N}$ such that $i \leqslant k$. If A is a set, $k \in \mathcal{N}$, then we set, as usual,

$$A^k \triangleq A^{\overline{1,k}};$$
$$(2.4.5)$$

relation (2.4.5) defines the Cartesian power of A; elements of A^k are all mappings of the "segment" $\overline{1, k}$ into A. If, in this case, $A = \mathbf{R}$ in (2.4.5), then this relation defines the k-dimensional arithmetic space.

The notion of Tychonoff product of topological spaces is very important for the subsequent constructions. For our purposes, it is sufficient to consider products (generally speaking, infinite) of copies of the same topological space, i.e., Tychonoff powers. Moreover, in almost all subsequent constructions the space-factors will be identified with a properly topologized real line $\mathbf{R}$. For the sake of completeness, we shall introduce this traditional notion in conformity with this case. Recall also the important notion of a base of a topological space.

If T is a nonempty set and β is a nonempty family of subsets of T, then β is called a (topological) base, or basis, of T if: (1) the union of all sets from β equals T; (2) $\forall\, U \in \beta,\, V \in \beta,\, x \in U \cap V$

$$\{W \in \beta \mid (x \in W)\&(W \subset U \cap V)\} \neq \varnothing.$$

We shall not consider now the most familiar examples of topological bases used in the theory of metric spaces but shall confine ourselves to a special construction, useful both for describing the Tychonoff product and for representating $*$-weak topologies, which are quite important for the sequel. However, let us first recall how a topology is generated by a base.

Given a nonempty set T, we denote by (BAS) $[T]$ the set of all topological bases on T; here $\forall\, \beta \in$ (BAS) $[T]$

$$\sqcup(\beta) \triangleq \left\{ \bigcup_{G \in \alpha} G \colon \alpha \in \mathcal{P}(\beta) \right\}$$

is the topology on the set T (generated by the base β) containing β as a subset. We shall also need some new symbols to denote the topological notions introduced in Sec. 2.2. Given a set H, we denote by (top) $[H]$ the (nonempty) family of all topologies H possessing the property that the intersection of any nonempty subfamily of (top) $[H]$ is its element, i.e., a topology on H. When some property is shared by some topologies but not by other topologies on H, this essential circumstance allows one to introduce, in some cases, the weakest topology on H among those sharing this property.

Given two nonempty topological spaces (U, τ) and (V, θ), we denote by $C(U, \tau, V, \theta)$ the set of all operators from V^U continuous in the sense of τ, θ. We are frequently interested in constructing the weakest topology for which all operators from a given set are still continuous, namely, if the sets U and V are nonempty, ϑ is a topology on V, and $\mathcal{H} \in \mathcal{P}(V^U)$, then the property of the family (top) $[U]$ stated above allows one to correctly define the unique topology $t^0(U, V, \vartheta, \mathcal{H}) \in$ (top) $[U]$ for which

$$(\mathcal{H} \subset C(U, t^0(U, V, \vartheta, \mathcal{H}), V, \vartheta))\&(\forall\, \tau \in \text{(top)}\,[U]\colon$$

$$(\mathcal{H} \subset C(U, \tau, V, \vartheta)) \Rightarrow (t^0(U, V, \vartheta, \mathcal{H}) \subset \tau));$$

this topology $t^0(U, V, \vartheta, \mathcal{H})$ is the weakest one in (top) $[U]$ with respect to which all operators from $\mathcal{H}$ remain continuous. Note that if $\mathcal{H}$ is a one-element set, i.e., $\mathcal{H} = \{h\}$ $(h \in V^U)$, then the following representation holds:

$$t^0(U, V, \vartheta, \mathcal{H}) = t^0(U, V, \vartheta, \{h\}) = \{h^{-1}(G) \colon G \in \vartheta\}. \tag{2.4.6}$$

If $\mathcal{H}$ is not a one-element set, then the construction of the above weakest topology becomes somewhat harder; nevertheless, it is not difficult to indicate a topological base generating this topology. Namely, $\underset{U}{\forall}\ \mathbf{S}[U \neq \varnothing]$ $\underset{V}{\forall}\ \mathbf{S}[V \neq \varnothing]\ \forall\,\vartheta \in (\mathrm{top})\,[V],\ \mathcal{H} \in \mathcal{P}(V^U),\ \mathcal{H} \neq \varnothing$:

$$\beta^0(U, V, \vartheta, \mathcal{H}) \triangleq \left\{ T \in \mathcal{P}(U) \,\middle|\, \exists\, m \in \mathcal{N},\ (h_i)_{i \in \overline{1,m}} \in \mathcal{H}^m, \right.$$

$$\left. (G_i)_{i \in \overline{1,m}} \in \vartheta^m \colon T = \bigcap_{i=1}^{m} h_i^{-1}(G_i) \right\} \in (\mathrm{BAS})\,[U] \qquad (2.4.7)$$

(the analogy between (2.4.6) and (2.4.7) is quite clear), and

$$\sqcup(\beta^0(U, V, \vartheta, \mathcal{H})) = t^0(U, V, \vartheta, \mathcal{H}). \qquad (2.4.8)$$

Relations (2.4.7) and (2.4.8) provide constructive means for constructing the weakest topology of the set U relative to which all operators from $\mathcal{H}$ are continuous. We shall use (2.4.7), (2.4.8) to construct the Tychonoff degree of $(\mathbf{R}, \tau)$, where $\tau \in (\mathrm{top})\,[\mathbf{R}]$, by making the parameters of (2.4.8) duly specific. To do this, we shall use as (V, ϑ) in (2.4.8) the space $(\mathbf{R}, \tau)$, as U a set of functionals with a fixed domain, and as $\mathcal{H}$ the set "composed" of all evaluation functionals of the elements of U at the points of their domain. Let $\underset{Z}{\forall}\ \mathbf{S}[Z \neq \varnothing]\ \forall\, z \in Z$:

$$(z - \mathrm{proj}\,)[Z] \triangleq (g(z))_{g \in \mathbf{R}^Z}.$$

This relation defines the functional on $\mathbf{R}^Z$, which we shall call, as usual [19], the projection or evaluation at z and interpret as

$$g \mapsto g(z) \colon \mathbf{R}^Z \to \mathbf{R}.$$

Now we can easily specify (2.4.7) and (2.4.8) because $\underset{Z}{\forall}\ \mathbf{S}[Z \neq \varnothing]\ \forall\, \tau \in (\mathrm{top})\,[\mathbf{R}]$:

$$\beta_{\otimes}^0(Z, \tau) \triangleq \beta^0(\mathbf{R}^Z, \mathbf{R}, \tau, \{(z - \mathrm{proj}\,)[Z] \colon z \in Z\}) \in (\mathrm{BAS})\,[\mathbf{R}^Z]. \qquad (2.4.9)$$

Of course, the topology that we want to introduce on $\mathbf{R}^Z$ can be defined as generated by some other bases [19, 54], but, from our point of view, it is representation (2.4.9) which is the most convenient for us, and we shall confine ourselves to it. Now, taking into account (2.4.8) and (2.4.9), we have $\underset{Z}{\forall}\ \mathbf{S}[Z \neq \varnothing]\ \forall \tau \in (\text{top})[\mathbf{R}]$:

$$\otimes^Z(\tau) \triangleq \sqcup(\beta_\otimes^0(Z, \tau)) = t^0(\mathbf{R}^Z, \mathbf{R}, \tau, \{(z - \text{proj})[Z] :\ z \in Z\}) \in (\text{top})[\mathbf{R}^Z]. \tag{2.4.10}$$

In the sequel, we shall be interested in only two copies of the topological space

$$(\mathbf{R}^Z, \otimes^Z(\tau)), \tag{2.4.11}$$

which we shall call the Tychonoff degree of $(\mathbf{R}, \tau)$. These copies are obtained by specifying the parameter τ in (2.4.10) and (2.4.11): (1) $\tau = \tau_\mathbf{R}$; (2) $\tau = \tau_\partial$ (see the notation in Sec. 2.2). We can take any nonempty sets as Z in (2.4.10) and (2.4.11) and, in particular, any nonempty families of sets. If Z is a nonempty set and τ is a Hausdorff topology in $\mathbf{R}$ (i.e., $(\mathbf{R}, \tau)$ is a Hausdorff space), then (2.4.11) is a Hausdorff space.

To conclude this section, we shall give, without detailed justification, some other properties of the space (2.4.11) in the case (1); the topology $\otimes^Z(\tau_\mathbf{R})$, where Z is a nonempty set, is usually called a topology of pointwise convergence on $\mathbf{R}^Z$. We shall be concerned with the familiar properties of this topology, given far from in full generality in the form of reference material for the subsequent chapters. The first fundamental property concerns compactness and is a corollary of the well-known Tychonoff theorem [19, 54]; the second one, with which we shall begin, concerns the representation of open sets. Namely, $\underset{T}{\forall}\ \mathbf{S}[T \neq \varnothing]\ \forall G \in \otimes^T(\tau_\mathbf{R}),\ g \in G\ \exists m \in \mathcal{N},\ \varepsilon \in (0, \infty),\ (t_i)_{i \in \overline{1,m}} \in T^m$:

$$\{h \in \mathbf{R}^T\ |\ \forall j \in \overline{1, m}:\ |g(t_j) - h(t_j)| < \varepsilon\} \subset G. \tag{2.4.12}$$

We shall make repeated use of (2.4.12) in the proofs of Chap. 5. The following important property has a rather general character: if Z is a nonempty set, $\tau \in (\text{top})[\mathbf{R}]$, and $(\mathbf{R}, \tau)$ is a Hausdorff space, then the family of all compact (see Sec. 2.2), in the sense of (2.4.11), subsets of $\mathbf{R}^Z$ and the family of all sets

$$F \in \mathcal{F}_{\otimes^Z(\tau)}, \tag{2.4.13}$$

for each of which $\forall z \in Z$ the set $\text{cl}\,(\{f(z) : f \in F\}, \tau)$ is compact in $(\mathbf{R}, \tau)$, coincide.

To formulate an important special case concerning the topology of pointwise convergence, we introduce the symbol $\leq$ to denote the pointwise ordering in the spaces of functionals with common domain, so that $\underset{Z}{\forall}\ \mathbf{S}[Z \neq \varnothing]$

$\forall g \in \mathbf{R}^Z$, $h \in \mathbf{R}^Z$ by definition:

$$(g \leq h) \Leftrightarrow (\forall z \in Z\colon g(z) \leq h(z)).$$

Then the assertion concerning the necessary and sufficient conditions for the set (2.4.13) to be compact implies the following property (here and henceforth we use the traditional notation for infinite Cartesian products): if U and V are nonempty sets and

$$(W_u)_{u\in U}\colon U \to \mathcal{P}(V),$$

then, as usual,

$$\prod_{u\in U} W_u \triangleq \{g \in V^U \,|\, \forall u \in U\colon g(u) \in W_u\};$$

by the axiom of choice, this product-set is nonempty provided that $\forall t \in U\colon W_t \neq \varnothing$. Namely, if Z is a nonempty set, $\alpha \in \mathbf{R}^Z$, $\beta \in \mathbf{R}^Z$, $\alpha \leq \beta$, then the product-set

$$\prod_{z\in Z} [\alpha(z), \beta(z)] = \{g \in \mathbf{R}^Z \,|\, \forall z \in Z\colon \alpha(z) \leqslant g(z) \leqslant \beta(z)\} \qquad (2.4.14)$$

is nonempty and compact in the topological space (2.4.11) for $\tau = \tau_{\mathbf{R}}$. We do not make direct use of this property of products (2.4.14) in the sequel; however, the statement of the Alaoglu theorem (for the special case of f.a. measures) on sufficiency, given in Sec. 3.4, is proved by an argument using the compactness inference (2.4.14) [9, p. 458]. Thus, this assertion, arising from the Tychonoff theorem, is, essentially, the basis of all subsequent constructions of compactification.

2.5. The Problem of Asymptotic Optimization in a Preordered Space and Its Extension

We shall consider one rather general extension scheme fixing for brevity: (1) two nonempty sets X and Ω, (2) an operator $s \in \Omega^X$, (3) a nonempty family $\mathcal{X}$ of subsets of X such that $\forall A \in \mathcal{X}$, $B \in \mathcal{X}$,

$$\{C \in \mathcal{X} \,|\, C \subset A \cap B\} \neq \varnothing$$

(the case $\varnothing \in \mathcal{X}$ is not excluded), (4) a topology $\theta \in (\text{top})[\Omega]$. The following general assertion is valid under these conditions:

$$\bigcap_{U\in\mathcal{X}} \mathrm{cl}\,(s^1(U), \theta) = \{\omega \in \Omega \,|\, \underset{T}{\exists}\, \mathrm{S}[T \neq \varnothing] \,\exists\, < \in (\mathrm{DIR})[T],\, h \in X^T\colon$$

$$(\forall L \in \mathcal{X}: \{m \in T \mid \forall t \in T: (m < t) \Rightarrow (h(t) \in L)\} \neq \varnothing)\&$$

$$((T, <, s \circ h) \xrightarrow{\theta} \omega)\}. \tag{2.5.1}$$

The proof of (2.5.1), which uses the construction of directed products (see Sec. 2.2 and (2.4.4)), is quite obvious, and therefore is omitted. This equality is important for us. On one hand, it bears on the problems of Chap. 6 connected with the study of the attainability of elements of Ω in the class of approximate solutions. On the other hand, it will play an important role in the constructions of asymptotic optimization; this second circumstance will be considered in this section.

If T is a nonempty set and $\ll$ is a preordering (see Sec. 2.2) on T, then we shall call the pair $(T, \ll)$ a preordered space. We shall set $\forall H \in \mathcal{P}(T)$

$$(\ll - \mathrm{MIN})[H] \triangleq \{u \in H \mid \forall v \in H: (v \ll u) \Rightarrow (u \ll v)\}, \tag{2.5.2}$$

obtaining the set of all minimal, in $(T, \ll)$, elements of H. We shall identify the process of minimization on H in the preordered space $(T, \ll)$ with the search for set (2.5.2); this setting is more "realistic" than the search for $\ll$-smallest elements of H, which can often be absent.

Given a topological space (Y, τ), $Y \neq \varnothing$, and a preordered space $(T, \ll)$, we set

$$\mathbf{C}_{\downarrow}(Y, \tau, T, \ll) \triangleq \{g \in T^Y \mid \forall t \in T: \{y \in Y \mid g(y) \ll t\} \in \mathcal{F}_\tau\}. \tag{2.5.3}$$

The elements of the set (2.5.3) will be called lower semicontinuous operators transforming (Y, τ) into $(T, \ll)$. The Zorn lemma [9, Chap. 1] implies the following theorem.

THEOREM 2.5.1. *Let* (Y, τ), $Y \neq \varnothing$, *be a compact space and* $(T, \ll)$ *a preordered space;* $g \in \mathbf{C}_{\downarrow}(Y, \tau, T, \ll)$, $F \in \mathcal{F}_\tau$, $F \neq \varnothing$. *Then*

$$(\ll - \mathrm{MIN})[g^1(F)] = \{z \in g^1(F) \mid \forall y \in F: (g(y) \ll z) \Rightarrow (z \ll g(y))\} \neq \varnothing.$$

In fact, Theorem 2.5.1 is a theorem of existence of an efficient solution; we note [1] among assertions of this kind for special cases of optimization problems.

We now complete the sequence $(X, \Omega, s, \mathcal{X}, \theta)$ by the preorder $\overline{\overline{<}}$ on Ω, thus obtaining, in particular, the space $(\Omega, \overline{\overline{<}}, \theta)$, which is simultaneously preordered and topological. Under these conditions, we may regard the set

$$(\overline{\overline{<}} - \mathrm{MIN})\left[\bigcap_{U \in \mathcal{X}} \mathrm{cl}\,(s^1(U), \theta)\right] \in \mathcal{P}(\Omega) \tag{2.5.4}$$

as an asymptotic "value" (here we must bear in mind (2.5.1)). The set (2.5.4) is regarded in the sequel as the value of the corresponding minimization problem. Clearly, efficient approximate solutions may be defined as nets $(T, \angle, h)$ in X for which: (1) $\forall U \in \mathcal{X}$ we have $h(t) \in U$ f.s.m.; (2) the net $(T, \angle, s \circ h)$ converges in the sense of (Ω, θ) to a point of the set on the left-hand side of (2.5.4). In this section, we shall concentrate our attention on the problem of determining the asymptotic value (2.5.4). To do this, we return to the problem of constructing asymptotically efficient approximate solutions in the case of optimization problems in pointwise ordering, which will be considered in the next section. Throughout the remainder of this section, we fix a nonempty compact space (Y, τ). Moreover, we assume, till the end of this section, that the topological space (Ω, θ) is Hausdorff, unless otherwise specified. The next statement is not connected with the introduction of $\overset{=}{<}$ and is valid without the assumption that Ω is endowed with the structure of a preordered space.

THEOREM 2.5.2. *Given $m \in Y^X$, let $g \in \Omega^Y$ be an operator continuous with respect to the topological spaces (Y, τ) and (Ω, θ) and satisfying the condition $s = g \circ m$. Then*

$$\bigcap_{U \in \mathcal{X}} \mathrm{cl}\left(s^1(U), \theta\right) = g^1\left(\bigcap_{U \in \mathcal{X}} \mathrm{cl}\left(m^1(U), \tau\right)\right).$$

In this theorem, the question is of the asymptotically attainable elements of Ω; in this sense it is, in fact, the basis for a number of assertions of Chap. 6. However, now we shall use this theorem for constructing generalized extremal problems. The proof of this theorem reduces to standard topological arguments. In particular, we use the procedure of thinning out an arbitrary net with values in a compact space to a convergent subnet. This argument is not complicated and is omitted in this exposition; in Sec. 6.2, we· shall prove some conceptually analogous statements, so that the interested reader can easily restore the scheme of the proof of Theorem 2.5.2.

COROLLARY. *Let all conditions of Theorem 2.5.2 hold. Moreover, let $g \in \mathbf{C}_{\downarrow}(Y, \tau, \Omega, \overset{=}{<})$, and the family $\mathcal{X}$ be a filter base on X. Then the asymptotic value (2.5.4) is nonempty.*

The proof of the corollary is obvious; its assertion provides a theorem of existence of asymptotically efficient solutions; this theorem has, however, a somewhat conditional character since in the hypothesis we assume the existence of a 4-tuple (Y, τ, m, g) satisfying a number of conditions. Note that we can combine the conditions of continuity of the operator g with respect to the spaces (Y,τ) and (Ω,θ) and of lower semicontinuity with respect to (Y,τ) and $(\Omega,\overset{=}{<})$ if we require a certain consistency in $(\Omega,\overset{=}{<},\theta)$. Indeed, the following theorem is true.

THEOREM 2.5.3. *Let* $(\Omega,\overset{=}{<},\theta)$ *be a preordered topological* (*not necessarily Hausdorff*) *space satisfying the following consistency condition:* $\forall\,\omega\in\Omega$

$$\{\nu\in\Omega|\ \nu\overset{=}{<}\omega\}\in\mathcal{F}_\theta. \tag{2.5.5}$$

Then $C(Y,\tau,\Omega,\theta)\subset \mathbf{C}_{\downarrow}(Y,\tau,\Omega,\overset{=}{<})$.

The proof follows directly from the definition of continuity and is therefore omitted. Note, as an obvious corollary, the natural theorem of existence of a solution for the generalized problem.

THEOREM 2.5.4. *Let* (1) $(\Omega,\overset{=}{<},\theta)$ *be a space not necessarily Hausdorff as* (Ω,θ) *is but consistent in the sense of* (2.5.5), (2) $g\in C(Y,\tau,\Omega,\theta)$, (3) $F\in\mathcal{F}_\tau$, $F\neq\varnothing$. *Then*

$$(\overset{=}{<}-\mathrm{MIN})[g^1(F)]\neq\varnothing.$$

The proof reduces to combining Theorems 2.5.1 and 2.5.3.

THEOREM 2.5.5. *Suppose that* $(\Omega,\overset{=}{<},\theta)$ *is Hausdorff as* (Ω,θ) *is and consistent in the sense of* (2.5.5). *Furthermore, suppose that* $m\in Y^X$, *the operator* $g\in C(Y,\tau,\Omega,\theta)$ *satisfies the condition* $s=g\circ m$, *and* $\mathcal{X}$ *is a filter base on* X. *Then the asymptotic value* (2.5.4) *is nonempty.*

To prove the theorem, it suffices to consider the corollary to Theorem 2.5.2 and Theorem 2.5.3. Theorem 2.5.5 provides a version of the conditional existence theorem which is more convenient for the sequel. We see that if $(\Omega,\overset{=}{<},\theta)$ is Hausdorff and consistent and $\mathcal{X}$ is a filter base on X, then the existence of a 4-tuple-compactificator (Y,τ,m,g), where (Y,τ), $Y\neq\varnothing$, is a compact space, g is a continuous operator $(Y,\tau)\to(\Omega,\theta)$ such that $s=g\circ m$, implies the existence of at least one asymptotically efficient approximate solution (see (2.5.1)).

To conclude the section, we shall consider a question that is not closely connected with the subsequent exposition but, at the same time, is essential from the standpoint of traditional settings of extremal problems. We shall speak about the so-called "conventional" value, or the value of an extremal problem in the class of "exact" solutions. We identify the latter with the elements of the intersection of all sets of the family $\mathcal{X}$ by setting

$$X_0 \triangleq \bigcap_{U\in\mathcal{X}} U$$

(X_0 is the set of all exact solutions). We encountered concepts of this kind in Sec. 1.4, where we considered an example. We shall now discuss these questions in the general case. Let us consider the problem

$$s(x) \to [\overline{\overline{<}} - \text{MIN}], \qquad x \in X_0,$$

understanding its solution as the minimization of values $s(x)$, $x \in X_0$, in the preordered space $(\Omega, \overline{\overline{<}})$. Till the end of this section we suppose that the topological space (Ω, θ) is Hausdorff and that the preordered topological space $(\Omega, \overline{\overline{<}}, \theta)$ is compatible in the sense of (2.5.5). Furthermore, we assume to be given a compact space (Y, τ), $Y \neq \varnothing$, and operators $g \in C(Y, \tau, \Omega, \theta)$ and $m \in Y^X$ for which $s = g \circ m$. Let us first consider the question of how to define a "conventional" solution. This definition really requires an explanation because the set X is not equipped with any "nice" topological structure, nor has the set X_0 properties that guarantee the existence of efficient solutions and minimal elements of the set $s^1(X_0)$. Let us consider, however, the set

$$Z_0 \triangleq \text{cl}\,(s^1(X_0), \theta)$$

as an "inessential" modification of the set $s^1(X_0)$ of all admissible (from the standpoint of the X_0-constraint) estimates. The elements of Z_0 can be regarded as estimates that are almost attainable in the class of exact solutions, or as almost admissible estimates. Defining the notion of conventional value, it is natural to require its nontriviality in the case of a nonempty X_0 (the consistency of a conventional problem); having replaced $s^1(X_0)$ by its closure, we have made a step in this direction. The nontriviality should be understood in the sense that when the "value" is somehow identified with a set, then this notion is naturally reduced to the condition of nonemptiness of this "value" (clearly, this question does not arise in the problem of scalar optimization, where, in case of nonemptiness of an admissible set, the "conventional" value is always defined as the greatest lower bound of the image of this admissible set). We shall show that this requirement is satisfied if we identify the "conventional" value with the set

$$(\overline{\overline{<}} - \text{MIN})[Z_0] \in \mathcal{P}(Z_0). \tag{2.5.6}$$

Note that under our assumptions

$$s^1(X_0) = (g \circ m)^1(X_0) = \{(g \circ m)(x)\colon x \in X_0\}$$

$$= \{g(m(x))\colon x \in X_0\} = \{g(y)\colon y \in m^1(X_0)\} = g^1(m^1(X_0)). \tag{2.5.7}$$

Let us now observe the following important fact: the map g is closed [19] like any continuous map from a compact space into a Hausdorff one. This property means that $\forall H \in \mathcal{F}_\tau: g^1(H) \in \mathcal{F}_\theta$. But in this case we have $\forall T \in \mathcal{P}(Y)$

$$\mathrm{cl}\,(g^1(T), \theta) = g^1(\mathrm{cl}\,(T, \tau)). \qquad (2.5.8)$$

Taking into account (2.5.7), we take the set $m^1(X_0)$ for T in relation (2.5.8), so that

$$Z_0 = \mathrm{cl}\,(g^1(m^1(X_0)), \theta) = g^1(\mathrm{cl}\,(m^1(X_0)), \tau). \qquad (2.5.9)$$

We see that the "conventional" value (2.5.6) coincides with the set

$$(\overline{\overline{<}} - \mathrm{MIN}\,)[g^1(\mathrm{cl}\,(m^1(X_0), \tau))],$$

which, in turn, can be regarded as the "value" of the minimization problem in $(\Omega, \overline{\overline{<}})$ of the elements $g(y)$, $y \in \mathrm{cl}\,(m^1(X_0), \tau)$. This determines the role of the set (2.5.9) in finding the value of a "conventional" problem. Of course, the value is identified with the corresponding set of minimal elements, which is typical of optimization problems in spaces with order. Let us note, however, that the set $\mathrm{cl}\,(m^1(X_0), \tau) \in \mathcal{F}_\tau$ is obviously nonempty if X_0 is nonempty. But in this case, as one can see from Theorem 2.5.4, the following implication holds:

$$(X_0 \neq \varnothing) \Rightarrow ((\overline{\overline{<}} - \mathrm{MIN}\,)[g^1(\mathrm{cl}\,(m^1(X_0), \tau))] \neq \varnothing). \qquad (2.5.10)$$

Now we obtain, combining (2.5.9) and (2.5.10),

$$(X_0 \neq \varnothing) \Rightarrow ((\overline{\overline{<}} - \mathrm{MIN}\,)[Z_0] \neq \varnothing). \qquad (2.5.11)$$

Relation (2.5.11) shows that the required nontriviality property is obtained (we mean the property of the set (2.5.6) assumed as a "conventional" value, i.e., the value of the problem of minimizing the elements $s(x)$, $x \in X_0$). Note that we have already used this construction in Sec. 1.4 when defining V_1; to be more precise, we used definition (2.5.6). Moreover, this question was discussed rather explicitly for the respective example of a control problem at the substantive level.

Let us also observe that $s^1(X_0) \subset s^1(U)$, $U \in \mathcal{X}$, by the definition of X_0. As a result $\forall U \in \mathcal{X}: Z_0 \subset \mathrm{cl}\,(s^1(U), \theta)$. Thus we get

$$Z_0 \subset \bigcap_{U \in \mathcal{X}} \mathrm{cl}\,(s^1(U), \theta). \qquad (2.5.12)$$

Let us take an arbitrary

$$z_0 \in (\overline{\overline{<}} - \mathrm{MIN}\,)[Z_0] \setminus (\overline{\overline{<}} - \mathrm{MIN}\,)\left[\bigcap_{U \in \mathcal{X}} \mathrm{cl}\,(s^1(U), \theta)\right].$$

Then, by (2.5.12), we can find an element ω_0 of the set (2.5.1) for which $\omega_0 \overline{\overline{<}} z_0$ but the condition $z_0 \overline{\overline{<}} \omega_0$ no longer holds. This means that the result z_0 can be improved in passing to the asymptotic setting. Actually, we can make a stronger inference. We ought to take into account the easily verifiable fact that $\forall K \in \mathcal{P}(\Omega)$, $k \in K$:

$$(\overline{\overline{<}} - \mathrm{MIN}\,)[\{t \in K \mid t \overline{\overline{<}} k\}] \subset (\overline{\overline{<}} - \mathrm{MIN}\,)[K]. \tag{2.5.13}$$

The following is an immediate consequence of (2.5.13):

$$\forall K \in \mathcal{F}_\tau,\ t \in g^1(K)\quad \exists r \in (\overline{\overline{<}} - \mathrm{MIN}\,)[g^1(K)]\colon r \overline{\overline{<}} t; \tag{2.5.14}$$

property (2.5.14) has the meaning of a certain "intrinsic" stability of $g^1(K)$. Now it is easy, again using Theorem 2.5.2, to verify the property having the meaning of improvability of the conventional value by the asymptotic one, namely, with due account of the definitions of X_0 and Z_0, we can show the validity of the following statement.

THEOREM 2.5.6. *Let us take an arbitrary*

$$t \in (\overline{\overline{<}} - \mathrm{MIN}\,)[Z_0].$$

Then we have

$$\left\{\tilde{t} \in (\overline{\overline{<}} - \mathrm{MIN}\,)\left[\bigcap_{U \in \mathcal{X}} \mathrm{cl}\,(s^1(U), \theta)\right] \,\middle|\, \tilde{t} \overline{\overline{<}} t\right\} \neq \varnothing.$$

The proof is an immediate consequence of (2.5.14) and is therefore omitted. Note that the meaning of this assertion consists in a certain stability of the asymptotic value and in its sufficiency in the sense of improvability of elements of the conventional value.

2.6. Extension of the Cone Optimization Problem

In this section, on the basis of the general extension constructions of the asymptotic optimization problem in a preordered space, we consider an important special case of this general problem. It is of great practical significance. Namely, we shall deal with problems of cone optimization in case of the cone corresponding to the pointwise ordering of the estimate space. We considered an example of such a setting in Sec. 1.4, where we discussed the control problem with integral constraints. We shall now discuss the general case of this problem. Throughout this section, we fix an arbitrary nonempty set Q. The space $\mathbf{R}^Q$ of all functionals on Q will be used as the set Ω from Sec. 2.5. The preorder $\leqq$ and the topology θ will be specified in terms of respective pointwise definitions; furthermore, we take for $\leqq$ a (partial) order, namely, we shall denote by $\leqq$ the pointwise ordering of $\mathbf{R}^Q$; in other words, $\leqq$ is the set of all ordered pairs (x, y), $x \in \mathbf{R}^Q$, $y \in \mathbf{R}^Q$, for each of which $x \leq y$. In fact, the ordering $\leqq$ is the "constriction" of $\leq$ to $\mathbf{R}^Q$, so that $\forall x \in \mathbf{R}^Q$, $y \in \mathbf{R}^Q$

$$(x \leqq y) \Leftrightarrow (\forall q \in Q\colon x(q) \leqslant y(q));$$

$(\mathbf{R}^Q, \leqq)$ is an ordered space. We shall use the ordering $\leqq$ as the basic relation $\underset{=}{<}$ of Sec. 2.5. Finally, we set $\theta \triangleq \otimes^Q(\tau_{\mathbf{R}})$. The space

$$(\mathbf{R}^Q, \leqq, \otimes^Q(\tau_{\mathbf{R}})) \tag{2.6.1}$$

is consistent in the sense of (2.5.5) and satisfies the separation condition of $(\mathbf{R}^Q, \otimes^Q(\tau_{\mathbf{R}}))$. The specification of the extremal problem of Sec. 2.6 leads to a setting connected with optimization over the cone

$$\mathbf{R}_+^Q \triangleq \{h \in \mathbf{R}^Q \mid \forall q \in Q\colon 0 \leqslant h(q)\}.$$

For such a problem of "order" optimization, which is sufficient for most applications, we intend to establish, under fairly general assumptions, the existence of a compactificator and thus to eliminate the somewhat conditional character of assertions of the preceding section, where we assumed the existence of the 4-tuple (Y, τ, m, g) with properties that are sufficient for constructing a well-defined extension. However, we shall postpone the realization of this aim till Chap. 3, where we shall introduce the necessary notions of the f.a. measure theory. Now we shall confine ourselves to the specification of assertions of Sec. 2.5 and also introduce one special construction of pointwise consistency. The corresponding conditions of such a consistency frequently admit a simple verification and therefore will be used in the main part (Chap. 8) when we study specific questions of the asymptotic insensitivity and stability in cone optimization problems under a perturbation of the system of constraints. Note that the set Q will be preserved as the main parameter of the extremal problems considered in Chaps. 8 and 9, so that the notation of this section will be largely preserved in these chapters. Let $\forall\limits_{M} \mathbf{S}[M \neq \varnothing]$:

$$\mathcal{Q}_M \triangleq H^Q\big|_{H=\mathbf{R}^M} \tag{2.6.2}$$

(the set M will be specified in Chaps. 3, 8, and 9; expression (2.6.2) is actively exploited in replacing M by the respective concrete "value"). Relation (2.6.2) defines, in particular, the set $\mathcal{Q}_X$ of all operators mapping from Q into $\mathbf{R}^X$ (see Sec. 2.5); they are nothing but parametrized families of functionals on X (the elements $q \in Q$ play the role of indices). Every element g of set (2.6.2) and every point $m \in M$ can be associated with a functional on Q, by defining the values of the latter as $g(q)(m)$, $q \in Q$. Namely, let, by definition, $\forall\limits_{M} \mathbf{S}[M \neq \varnothing]$, $\forall g \in \mathcal{Q}_M$,

$$(\text{term})\,[g]\colon M \to \mathbf{R}^Q \tag{2.6.3}$$

be an operator such that $\forall \mu \in M$

$$(\text{term})\,[g](\mu) \triangleq (g(q)(\mu))_{q \in Q}. \tag{2.6.4}$$

It stands to reason that (2.6.3) and (2.6.4) is an inessential transformation of the initial operator g and nothing else. However, it provides some convenience. It is connected with the condition of superposition used in Theorem 2.5.2 and in a number of other assertions of the preceding section. In this connection, let us observe that $\forall\limits_{U} \mathbf{S}[U \neq \varnothing]\,\forall\limits_{V} \mathbf{S}[V \neq \varnothing]\,\forall \alpha \in \mathcal{Q}_U,\ \beta \in \mathcal{Q}_V,$ $h \in V^U$:

$$((\text{term})\,[\alpha] = (\text{term})\,[\beta] \circ h) \Leftrightarrow (\forall q \in Q\colon \alpha(q) = \beta(q) \circ h). \tag{2.6.5}$$

Assertion (2.6.5) can, in particular, be realized under the conditions in which we use for U the set X of the previous section, and for V the set Y endowed with a compact topology. In connection with this last case, we introduce the following definition, namely, $\forall\limits_{T} \mathbf{S}[T \neq \varnothing]\,\forall \nu \in (\text{top})\,[T]$:

$$\mathcal{Q}^0_T(\nu) \triangleq H^Q\big|_{H=\mathbf{C}(T,\nu)} \in \mathcal{P}(\mathcal{Q}_T); \tag{2.6.6}$$

we introduce in (2.6.6) the set of all operators on Q whose values are ν-continuous functionals on T. From the definition of the topology of pointwise convergence it easily follows that $\forall\limits_{T} \mathbf{S}[T \neq \varnothing]\,\forall \nu \in (\text{top})\,[T],\ g \in \mathcal{Q}^0_T(\nu)$:

$$(\text{term})\,[g] \in C(T, \nu, \mathbf{R}^Q, \otimes^Q(\tau_\mathbf{R})). \tag{2.6.7}$$

Relation (2.6.7) will be needed to specify (T, ν) as (Y, τ) (Sec. 2.5). Now it is natural to use (2.6.5) and (2.6.7) in conjunction. In particular, it is useful to exploit this circumstance when checking the conditions of Theorem 2.5.2, if we specify $(\Omega, \overline{\overline{<}}, \theta)$ of Sec. 2.5 in the form (2.6.1). Recall that X is a nonempty set; $\mathcal{X}$ is a nonempty family of subsets of X, which contains, together with every pair of its sets, a subset of its intersection; $\Omega = \mathbf{R}^Q$, $\theta = \otimes^Q(\tau_{\mathbf{R}})$. Suppose that throughout the remainder of this section (Y, τ), $Y \neq \varnothing$, is a fixed compact space and $m \in Y^X$. Finally, we fix in this section $s \in \mathcal{Q}_X$; this modification of the interpretation of the objective operator is inessential from the conceptual point of view, but, as we have already noted, it provides some convenience. We shall now formulate a corollary of Theorem 2.5.2 using (2.6.5) and (2.6.7).

THEOREM 2.6.1. *Let* $g \in \mathcal{Q}_Y^0(\tau)$ *and assume that* $\forall\, q \in Q : s(q) = g(q) \circ m$. *Moreover, let*

$$(S \triangleq (\text{term})\,[s])\,\&\,(G \triangleq (\text{term})\,[g]). \tag{2.6.8}$$

Then

$$\bigcap_{U \in \mathcal{X}} \mathrm{cl}\,(S^1(U), \otimes^Q(\tau_{\mathbf{R}})) = G^1\left(\bigcap_{U \in \mathcal{X}} \mathrm{cl}\,(m^1(U), \tau)\right).$$

The proof is obvious. Note that it is the operator S in the hypothesis of the theorem, which maps X into $\mathbf{R}^Q$, which plays the role of an objective operator in the constructions of Sec. 2.5 (in the statement of Theorem 2.5.2 s must be replaced by S and g by G). The corollary of Theorem 2.5.2 is specified analogously; in this case the preorder $\overline{\overline{<}}$ of Sec. 2.5 is identified with $\leqq$, so that in the form of (2.6.1) we realize here the space $(\Omega, \overline{\overline{<}}, \theta)$ from Sec. 2.5.

THEOREM 2.6.2. *Suppose that all conditions of Theorem 2.6.1 are satisfied and let* $\mathcal{X}$ *be a filter base on* X, *while* S *and* G *correspond to* (2.6.8). *Then*

$$(\leqq - \mathrm{MIN})\left[\bigcap_{U \in \mathcal{X}} \mathrm{cl}\,(S^1(U), \otimes^Q(\tau_{\mathbf{R}}))\right] \neq \varnothing.$$

The proof obviously follows from Theorem 2.5.5 with due account of (2.6.5) and (2.6.7).

Regarding the last two assertions, we note that as one can see from Theorems 2.6.1 and 2.6.2, under the conditions of pointwise consistency (see Theorem 2.6.1) the search for the asymptotic value of the minimization problem of $S(x)$, $x \in X$, in the sense of $\leqq$ under the "constraints" $\mathcal{X}$, can be replaced by solving the generalized problem

$$G(y) \to [\leqq - \mathrm{MIN}], \qquad y \in \bigcap_{U \in \mathcal{X}} \mathrm{cl}\,(m^1(U), \otimes^Q(\tau_{\mathbf{R}})), \qquad (2.6.9)$$

which is always solvable if $\mathcal{X}$ is a filter base on X (S and G are connected with the initial data s and g, respectively, by relations (2.6.8)).

To conclude the section, we shall now briefly discuss the question of constructing asymptotically efficient approximate solutions. We mean relation (2.5.1), where we must make some minor changes that take into account the fact that (under the same X and $\mathcal{X}$)

$$(\Omega, \theta) = (\mathbf{R}^Q, \otimes^Q(\tau_{\mathbf{R}}))$$

and $S =$ (term) $[s]$ of (2.6.8) is used as an objective operator, where $s \in \mathcal{Q}_X$; namely

$$\bigcap_{U \in \mathcal{X}} \mathrm{cl}\,(S^1(U), \otimes^Q(\tau_{\mathbf{R}})) = \Big\{ \omega \in \mathbf{R}^Q \mid \underset{T}{\exists}\, \mathbf{S}[T \neq \varnothing]\, \exists \angle \in (\mathrm{DIR})\,[T],\, h \in X^T$$

$$(\forall L \in \mathcal{X}\colon \{l \in T \mid \forall t \in T\colon (l \angle t) \Rightarrow (h(t) \in L)\} \neq \varnothing)\&$$

$$((T, \angle, S \circ h) \xrightarrow{\ \otimes^Q(\tau_{\mathbf{R}})\ } \omega) \Big\}. \qquad (2.6.10)$$

Asymptotically efficient approximate solutions are defined as $\mathcal{X}$-admissible nets in X (we introduced this notion in Sec. 2.2 for the case where $\mathcal{X}$ is a filter base. However, this definition can be easily extended to more general cases; the required form of this condition is actually contained in (2.6.10)), on which the values of the operator S converge to elements of the asymptotic value. Theorem 2.6.1 shows that in the process of constructing these solutions, problem (2.6.9) may play a major role. In this connection, we shall now discuss some questions related to the efficient solutions of "standard" cone optimization problems. If T is a nonempty set, $H \in \mathcal{P}(T)$, and

$$l\colon T \to \mathbf{R}^Q,$$

then the set

$$(l - \min)[H] \triangleq \{h \in H \mid l(h) \in (\leqq - \mathrm{MIN}\,)[l^1(H)]\}$$

$$= H \cap l^{-1}((\leqq - \mathrm{MIN}\,)[l^1(H)]) \in \mathcal{P}(H) \qquad (2.6.11)$$

obviously possesses the property

$$l^1((l - \min)[H]) = (\leqq - \mathrm{MIN}\,)[l^1(H)]. \qquad (2.6.12)$$

Relation (2.6.11) defines the set of all efficient solutions in the traditional sense; (2.6.12) provides the natural "completeness" property of these solutions in the sense of realization of the value (the set of all minimal elements of the image of H). Modification (2.6.11) is more convenient for our purposes; it is oriented to a parametric specification of the objective operator. Namely, we set $\underset{T}{\forall}\, \mathbf{S}[T \neq \varnothing]\ \forall f \in \mathcal{Q}_T,\ \Lambda \in \mathcal{P}(T)$:

$$(f - \mathrm{Min}\,)[\Lambda] \triangleq ((\mathrm{term})\,[f] - \min)[\Lambda].$$

Then, as one can easily check, $\forall g \in \mathcal{Q}_Y^0(\tau),\ F \in \mathcal{F}_\tau \setminus \{\varnothing\}$

$$(g - \mathrm{Min}\,)[F] \in 2^F. \tag{2.6.13}$$

Observe that in our settings of the asymptotic optimization problem, we must take for F, as one can see from Theorem 2.6.1 (under the condition that $\mathcal{X}$ is a filter base on X), the admissible set of problem (2.6.9). In other words, in the most important case, where $g \in \mathcal{Q}_Y^0(\tau)$ satisfies the condition $s(q) = g(q) \circ m,\ q \in Q$, we must specify (2.6.13) as the set

$$(g - \mathrm{Min}\,)\left[\bigcap_{U \in \mathcal{X}} \mathrm{cl}\,(m^1(U), \tau) \right]. \tag{2.6.14}$$

If $\mathcal{X}$ is a filter base on X, then, as one can see from (2.6.13), the set (2.6.14) is nonempty. The set (2.6.14) consists of generalized efficient solutions of problem (2.6.9). It turns out that every element of the set (2.6.14) can be associated with an asymptotically efficient approximate solution, which is determined, in the general case, by a rather complicated, but quite definite, law which we shall consider now. Here we shall use the familiar construction of directed product (see Sec. 2.2). Throughout the remainder of this section, we suppose that $\forall y \in Y$: $\mathbf{N}_\tau(y)$ is, by definition, the family of all neighborhoods of y in the (compact) topological space (Y, τ); see the respective definitions in Sec. 2.2. Then every family of this kind provides a filter and, in particular, a filter base on Y.

Given an arbitrary family of sets $\mathcal{G}$, we denote by $\dashv (\mathcal{G})$ the binary relation on $\mathcal{G}$ such that

$$\forall U \in \mathcal{G}, \quad V \in \mathcal{G} \text{ def}: \qquad (U \dashv (\mathcal{G})V) \Leftrightarrow (V \subset U).$$

If T is a nonempty set and $\mathcal{G}$ a filter base on T, then $\dashv (\mathcal{G})$ is a direction on $\mathcal{G}$. We can easily apply the construction of products from Sec. 2.2 to directed sets of this kind. Namely, given nonempty sets U and V, let $\mathcal{U}$ be a filter base on U, and $\mathcal{V}$ be a filter base on V. Then

$$\odot[\mathcal{U}, V] \triangleq \dashv (\mathcal{U}) \odot \dashv (\mathcal{V}) \qquad (2.6.15)$$

is a direction on $\mathcal{U} \times \mathcal{V}$, so that

$$(\mathcal{U} \times \mathcal{V}, \odot[\mathcal{U}, V])$$

is a nonempty directed set; in this case $\forall A_1 \in \mathcal{U},\, B_1 \in \mathcal{V},\, A_2 \in \mathcal{U},\, B_2 \in \mathcal{V}$:

$$((A_1, B_1) \odot [\mathcal{U}, V](A_2, B_2)) \Leftrightarrow ((A_2 \subset A_1)\&(B_2 \subset B_1)).$$

Till the end of this section, let $\mathcal{X}$ be a filter base on X. Then $\forall y \in Y$, in particular, we have a nonempty directed set

$$(\mathcal{X} \times \mathbf{N}_\tau(y), \odot[\mathcal{X}, \mathbf{N}_\tau(y)]), \qquad (2.6.16)$$

constructed by formula (2.6.15). In particular, in (2.6.16) we can, in the case of $g \in \mathcal{Q}_Y^0(\tau)$, choose y from the nonempty set (2.6.14). However, $\forall y \in \cap_{U \in \mathcal{X}} \mathrm{cl}\,(m^1(U), \tau),\, A \in \mathcal{X},\, B \in \mathbf{N}_\tau(y)$:

$$A \cap m^{-1}(B) \neq \varnothing. \qquad (2.6.17)$$

We omit the verification of (2.6.17) because it reduces to using the standard definition of closure in (Y, τ). This implies that

$$\forall y \in \bigcap_{U \in \mathcal{X}} \mathrm{cl}\,(m^1(U), \tau)$$

the operator

$$(A, B) \to A \cap m^{-1}(B)\colon \mathcal{X} \times \mathbf{N}_\tau(y) \to 2^X \qquad (2.6.18)$$

is defined for which, by the choice axiom,

$$(\mathrm{Sel})\,[\mathcal{X}, \tau, y] \triangleq \prod_{(A,B) \in \mathcal{X} \times \mathbf{N}_\tau(y)} (A \cap m^{-1}(B)) \neq \varnothing; \qquad (2.6.19)$$

the set (2.6.19) consists of the selectors of the (multivalued) map (2.6.18). If we complete (2.6.16) by selector (2.6.18), i.e., by an element of the set (2.6.19), the result will be a net in X.

THEOREM 2.6.3. *Let $g \in \mathcal{Q}_Y^0(\tau)$ satisfy the condition of pointwise consistency with $s \in \mathcal{Q}_X$, i.e., $\forall q \in Q\colon s(q) = g(q) \circ m$. Let, moreover, y be an element of the set (2.6.14). Then, in the notation of (2.6.8), we have*

$$G(y) \in (\leqq - \mathrm{MIN})\left[\bigcap_{U \in \mathcal{X}} \mathrm{cl}\,(S^1(U), \otimes^Q(\tau_{\mathbf{R}}))\right]. \qquad (2.6.20)$$

If $\rho \in$ (Sel) $[\mathcal{X}, \tau, y]$, then

$$(\mathcal{X} \times \mathbf{N}_\tau(y), \odot[\mathcal{X}, \mathbf{N}_\tau(y)], \rho)$$

is an $\mathcal{X}$-admissible net in X $(\forall\, U_0 \in \mathcal{X} \,\exists\, \zeta \in \mathcal{X} \times \mathbf{N}_\tau(y)\, \forall\, \xi \in \mathcal{X} \times \mathbf{N}_\tau(y):$

$$(\zeta \odot [\mathcal{X}; \mathbf{N}_\tau(y)]\xi) \Rightarrow (\rho(\xi) \in U_0);$$

for more details see Secs. 2.2 and 2.3), for which

$$(\mathcal{X} \times \mathbf{N}_\tau(y), \odot[\mathcal{X}; \mathbf{N}_\tau(y)], S \circ \rho)$$

is a net in $\mathbf{R}^Q$ converging to $G(y)$ in the sense of topology $\otimes^Q(\tau_\mathbf{R})$.

The proof is practically obvious. Relation (2.6.20) follows from Theorem 2.6.1 and the definition of sets (2.6.13) and (2.6.14); in particular, we have to take into account (2.6.11) and (2.6.12). Next, the structure of the net (see (2.6.19))

$$(\mathcal{X} \times \mathbf{N}_\tau(y), \odot[\mathcal{X}; \mathbf{N}_\tau(y)], m \circ \rho) \tag{2.6.21}$$

clearly yields the property of convergence to y in the sense of topology τ. However, G (2.6.8) possesses the property of continuity as an operator

$$(Y, \tau) \to (\mathbf{R}^Q, \otimes^Q(\tau_\mathbf{R}))$$

(see, in particular, (2.6.7)). Therefore, the convergence of (2.6.21) implies the convergence of the net

$$(\mathcal{X} \times \mathbf{N}_\tau(y), \odot[\mathcal{X}; \mathbf{N}_\tau(y)], G \circ m \circ \rho)$$

to $G(y)$ in the sense of $\otimes^Q(\tau_\mathbf{R})$. However, we have $G \circ m \circ \rho = S \circ \rho$ (see (2.6.5)). The other arguments are obvious.

In Theorem 2.6.3, we were concerned with the asymptotic realization of one point (2.6.20) of the asymptotic value, so that in this theorem we have the technique of constructing an "individual" asymptotically efficient approximate solution. We shall now use relation (2.6.12) for describing the technique of asymptotic realization of the entire asymptotic value. For this purpose, it is expedient to use the appropriate specification, assuming, throughout the remainder of this section, that $g \in \mathcal{Q}_Y^0(\tau)$ satisfies the condition of pointwise consistency with s (see Theorem 2.6.3), and the operators S and G are defined by (2.6.8). By virtue of (2.6.12), under these conditions we have the relation

$$G^1\!\left((g - \mathrm{Min}\,)\!\left[\bigcap_{U \in \mathcal{X}} \mathrm{cl}\,(m^1(U), \tau)\right]\right)$$

$$= \left\{ G(y) \colon y \in (g - \text{Min}) \left[\bigcap_{U \in \mathcal{X}} \text{cl}\,(m^1(U), \tau) \right] \right\}$$

$$= (\leqq - \text{MIN}) \left[G^1 \left(\bigcap_{U \in \mathcal{X}} \text{cl}\,(m^1(U), \tau) \right) \right]$$

$$= (\leqq - \text{MIN}) \left[\bigcap_{U \in \mathcal{X}} \text{cl}\,(S^1(U), \otimes^Q(\tau_{\mathbf{R}})) \right]. \tag{2.6.22}$$

One can see from (2.6.22) that all points of the asymptotic value can be realized as elements of $G(y)$ when y runs over the set (2.6.14). But for any element y we already have a construction of an (asymptotically efficient) approximate net-solution $(D, \angle, h)$ in X, along which the net $(D, \angle, \mathbf{S} \circ h)$ converges to $G(y)$. Hence, from the conceptual point of view, it is the set (2.6.14) that contains all the necessary information for the "complete" realization of the asymptotic value in the class of approximate solutions; moreover, in Theorem 2.6.3 we indicate a quite specific way of associating the generalized solutions (points of the set (2.6.14)) with the asymptotically efficient approximate solutions. There are other ways of treating generalized solutions (and we shall use them in the sequel); they may turn out to be simpler in some cases. Some of these possibilities were considered in Sec. 2.3; they require, of course, some additional conditions that we shall not consider now.

Chapter 3

Finitely Additive Measures on a Semialgebra of Sets

3.1. Introduction

In this chapter, we provide some necessary material from the f.a. measure theory which we shall need in the sequel. This is connected with the use of f.a. measures as a sort of material for constructing extensions, which, in turn, is motivated, on one hand, by the possible presence of various kinds of discontinuous dependences in the settings of extremal problems under consideration and, on the other hand, by the existence of nice conditions of compactness following from Alaoglu's theorem [9]. However, such an approach to the problem of extension is not traditional and requires some explanation. To some extent, the need for other (nontraditional) extension of constructions was shown by the last example of Sec. 1.2, whose conditions included the discontinuous functions. It is possible to construct more substantive examples of this kind, by considering, for instance, the control problem of Sec. 1.3, but we will not do so now. Instead, we shall try to characterize more general properties of f.a. measures that will allow us to realize compactifications with due account of possible discontinuous dependences in the conditions of the problem. Let us, however, first, briefly consider some other approaches to extensions, which use measures. First of all, we should mention the so-called Varga controls [4] that close, in a required sense, the space of conventional controls, on which geometric constraints are imposed. The concrete image of these "controls" generating sliding modes in the respective (controlled) system can differ: the controls can be represented as in [4, 6, 20, 21, 55] as functions of time taking values in sets of normalized Borel measures on the set specifying the geometric constraints imposed on the choice of the control vector at any time moment, but they can also be interpreted as measures on the Cartesian product of measurable spaces as in [32, p. 160]. In the latter case (differing from the first representation only in form but not in essence), the generalized controls are identified with linear continuous functionals on $C(K) = C(K, \tau_K)$, where K is a finite-dimensional compact space with the natural topology τ_K of coordinatewise convergence, having the form of a product $K = [t_0, \vartheta_0] \times P$ (here P is a set specifying the geometric constraints on the choice of "instantaneous" controls, $t_0 < \vartheta_0$). We used a similar representation, with an appropriate topologization of the space of generalized controls, in the last example of Sec. 1.2, where we considered the convergence of type (1.2.15); in this case, however, we considered a controllable system with a quite different type of constraint, so that, in fact, the approach of [4, 6, 55] can be used in other situations as well (we mention here an important characterization [53] of extension procedures in terms of convexations).

An essential role in these representations is played by the well-known Riesz theorem on the representation of the space $C^*(K)$ topologically dual to $C(K)$, and also by similar assertions concerning the Caratheodory function space (see [4]). In the sequel, we use somewhat different theorems on the integral representation, although it is easy to trace some analogy with the approach mentioned above. The main difference is in using, instead of $C(K)$, an appropriate space of discontinuous functions [9, Chap. 4], which should, however, meet some requirements connected, in particular, with the so-called universal integrability [35, 58]. We must ensure such an integrability within the corresponding class of measures because they will "replace" the control functions whose choice must be at our disposal. As a result, we arrive at the space of so-called stratum functions that can be uniformly approximated by step (relative to a corresponding space) functions. Let us observe in passing that the elements of $C(K)$ automatically have the property of universal integrability with respect to Radon measures, so that in this problem there is also a certain analogy with the approach of [9, 19, 53]. In the sequel, in the study of f.a. measures, we shall devote much attention to issues connected with the approximation of these measures by indefinite integrals. Essentially, we shall be concerned with constructing some new approximate analogs of the Radon–Nicodym property, which does not hold, in its precise form, in the case of f.a. measures, as is shown in [59]; these issues are considered in the next chapter. This chapter contains a summary of general concepts and the definitions of integrals and of *-weak topology.

3.2.　Semialgebras and Algebras of Sets

In what follows, we shall fix an arbitrary nonempty set E. We shall consider subsets of E and families of subsets of E possessing some special properties. We shall need to use partitions of subsets of E; a special role will be played in the sequel by finite partitions of a subset of E by elements of a given family. Namely, if $m \in \mathcal{N}$, H is a subset of E, and $\mathcal{H}$ is a nonempty family of subsets of E, then we shall denote by $\Delta_m(H, \mathcal{H})$ the set of all maps

$$(H_i)_{i \in \overline{1,m}} \colon \overline{1, m} \to \mathcal{H}, \tag{3.2.1}$$

for each of which we have

$$\left(H = \bigcup_{i=1}^{m} H_i \right) \& (\forall\, p \in \overline{1, m},\ q \in \overline{1, m} \colon (p \neq q) \Rightarrow (H_p \cap H_q = \varnothing)).$$

We have thus introduced the sets of "ordered partitions" (here and henceforth we mean finite partitions), which is emphasized in (3.2.1) by means of a specified numbering. It is also convenient to introduce "unordered" partitions, assuming that for any set (and, in particular, for any family of sets) $\mathbf{M}$: Fin $(\mathbf{M})$ is, by definition, the family of all nonempty finite subsets of $\mathbf{M}$. Namely, if H is a subset of E, $\mathcal{H}$ is a nonempty family of subsets of E, then we can denote by $\mathbf{D}(H, \mathcal{H})$ the family of all $\mathcal{K} \in$ Fin $(\mathcal{H})$ such that

$$\left(\bigcup_{K \in \mathcal{K}} K = H \right) \& (\forall\, A \in \mathcal{K},\ B \in \mathcal{K} \colon (A \cap B \neq \varnothing) \Rightarrow (A = B)).$$

The relationship between these two definitions is obvious, and we shall not dwell on it in greater detail. As usual [25], we regard as a semialgebra (of subsets of E) any family $\mathcal{I}$ of subsets of E such that: (1) $\varnothing \in \mathcal{I}$, (2) $E \in \mathcal{I}$, (3) $\forall\, A \in \mathcal{I}$, $B \in \mathcal{I}$ we have $A \cap B \in \mathcal{I}$, (4) $\forall\, \Lambda \in \mathcal{I}\ \exists\, m \in \mathcal{N}$:

$$\Delta_m(E \setminus \Lambda, \mathcal{I}) \neq \varnothing.$$

The algebra (of subsets of E) is any semialgebra $\mathcal{A}$ of subsets of E such that $\forall\, A \in \mathcal{A}$: $E \setminus A \in \mathcal{A}$. Actually, the difference between these two types of subsets of E is inessential; the transition from the semialgebra $\mathcal{I}$ to the smallest algebra $\mathcal{A}$ still containing $\mathcal{I}$ is very easy and can be described in terms of finite partitions: this algebra $\mathcal{A}$ (generated by $\mathcal{I}$) is the family of all sets A, $A \subset E$, for each of which $\exists\, m \in \mathcal{N}$:

$$\Delta_m(A, \mathcal{I}) \neq \varnothing.$$

Of course, here and in the sequel we are concerned with algebras and semialgebras of subsets of the set E, since the latter is fixed. As usual, the algebra $\mathcal{A}$ is called a σ-algebra if the union of any sequence of sets from $\mathcal{A}$ is an element of $\mathcal{A}$; in this case, we call the pair $(E, \mathcal{A})$ a standard measurable space. Given an algebra $\mathcal{A}$ (of subsets of E) and $m \in \mathcal{N}$,

$$(A_i)_{i \in \overline{1,m}} \colon \ \overline{1, m} \to \mathcal{A},$$

we clearly have

$$\left(\bigcup_{i=1}^{m} A_i \in \mathcal{A} \right) \& \left(\bigcap_{i=1}^{m} A_i \in \mathcal{A} \right).$$

We shall not confine our consideration to the case of standard measurable spaces, nor to the spaces with an algebra of sets. This is due to the use of f.a. measures, for which the procedures of additive extension to the naturally generated standard measurable structure are rather complicated, although the possibility of such an extension is provided by the well-known Hahn–Banach theorem. The difficulties of constructing even individual examples of f.a., but not countably additive, measures on standard measurable spaces are well known. Note, in particular, that in [9], containing, probably, the most extensive summary of basic statements (and their proofs) of the f.a. measure theory, the question is of measures with the domain of definition in the form of an algebra (but, in general, not a σ-algebra) of sets. The procedure of extending a f.a. measure from a semialgebra to the algebra generated by this semialgebra causes no significant difficulties [25]; however, the corresponding concrete constructions may turn out to be rather unwieldy, and, what is most important, often simply unnecessary. The corresponding constructions of the f.a. integration on measurable spaces with semialgebras of sets are rather easy and, in a number of cases, more suitable for practical use. Let us note, as an example, that in Secs. 1.3 and 1.4, for the purposes of exhaustive description of the measurable structure, it suffices to assume that the set E coincides with I, and to identify the corresponding semialgebra (of subsets of $E = I$) with the family of all intervals

$$[\alpha, \beta), \qquad t_0 \leqslant \alpha \leqslant \vartheta_0, \qquad t_0 \leqslant \beta \leqslant \vartheta_0,$$

including the empty one. A more extensive measurable structure may be needed in the examples of Secs. 1.3 and 1.4 if the functions $b(\cdot)$, $s_1(\cdot), \ldots,$ $s_n(\cdot)$, $B(\cdot)$, and $S(\cdot)$ go beyond the scope of the class considered in Chap. 1 (recall that we assumed these functions to be pw.c. and r.c.). At the same time, there may arise concrete situations where standard measurable spaces are needed essentially, in particular, in some probabilistic settings of extremal problems. At any rate, the case of a standard measurable space can be "embedded" into the standard setting, and we shall not isolate it into a special case.

3.3. Finitely Additive Measures

In this section, we shall introduce some initial notions of f.a. measure theory. The definitions of f.a. measures on a measurable space with a semialgebra of sets are conceptually equivalent to those of [25, Sec. 1.6], they are somewhat simpler in the case of a space with an algebra of sets. We denote by $\Pi[E]$ the set of all semialgebras of subsets of E; $\forall \mathcal{I} \in \Pi[E]$:

$$(\text{add})\,[\mathcal{I}] \triangleq \left\{ \mu \in \mathbf{R}^{\mathcal{I}} \,\middle|\, \forall \Lambda \in \mathcal{I},\ m \in \mathcal{N},\ (\Lambda_i)_{i \in \overline{1,m}} \right.$$

$$\left. \in \Delta_m(\Lambda, \mathcal{I}):\ \mu(\Lambda) = \sum_{i=1}^{m} \mu(\Lambda_i) \right\},$$

$$(\text{add})_+ [\mathcal{I}] \triangleq \{\mu \in (\text{add})\,[\mathcal{I}]|\ \forall \Lambda \in \mathcal{I}:\ 0 \leqslant \mu(\Lambda)\}, \tag{3.3.1}$$

$$\mathbf{P}(\mathcal{I}) \triangleq \{\mu \in (\text{add})_+ [\mathcal{I}]|\ \mu(E) = 1\}, \tag{3.3.2}$$

$$\mathbf{T}(\mathcal{I}) \triangleq \{\mu \in \mathbf{P}(\mathcal{I})|\ \forall \Lambda \in \mathcal{I}:\ (\mu(\Lambda) = 0) \vee (\mu(\Lambda) = 1)\}. \tag{3.3.3}$$

If $\mathcal{A}$ is an algebra (of subsets of E), then

$$(\text{add})\,[\mathcal{A}] \triangleq \{\mu \in \mathbf{R}^{\mathcal{A}}|\ \forall U \in \mathcal{A},\ V \in \mathcal{A}:$$

$$(U \cap V = \varnothing) \Rightarrow (\mu(U \cup V) = \mu(U) + \mu(V))\}.$$

The procedure of additive extension of f.a. measures from the semialgebra $\mathcal{I} \in \Pi[E]$ to the algebra $\mathcal{A}$ generated by this semialgebra is simple (we should certainly take into account the representation of $\mathcal{A}$ given in Sec. 3.2): $\forall \mu \in (\text{add})\,[\mathcal{I}]$ the unique f.a. measure $\nu \in (\text{add})\,[\mathcal{A}]$, for which $\mu = (\nu|\mathcal{I})$ is such that $\forall A \in \mathcal{A}$,

$$m \in \mathcal{N},\qquad (\Lambda_i)_{i \in \overline{1,m}} \in \Delta_m(A, \mathcal{I}):\ \sum_{i=1}^{m} \mu(\Lambda_i) = \nu(A). \tag{3.3.4}$$

It is expedient to resort to the procedure of extension defined by (3.3.4), in order to check some properties of f.a. measures on the initial semialgebra of sets (for more details see, e.g., [36, pp. 85–104]). Note, by the way, that this extension preserves the property of positiveness (to be precise, nonnegativity) of the corresponding f.a. measure by moving cone (3.3.1) into $(\mathrm{add})_+ [\mathcal{A}]$. The use of procedure (3.3.4) allows us to check the monotonicity of measures from (3.3.1) on $\mathcal{I}$. Another, no less useful property, connected with the Jordan decomposition, can also be checked by using the scheme of [36, p. 103] with the aid of the indicated extension (3.3.4). We shall not realize these quite obvious constructions now in detail. Let us set $\forall\, \mathcal{I} \in \Pi[E]$:

$$\mathbf{A}(\mathcal{I}) \triangleq \{\mu - \nu \colon (\mu, \nu) \in (\mathrm{add})_+ [\mathcal{I}] \times (\mathrm{add})_+ [\mathcal{I}]\}. \tag{3.3.5}$$

By virtue of the above circumstance concerning the Jordan decomposition, (3.3.5) is the set of all f.a. measures on $\mathcal{I}$ having bounded variation. Clearly, (3.3.5) provides a linear subspace of $\mathbf{R}^{\mathcal{I}}$ generated by cone (3.3.1). In turn, (3.3.2) and (3.3.3) are subsets of this cone, essential for some extension constructions for extremal problems. The elements of (3.3.2) are normalized f.a. measures on $\mathcal{I}$, or f.a. "probabilities." The elements of (3.3.5) are two-valued $(0, 1)$-measures on the semialgebra $\mathcal{I}$. For a number of reasons, these objects can be used as the material for creating rather universal extension constructions (see, for instance, [34]). One of the simplest examples of a f.a. measure, an element of (3.3.3), can be provided by Dirac's measure concentrated at a point $x \in E$ and restricted to $\mathcal{I}$; the indicated (countably additive) measures do not exhaust, generally speaking, (3.3.3) but possess some density properties on this set. To describe these properties, we shall need an adequate topologization of (3.3.5) corresponding to the natural duality (similar to that of [9, Chap. 4]) with respect to the space of stratum functionals. We shall consider these definitions later; now let us recall some basic notions concerning the (pointwise) ordering in $\mathbf{A}(\mathcal{I})$, where $\mathcal{I} \in \Pi[E]$. The question is the specification of well-known concepts of the theory of vector lattices, so that the relevant summary of basic notions will be quite short; we refer the reader for details to [18, 52]. Recall that $\forall\, \mathcal{I} \in \Pi[E]$, $\mu \in \mathbf{A}(\mathcal{I})$, $\nu \in \mathbf{A}(\mathcal{I})$:

$$(\mu \leq \nu) \Leftrightarrow (\forall\, \Lambda \in \mathcal{I} \colon \mu(\Lambda) \leqslant \nu(\Lambda)). \tag{3.3.6}$$

In (3.3.6), we defined the pointwise ordering in $\mathbf{A}(\mathcal{I})$, $\mathcal{I} \in \Pi[E]$. The following two circumstances are essential for the sequel: the agreement of the linear (induced from the space of all functionals on the initial semialgebra of sets, the space being endowed with pointwise linear operations) and the order structures in sets of the form (3.3.5), and the existence of the least upper and greatest lower bounds for the subsets of (3.3.5) satisfying natural conditions of order boundedness, above and below respectively. The first circumstance is clarified by obvious arguments following directly from the definitions, so that (3.3.5) is actually an ordered vector space. As for the second circumstance, we ought to provide some explanations, the more so since it is connected with the familiar Jordan decomposition, which it seems reasonable to recall in terms of the theory of vector lattices.

We suppose that $\forall \mathcal{I} \in \Pi[E]$, $\mu \in (\text{add})[\mathcal{I}]$, $\Lambda \in \mathcal{I}$:

$$(\underset{\Lambda}{\text{VAR}})[\mu] \triangleq \bigcup_{m \in \mathcal{N}} \left\{ \sum_{i=1}^{m} |\mu(\Lambda_i)| : (\Lambda_i)_{i \in \overline{1,m}} \in \Delta_m(\Lambda, \mathcal{I}) \right\}, \qquad (3.3.7)$$

thus obtaining every time a nonempty subset of $\mathbf{R}$. Relation (3.3.5) singles out from $(\text{add})[\mathcal{I}]$ the set of all f.a. measures on $\mathcal{I}$ possessing the property of boundedness of (3.3.7) for $\Lambda = E$; this assertion can easily be checked with the use of the "conventional" Jordan decomposition [9] (in a form specially adjusted to the case of f.a. measures on a semialgebra of sets; the relevant constructions are given in [36, p. 103]). It is easy to verify that the elements of (3.3.5), the f.a. measures of bounded variation, possess the property of boundedness of (3.3.7) as functions of sets on a corresponding semialgebra. This allows us to introduce $\forall \mathcal{I} \in \Pi[E]$, $\mu \in \mathbf{A}(\mathcal{I})$ the functional

$$v_\mu \triangleq (\sup ((\underset{\Lambda}{\text{VAR}})[\mu]))_{\Lambda \in \mathcal{I}} \in \mathbf{R}^{\mathcal{I}}$$

(the variation of μ as a function of sets). If $\mathcal{A}$ is an algebra of subsets of E, then, using a simpler form of representing the additivity for the elements of $(\text{add})[\mathcal{A}]$, mentioned earlier, it is easy to verify [36, pp. 98–99] that $v_\mu \in (\text{add})_+[\mathcal{A}]$ for $\mu \in \mathbf{A}(\mathcal{A})$. Of course, this can be justified by using different methods, and it is convenient to use definitions that are typical of the case of an algebra of sets. Now it is quite easy to go back to measures on a semialgebra, treating them as restrictions of similar measures on the generated algebra of sets, so that, in reality, $\forall \mathcal{I} \in \Pi[E]$, $\mu \in \mathbf{A}(\mathcal{A})$:

$$v_\mu \in (\text{add})_+ [\mathcal{I}], \qquad \mu^+ \triangleq \frac{1}{2}(v_\mu + \mu) \in (\text{add})_+ [\mathcal{I}],$$

$$\mu^- \triangleq \frac{1}{2}(v_\mu - \mu) \in (\text{add})_+ [\mathcal{I}], \qquad \mu = \mu^+ - \mu^-, \qquad v_\mu = \mu^+ + \mu^-.$$
$$(3.3.8)$$

In (3.3.8), we introduced the Jordan decomposition of the f.a. measure μ. Now it is appropriate to construe this representation in terms of ordering (3.3.6). However, we first ought to observe that $\forall \mathcal{I} \in \Pi[E]$ the functional

$$\mu \mapsto v_\mu(E)\colon \mathbf{A}(\mathcal{I}) \to [0,\infty) \tag{3.3.9}$$

is a norm, called in the sequel a (strong) variation-norm. In the sequel, we shall use norm (3.3.9) in constructing the canonical duality with respect to the spaces of stratum functionals on E. Returning to order relations, we first of all give the well-known interpretation of the f.a. measures used in (3.3.8), in terms of the least upper bounds of two-element subsets of (3.3.5). Given $\mathcal{I} \in \Pi[E]$, we denote by $\mathcal{O}_\mathcal{I}$ the measure from $\mathbf{A}(\mathcal{I})$ identically zero on $\mathcal{I}$. Now observe that the operation of extension, characterized by (3.3.4) and transferring the set $(\mathrm{add})[\mathcal{I}]$, $\mathcal{I} \in \Pi[E]$, onto $(\mathrm{add})[\mathcal{A}]$, where $\mathcal{A}$ is the algebra generated by $\mathcal{I}$, preserves linear operations and has the following property: if

$$\mu_1 \in (\mathrm{add})[\mathcal{I}], \quad \mu_2 \in (\mathrm{add})[\mathcal{I}], \quad \nu_1 \in (\mathrm{add})[\mathcal{A}], \quad \nu_2 \in (\mathrm{add})[\mathcal{A}]$$

and the measure ν_1 (measure ν_2) is defined via μ_1 (via μ_2) by a relation similar to (3.3.4), so that $\mu_1 = (\nu_1|\mathcal{I})$ and $\mu_2 = (\nu_2|\mathcal{I})$, then

$$(\mu_1 \leq \mu_2) \Leftrightarrow (\nu_1 \leq \nu_2).$$

We have characterized the extension operator as an order isomorphism. Now we have the opportunity to study the order properties of f.a. measures of bounded variation defined on a semialgebra of sets by invoking their analogs for similar measures on the algebra generated by this semialgebra. These analogs can be established more easily, and their subsequent transfer to the case of measurable spaces, of interest to us, with semialgebras of sets is thus quite simple (observe that this case was considered in detail in [37, 38], to which the interested reader is referred). Among the circumstances simplifying similar constructions in a more traditional case of measurable spaces with an algebra of sets (see, in particular, [9, Chap. 3]), we mention only the following familiar representation: if $\mathcal{A}$ is an algebra of subsets of E, $\mu \in \mathbf{A}(\mathcal{A})$, $A \in \mathcal{A}$, then the nonempty set

$$\{\mu(\tilde{A})\colon \tilde{A} \in \mathcal{A}, \ \tilde{A} \subset A\}$$

is bounded from above in $\mathbf{R}$ and its least upper bound coincides with $\mu^+(A)$.

By virtue of the above arguments, we come, after elementary transformations, to the important conclusion that $\forall \mathcal{I} \in \Pi[E]$, $\mu \in \mathbf{A}(\mathcal{I})$:

$$(\mu \leq \mu^+) \& (\forall \nu \in (\mathrm{add})_+[\mathcal{I}]\colon (\mu \leq \nu) \Rightarrow (\mu^+ \leq \nu)).$$

This relation means that μ^+ is the least (in the sense of $\leq$) majorant of the two-element set $\{\mu; \mathcal{O}_{\mathcal{I}}\}$, i.e., is its least upper bound. Now, using the agreement between the order (3.3.6) and the linear operations in $\mathbf{A}(\mathcal{I})$, it is easy to verify that $\forall \mathcal{I} \in \Pi[E]$, $\mu \in \mathbf{A}(\mathcal{I})$, $\nu \in \mathbf{A}(\mathcal{I})$ we have

$$\mu \vee \nu \triangleq \mu + (\nu - \mu)^+ \in \mathbf{A}(\mathcal{I}):$$

$$(\mu \leq \mu \vee \nu) \& (\nu \leq \mu \vee \nu) \& (\forall \gamma \in \mathbf{A}(\mathcal{I}): ((\mu \leq \gamma) \& (\nu \leq \gamma)) \Rightarrow (\mu \vee \nu) \leq \gamma).$$
$$(3.3.10)$$

In (3.3.10), we have the property of existence (and the concrete structure) of the least upper bound of a two-element subset of $\mathbf{A}(\mathcal{I})$. Now it is quite easy to establish the existence of the greatest lower bound of these sets. It is easier to establish this fact in the case where one of the elements of the unordered pair $\{\mu, \nu\}$ coincides with $\mathcal{O}_{\mathcal{I}}$. Let us first observe that, according to (3.3.7), we have that $\forall \mathcal{I} \in \Pi[E]$, $\mu \in \mathbf{A}(\mathcal{I})$:

$$v_\mu = v_\nu\big|_{\nu=-\mu}, \qquad \mu^- = (-\mu)^+. \tag{3.3.11}$$

Relations (3.3.11) imply, in particular, the relation $\mu^- = (-\mu) \vee \mathcal{O}_{\mathcal{I}}$. Under the conditions defining (3.3.11), this means that the f.a. measure

$$-\nu\big|_{\nu=\mu^-} \in \mathbf{A}(\mathcal{I})$$

is the greatest (in the sense of $\leq$) minorant of $\{\mu, \mathcal{O}_{\mathcal{I}}\}$, i.e., the greatest lower bound of the above two-element set. Now it is easy to indicate the greatest lower bound of an arbitrary two-element subset of (3.3.5). Namely, $\forall \mathcal{I} \in \Pi[E]$, $\mu \in \mathbf{A}(\mathcal{I})$, $\nu \in \mathbf{A}(\mathcal{I})$ we have

$$\mu \wedge \nu \triangleq \mu - (\mu - \nu)^+ \in \mathbf{A}(\mathcal{I}): (\mu \wedge \nu \leq \mu) \& (\mu \wedge \nu \leq \nu) \& (\forall \gamma \in \mathbf{A}(\mathcal{I}):$$

$$((\gamma \leq \mu) \& (\gamma \leq \nu)) \Rightarrow (\gamma \leq \mu \wedge \nu)). \tag{3.3.12}$$

Thus, $\mathbf{A}(\mathcal{I})$ with the pointwise ordering is a (vector) lattice. Moreover, it possesses the important completeness property. This means that any nonempty subset of $\mathcal{A}(\mathcal{I})$ with a majorant (minorant) has the least upper (greatest lower) bound. This is an easy consequence of a similar assertion for a positive cone, which can be checked by analogy with [25], where the case of countably additive measures was considered. Namely, if

$$\mathcal{I} \in \Pi[E], \qquad \varnothing \neq G \subset (\mathrm{add})_+[\mathcal{I}], \tag{3.3.13}$$

and

$$(\forall \mu \in G, \nu \in G: \mu \vee \nu \in G) \& (\{\gamma \in (\mathrm{add})_+[\mathcal{I}]| \forall \zeta \in G: \zeta \leq \gamma\} \neq \varnothing),$$
$$(3.3.14)$$

then $\forall \Lambda \in \mathcal{I}$ the (nonempty) set $\{\xi(\Lambda): \xi \in G\}$ is bounded from above; moreover, (under the conditions of (3.3.13) and (3.3.14)) the functional

$$\Lambda \mapsto \sup(\{\tilde{\mu}(\Lambda): \tilde{\mu} \in G\}): \mathcal{I} \to \mathbf{R}$$

is an element of $(\mathrm{add})_+[\mathcal{I}]$, and, moreover, is a $\leq$-majorant of G. Let us observe that a simple example of set G (3.3.13) immediately satisfying condition (3.3.14) can be obtained by defining G as the set of all nonnegative minorants of the arbitrary nonempty set K, $K \subset (\mathrm{add})_+[\mathcal{I}]$. This is related to the now easily verifiable property of completeness that we spoke about earlier. Given $\mathcal{I} \in \Pi[E]$, we denote by $\mathcal{B}_{\leq}^{\downarrow}(\mathcal{I})$ (by $\mathcal{B}_{\leq}^{\uparrow}(\mathcal{I})$) the family of all nonempty subsets of $\mathbf{A}(\mathcal{I})$ possessing minorants (majorants) in $\mathbf{A}(\mathcal{I})$ with respect to $\leq$. Then: (1) $\forall H \in \mathcal{B}_{\leq}^{\downarrow}(\mathcal{I})$ one can correctly define the unique f.a. measure $\mathrm{Inf}\,(H) \in \mathbf{A}(\mathcal{I})$ such that

$$(\forall \mu \in H: \mathrm{Inf}\,(H) \leq \mu) \& (\forall \nu \in \mathbf{A}(\mathcal{I}): (\forall \zeta \in H: \nu \leq \zeta) \Rightarrow (\nu \leq \mathrm{Inf}\,(H)));$$
$$(3.3.15)$$

(2) $\forall \tilde{H} \in \mathcal{B}_{\leq}^{\uparrow}(\mathcal{I})$ one can correctly define the unique f.a. measure $\mathrm{Sup}\,(\tilde{H}) \in \mathbf{A}(\mathcal{I})$ such that

$$(\forall \mu \in \tilde{H}: \mu \leq \mathrm{Sup}\,(\tilde{H})) \& (\forall \gamma \in \mathbf{A}(\mathcal{I}): (\forall \xi \in \tilde{H}: \xi \leq \gamma) \Rightarrow (\mathrm{Sup}\,(\tilde{H}) \leq \gamma)).$$
$$(3.3.16)$$

In the case of two-element subsets of $\mathbf{A}(\mathcal{I})$, $\mathcal{I} \in \Pi[E]$, which simultaneously are, as one can easily check, sets from $\mathcal{B}_{\leq}^{\downarrow}(\mathcal{I})$ and $\mathcal{B}_{\leq}^{\uparrow}(\mathcal{I})$, definitions (3.3.15) and (3.3.16) reduce to (3.3.12) and (3.3.10) respectively. Let $\forall \mathcal{I} \in \Pi[E]$, $\mu \in \mathbf{A}(\mathcal{I})$, $\nu \in \mathbf{A}(\mathcal{I})$:

$$[\mu, \nu]^{(0)} \triangleq \{\zeta \in \mathbf{A}(\mathcal{I}) \mid (\mu \leq \zeta) \& (\zeta \leq \nu)\}; \qquad (3.3.17)$$

expression (3.3.17) defines the order interval with endpoints μ and ν. Using (3.3.15)–(3.3.17), we can introduce the important notion of a band (component) of $\mathbf{A}(\mathcal{I})$. In this connection, recall that a linear subspace of $\mathbf{A}(\mathcal{I})$, where $\mathcal{I} \in \Pi[E]$, is any nonempty set M, $M \subset \mathbf{A}(\mathcal{I})$, such that

$$(\forall \alpha \in \mathbf{R}, \mu \in M: \alpha\mu \in M) \& (\forall \mu \in M, \nu \in M: \mu + \nu \in M).$$

The components are linear subspaces of $\mathbf{A}(\mathcal{I})$ possessing the properties of solidity and order completeness. The first property is related to intervals of the form (3.3.17) and the second one is related to (3.3.15) and (3.3.16).

Namely, if $\mathcal{I} \in \Pi[E]$ and the set H is a linear subspace of $\mathbf{A}(\mathcal{I})$, then H is a band of $\mathbf{A}(\mathcal{I})$ provided that: (1) $\forall\,\mu \in H$

$$[-v_\mu, v_\mu]^{(0)} \subset H, \tag{3.3.18}$$

(2) $\forall\, M \in \mathcal{B}^\uparrow_{\leq}(\mathcal{I})$ we have

$$(M \subset H) \Rightarrow (\mathrm{Sup}\,(M) \in H). \tag{3.3.19}$$

Property (3.3.18) is called an order solidity and (3.3.19) is called a completeness (we could replace property (2) by requiring the fulfillment of the condition $\mathrm{Inf}\,(\widetilde{M}) \in H$ for $\widetilde{M} \in \mathcal{B}^\downarrow_{\leq}(\mathcal{I}) \cap 2^H$). If H is a band of $\mathbf{A}(\mathcal{I})$, where $\mathcal{I} \in \Pi[E]$, then H, together with the band

$$H_d \triangleq \{\mu \in \mathbf{A}(\mathcal{I})\,|\, \forall\,\nu \in H\colon v_\mu \wedge v_\nu = \mathcal{O}_{\mathcal{I}}\}, \tag{3.3.20}$$

forms a decomposition of $\mathbf{A}(\mathcal{I})$ into an ordered direct sum (observe that (3.3.20) is a band if H in this relation is any nonempty subset of $\mathbf{A}(\mathcal{I})$): every element of $\mathbf{A}(\mathcal{I})$ is uniquely representable as the sum $\mu + \nu$, where $\mu \in H$, $\nu \in H_d$, and $\forall\,\mu_1 \in H$, $\nu_1 \in H_d$, $\mu_2 \in H$, $\nu_2 \in H_d$

$$(\mu_1 + \nu_1 \leq \mu_2 + \nu_2) \Rightarrow ((\mu_1 \leq \mu_2)\&(\nu_1 \leq \nu_2)).$$

As the best-known representation of this kind, recall the Hewitt–Yosida decomposition [9, 18, 57], where H is identified with a band of countably additive measures of bounded variation; in this case, H_d is a band of purely f.a. measures. Another representation of this kind will be considered in the next chapter. We conclude this brief summary of the most elementary properties concerning the order structure of the space of f.a. measures of bounded variation on a semialgebra of sets by observing that $\forall\,\mathcal{I} \in \Pi[E]$, $\mu \in \mathbf{A}(\mathcal{I})$:

$$v_\mu = \mu \vee (-\mu).$$

In the sequel, we shall augment, when necessary, the summary of these properties. Now we shall be concerned with some issues of characterization of $\mathbf{A}(\mathcal{I})$, $\mathcal{I} \in \Pi[E]$, as a topological vector space.

3.4. Measures and Integral Representation of Linear Functionals

We shall now discuss a very important problem of representing $\mathbf{A}(\mathcal{I})$, $\mathcal{I} \in \Pi[E]$, as a topological conjugate of the space of stratum (more precisely,

$\mathcal{I}$-stratum) functionals on E. This representation (see [9, Chap. 4] for a measurable space with an algebra of sets) admits a natural analogy with the theorem of Riesz which was discussed in Sec. 3.1. We shall now give exact definitions using an extremely short manner of presentation; the main attention will be given to the topological structure related to the representation of $\mathbf{A}(\mathcal{I})$, which is important in the sequel.

Given a subset L of E, we denote by χ_L the characteristic function of L: $\chi_L \in \mathbf{B}(E)$, $\chi_L(x) \triangleq 1$ for $x \in L$, $\chi_L(y) \triangleq 0$ for $y \in E \setminus L$. Then $\forall \mathcal{I} \in \Pi[E]$:

$$B_0(E,\mathcal{I}) \triangleq \left\{ g \in \mathbf{R}^E \,\middle|\, \exists\, m \in \mathcal{N},\ (\alpha_i)_{i \in \overline{1,m}} \in \mathbf{R}^m, \right.$$

$$\left. (\Lambda_i)_{i \in \overline{1,m}} \in \Delta_m(E,\mathcal{I}) \colon g = \sum_{i=1}^{m} \alpha_i \chi_{\Lambda_i} \right\} \subset \mathbf{B}(E). \tag{3.4.1}$$

The set on the left-hand side of (3.4.1) is the linear span of $\{\chi_\Lambda \colon \Lambda \in \mathcal{I}\}$ in $\mathbf{R}^E$ with linear operations (and multiplication) defined pointwise, and it plays an important role in the sequel; we shall denote by $B_0^+(E,\mathcal{I})$ the $\leq$-positive cone $B_0(E,\mathcal{I})$. Elements of (3.4.1) will be called $\mathcal{I}$-step functionals on E, or step-functionals on $(E,\mathcal{I})$.

In the sequel, we denote by $\|\cdot\|$ the natural sup-norm in $\mathbf{B}(E)$, so that $\forall g \in \mathbf{B}(E)$: $\|g\| \triangleq \sup(\{|g(x)| \colon x \in E\})$. It turns $(\mathbf{B}(E), \|\cdot\|)$ into a Banach space. Given $\mathcal{I} \in \Pi[E]$, we denote by $B(E,\mathcal{I})$ the closure of $B_0(E,\mathcal{I})$ in $(\mathbf{B}(E), \|\cdot\|)$. In other words, if $\mathcal{U}$ is the $\|\cdot\|$-topology on $\mathbf{B}(E)$, then $\forall \mathcal{I} \in \Pi[E]$:

$$B(E,\mathcal{I}) \triangleq \operatorname{cl}(B_0(E,\mathcal{I}),\mathcal{U}). \tag{3.4.2}$$

Elements of the set on the left-hand side of (3.4.2) will be called $\mathcal{I}$-stratum functionals on E, or stratum functionals on $(E,\mathcal{I})$. In this case, $\forall \mathcal{I} \in \Pi[E]$ the set $B(E,\mathcal{I})$ is a linear subspace of $\mathbf{B}(E)$ with pointwise executed linear operations; inducing the norm from $(\mathbf{B}(E), \|\cdot\|)$ into $B(E,\mathcal{I})$, we get a Banach space whose topological dual will be denoted by $B^*(E,\mathcal{I})$ and positive cone will be denoted by $B^+(E,\mathcal{I})$. The fact that $\mathbf{A}(\mathcal{I})$ and $B^*(E,\mathcal{I})$, where $\mathcal{I} \in \Pi[E]$, are topologically isomorphic is very important for the subsequent exposition; a corresponding isometric isomorphism can be defined by means of the simplest operation of integration [35, p. 75; 36], which we shall briefly explain somewhat later. Now we observe that

$$\mathbf{B}(E) = B(E, \mathcal{P}(E)), \tag{3.4.3}$$

so that the entire $\mathbf{B}(E)$ can be represented as a space of stratum functionals. Relation (3.4.3) defines a certain extreme case, or, to be more precise, the case of the larger space of stratum functionals on E. In the sequel, we shall need some intermediate cases; more precisely, we shall be interested in considering sets (3.4.2) for different values of $\mathcal{I}$. This is mainly due to the use of compactificators from Sec. 2.3, in order to select them for studying the problems of asymptotic optimization with bounded goal functionals. Now we shall confine ourselves to a simple remark about a relationship for sets (3.4.2), which follows from the constructions of Sec. 3.2. Namely, if $\mathcal{I} \in \Pi[E]$ and the family $\mathcal{A}$ is an algebra generated by $\mathcal{I}$ (see Sec. 3.2), then

$$(B_0(E,\mathcal{A}) = B_0(E,\mathcal{I}))\&(B(E,\mathcal{A}) = B(E,\mathcal{I})), \qquad (3.4.4)$$

because from the description of $\mathcal{A}$ in terms of finite $\mathcal{I}$-partitions (see Sec. 3.2) it follows that $\forall\, A \in \mathcal{A}$: $\chi_A \in B_0(E,\mathcal{I})$. Relation (3.4.4) allows us not to distinguish, in essence, the measurable spaces $(E,\mathcal{I})$ and $(E,\mathcal{A})$ from the standpoint of all our subsequent purposes. To this we must add the fact that the case of the measurable space $(E,\mathcal{A})$, where $\mathcal{A}$ is an algebra of subsets of E, is more often treated in the literature. Let us also observe that in the case where $\mathcal{A}$ is not merely an algebra but also a σ-algebra of subsets of E, $B(E,\mathcal{A})$ is the set of all $g \in \mathbf{B}(E)$ such that $\forall\, c_0 \in \mathbf{R}$: $g^{-1}((-\infty,c_0)) \in \mathcal{A}$; in other words, in the case of a standard measurable space $(E,\mathcal{A})$, we have for $B(E,\mathcal{A})$ the set of all bounded $\mathcal{A}$-measurable functionals on E.

Now we shall briefly recall the main elements of the simplest integration construction with respect to a f.a. measure (for more details, see [35, pp. 75–76; 36]). We should bear in mind the equivalence of the definition given below to that of the Darboux integral [35], so that the indicated construction can be easily reduced to constructions using the notions of lower and upper integrals.

Let us fix, till the end of this section, a semialgebra $\mathcal{I} \in \Pi[E]$. Then $\forall\, m \in \mathcal{N}$, $(\alpha_i)_{i \in \overline{1,m}} \in \mathbf{R}^m$, $(P_i)_{i \in \overline{1,m}} \in \Delta_m(E,\mathcal{I})$, $n \in \mathcal{N}$, $(\beta_j)_{j \in \overline{1,n}} \in \mathbf{R}^n$, $(Q_j)_{j \in \overline{1,n}} \in \Delta_n(E,\mathcal{I})$:

$$\left(\sum_{i=1}^{m} \alpha_i \chi_{P_i} = \sum_{j=1}^{n} \beta_j \chi_{Q_j} \right) \Rightarrow$$

$$\left(\forall\, \mu \in (\text{add})\,[\mathcal{I}]: \sum_{i=1}^{m} \alpha_i \mu(P_i) = \sum_{j=1}^{n} \beta_j \mu(Q_j) \right); \qquad (3.4.5)$$

the proof of (3.4.5) is a simple consequence of additivity used for partitions of P_i and Q_j by the sets

$$P_i \cap Q_j \in \mathcal{I}, \qquad (i,j) \in \overline{1,m} \times \overline{1,n}.$$

With due account of (3.4.5), the following definition is correct. Namely, $\forall \mu \in (\mathrm{add})\,[\mathcal{I}]$, $g \in B_0(E,\mathcal{I})$, the number (elementary μ-integral of g)

$$\overset{(el)}{\int_E} g(x)\,\mu(dx) \in \mathbf{R} \tag{3.4.6}$$

is, by definition, such that $\forall k \in \mathcal{N}$, $(a_i)_{i\in\overline{1,k}} \in \mathbf{R}^k$, $(\Lambda_i)_{i\in\overline{1,k}} \in \Delta_k(E,\mathcal{I})$ we have

$$\left(g = \sum_{i=1}^k a_i \chi_{\Lambda_i}\right) \Rightarrow \left(\overset{(el)}{\int_E} g(x)\,\mu(dx) = \sum_{i=1}^k a_i \mu(\Lambda_i)\right).$$

As a result of this construction, we get a linear functional on $B_0(E,\mathcal{I})$, associating each element of this set with the number (3.4.6). If $\mu \in \mathbf{A}(\mathcal{I})$, then this functional is uniformly continuous on $B_0(E,\mathcal{I})$ with respect to $\|\cdot\|$, and this allows us to prove the following simple property: if $g \in B(E,\mathcal{I})$,

$$(h_i)_{i\in\mathcal{N}}\colon \mathcal{N} \to B_0(E,\mathcal{I}), \qquad (s_i)_{i\in\mathcal{N}}\colon \mathcal{N} \to B_0(E,\mathcal{I}),$$

then the following implication is true:

$$((\|h_i - g\| \to 0)\,\&\,(\|s_i - g\| \to 0)) \Rightarrow$$

$$\left(\left|\overset{(el)}{\int_E} h_i(x)\,\mu(dx) - \overset{(el)}{\int_E} s_i(x)\,\mu(dx)\right| \to 0\right). \tag{3.4.7}$$

In turn, taking account of the completeness of $\mathbf{R}$, (3.4.7), in fact, provides the following definition: $\forall \mu \in \mathbf{A}(\mathcal{I})$, $g \in B(E,\mathcal{I})$ we set, by definition, the number (called in the sequel the definite integral of the function g with respect to the f.a. measure μ)

$$\int_E g(x)\,\mu(dx) \in \mathbf{R} \tag{3.4.8}$$

such that $\forall\,(g_i)_{i\in\mathcal{N}}$:

$$(g_i)_{i\in\mathcal{N}}\colon \mathcal{N} \to B_0(E,\mathcal{I}),$$

we have the implication

$$(\|g_i - g\| \to 0) \Rightarrow \left(\overset{(el)}{\int_E} g_i(x)\,\mu(dx) \to \int_E g(x)\,\mu(dx)\right).$$

It is essential for the sequel that $B(E,\mathcal{I})$ is an algebra: $\forall\, u \in B(E,\mathcal{I})$, $v \in B(E,\mathcal{I})$

$$uv \triangleq (u(x)v(x))_{x\in E} \in B(E,\mathcal{I}).$$

Taking this into account, we can easily define an indefinite integral with respect to a f.a. measure of bounded variation on the space $B(E,\mathcal{I})$, namely, $\forall\, \mu \in \mathbf{A}(\mathcal{I})$, $g \in B(E,\mathcal{I})$, $\Lambda \in \mathcal{I}$:

$$\int_{\Lambda} g\, d\mu \triangleq \int_{E} (g\chi_{\Lambda})(x)\, \mu(dx) \in \mathbf{R}. \tag{3.4.9}$$

Of course, if $\Lambda = E$, then integrals (3.4.8) and (3.4.9) coincide. Similarly, in the case of $\mu \in \mathbf{A}(\mathcal{I})$, $g \in B_0(E,\mathcal{I})$, $\Lambda = E$, we have a natural coincidence of integrals (3.4.6) and (3.4.9). Everywhere in the sequel, we shall understand the indefinite integral of an element $g \in B(E,\mathcal{I})$ with respect to the f.a. measure $\mu \in \mathbf{A}(\mathcal{I})$ as the functional

$$\Lambda \mapsto \int_{\Lambda} g\, d\mu\colon \mathcal{I} \to \mathbf{R}, \tag{3.4.10}$$

which is obviously a f.a. measure from $\mathbf{A}(\mathcal{I})$; we shall use the symbol $g * \mu$ to denote measure (3.4.10). Observe the following natural and easily verifiable property: $\forall\, \mu \in \mathbf{A}(\mathcal{I})$, $u \in B(E,\mathcal{I})$, $v \in B(E,\mathcal{I})$, $\Lambda \in \mathcal{I}$,

$$\int_{\Lambda} uv\, d\mu = \int_{\Lambda} u\, dv * \mu. \tag{3.4.11}$$

If we assume that Λ is nonfixed in (3.4.11), then we get the equality

$$u * (v * \mu) = (uv) * \mu = (vu) * \mu = v * (u * \mu),$$

where μ, u, v conform to conditions (3.4.11). All the above-mentioned properties are well known and are given for the sake of completeness of the exposition. We do not use more general definitions of the integral with respect to a f.a. measure (see, e.g., [9, Chap. 3]) because they are not needed in the sequel owing to the specific character of the space $B(E,\mathcal{I})$. Note that the simplest estimates of the proximity for integrals (3.4.9) are connected with the above-mentioned property of isometric isomorphism of $\mathbf{A}(\mathcal{I})$ and $B^*(E,\mathcal{I})$. In this case, the isometric isomorphism is defined as the operator associating the measure $\mu \in \mathbf{A}(\mathcal{I})$ with the functional

$$g \mapsto \int_{E} g\, d\mu\colon B(E,\mathcal{I}) \to \mathbf{R}.$$

Thus, the pair

$$(B(E,\mathcal{I}), \mathbf{A}(\mathcal{I})) \tag{3.4.12}$$

provides us with a duality; in this case, the canonical bilinear form is given by the functional

$$(g,\mu) \mapsto \int_E g\,d\mu \colon B(E,\mathcal{I}) \times \mathbf{A}(\mathcal{I}) \to \mathbf{R}.$$

Duality (3.4.12) is related to the natural definition of the $*$-weak topology $\tau_*(\mathcal{I})$ of the set $\mathbf{A}(\mathcal{I})$ as its weakest topology with respect to which every functional

$$\mu \mapsto \int_E g\,d\mu \colon \mathbf{A}(\mathcal{I}) \to \mathbf{R}$$

is continuous for $g \in B(E,\mathcal{I})$. Thus, $\tau_*(\mathcal{I})$ is the family of all sets G, $G \subset \mathbf{A}(\mathcal{I})$ such that $\forall\,\mu \in G\ \exists\,K \in \mathrm{Fin}\,(B(E,\mathcal{I})), \varepsilon \in (0,\infty)$:

$$\left\{\nu \in \mathbf{A}(\mathcal{I})\,\Big|\,\forall\,g \in K\colon \left|\int_E g\,d\mu - \int_E g\,d\nu\right| < \varepsilon\right\} \subset G.$$

In this case, the pair

$$(\mathbf{A}(\mathcal{I}), \tau_*(\mathcal{I})) \tag{3.4.13}$$

is a Hausdorff locally convex space, in which the conditions of compactness are specified by the well known Alaoglu theorem [9, p. 459]: the set F, $F \subset \mathbf{A}(\mathcal{I})$, is compact in (3.4.13) if and only if

$$(F \in \mathcal{F}_{\tau_*(\mathcal{I})})\,\&(\{c \in [0,\infty]|\ \forall\,\mu \in F\colon v_\mu(E) \leqslant c\} \neq \varnothing). \tag{3.4.14}$$

In other words (according to (3.4.14)), the set F is $*$-weakly compact if and only if it is $*$-weakly closed and strongly bounded. In the sequel, we shall use (3.4.14) in extension constructions of extremal problems when constructing the main scheme of compactification of the space of solutions. To conclude this section, we observe one obvious property of the spaces $B_0(E,\mathcal{I})$ and $B(E,\mathcal{I})$.

Given $s \in \mathbf{R}^E$, we denote, as usual, by $|s|$ the functional

$$x \mapsto |s(x)|\colon E \to [0,\infty).$$

Then, as one can easily see,

$$(\forall\,g \in B_0(E,\mathcal{L})\colon |g| \in B_0^+(E,\mathcal{I}))\,\&(\forall\,h \in B(E,\mathcal{I})\colon |h| \in B^+(E,\mathcal{I})).$$

3.5. Compactification in the Class of Two-Valued Measures

We now present a natural extension construction for the solution space
using directly (3.4.14). The question is of imbedding this space into one of
the sets (3.3.3) with a suitable choice of the semialgebra $\mathcal{I}$ of subsets of E. In
this section, however, we shall fix the semialgebra $\mathcal{I} \in \Pi[E]$. This allows us
to shorten the notation and to introduce the main notions in a more compact
form. At the same time, we notice that this extension construction, based
on the imbedding of the set E into $\mathbf{T}(\mathcal{I})$, will be seldom used in the sequel; it
will be used only in formulating one fairly obvious existence theorem for the
problem of asymptotic cone optimization (see the preceding chapter), and for
this reason we shall not consider it separately, referring the reader to [34, 38,
39] and confining ourselves to a summary of the main definitions and simplest
assertions given without proofs. (The latter, using the natural relation of
elements of (3.3.3) and ultrafilters of the corresponding semialgebra of sets,
are quite standard from the standpoint of Stone representation constructions
and can be easily restored by the interested reader without assistance; see
[5, 31]). Given $x \in E$, we denote by $\delta_x \in \mathbf{T}(\mathcal{P}(E))$ the Dirac measure
concentrated at x (and defined on the family of all subsets of E), so that

$$(\delta_x|\mathcal{I}) \in \mathbf{T}(\mathcal{I}) \tag{3.5.1}$$

is the Dirac x-measure on $\mathcal{I}$ obtained by the constriction of δ_x; the (countably) additive measures (3.5.1) provide the simplest example of a $(0,1)$-
measure and also a concrete operator $\mathcal{J}[\mathcal{I}]$ of the form

$$x \mapsto (\delta_x|\mathcal{I})\colon E \to \mathbf{T}(\mathcal{I}), \tag{3.5.2}$$

characterizing the imbedding of E into $\mathbf{T}(\mathcal{I})$. In the subsequent constructions of this section, we proceed from the idea of closing the image of E
by virtue of $\mathcal{J}[\mathcal{I}]$ (3.5.2). Let us, however, first recall one useful (and mentioned above) representation of elements of (3.3.3) in terms of characteristic
functions of ultrafilters of a measurable space. Thus, let

$$\tilde{\mathbf{F}}_{\mathcal{I}} \triangleq \{\mathcal{H} \in 2^{\mathcal{I}}| \,(\varnothing \notin \mathcal{H})\&(\forall U \in \mathcal{H},\, V \in \mathcal{H}\colon U \cap V \in \mathcal{H})\&$$

$$(\forall H \in \mathcal{H}\colon \{\Lambda \in \mathcal{I}| \, H \subset \Lambda\} \subset \mathcal{H})\}. \tag{3.5.3}$$

$$\tilde{\mathbf{F}}_{\mathcal{I}}^{0} \triangleq \{\mathcal{H} \in \tilde{\mathbf{F}}_{\mathcal{I}}| \,\forall \mathcal{G} \in \tilde{\mathbf{F}}_{\mathcal{I}}\colon (\mathcal{H} \subset \mathcal{G}) \Rightarrow (\mathcal{G} = \mathcal{H})\}. \tag{3.5.4}$$

The elements of (3.5.3) are filters on $\mathcal{I}$ and the elements of (3.5.4) are the
maximal filters from (3.5.3), or ultrafilters on $\mathcal{I}$ (in the literature, these

definitions are usually given for the case of measurable spaces with algebras of sets). Given $\mathcal{E} \in \mathcal{P}(\mathcal{I})$, the functional $\mathbf{X}_{\mathcal{E}} \colon \mathcal{I} \to \mathbf{R}$ is such that it is natural to call

$$(\forall \Lambda \in \mathcal{I} \setminus \mathcal{E} \colon \mathbf{X}_{\mathcal{E}}(\Lambda) \triangleq 0) \& (\forall \Sigma \in \mathcal{E} \colon \mathbf{X}_{\mathcal{E}}(\Sigma) \triangleq 1) \qquad (3.5.5)$$

the characteristic function of $\mathcal{E}$. The set (3.3.3) admits the representation

$$\mathbf{T}(\mathcal{I}) = \{ \mathbf{X}_{\mathcal{H}} \colon \mathcal{H} \in \widetilde{\mathbf{F}}_{\mathcal{I}}^{0} \} \qquad (3.5.6)$$

in terms of characteristic functions (3.5.5) (for details, see [39]). On the basis of (3.5.6), we can obtain, in some cases, a complete description of set (3.3.3); for details, see [39, p. 76]. As another consequence of (3.5.6), note the theorem on the density of all Dirac measures, namely

$$\mathbf{T}(\mathcal{I}) \triangleq \mathrm{cl}\,(\{ (\delta_x | \mathcal{I}) \colon x \in E \}, \tau_*(\mathcal{I})). \qquad (3.5.7)$$

When justifying (3.5.7), we use the Birkhoff theorem [19] and the natural construction of the $\mathcal{I}$-filter associated with a net, which allows us to use representation (3.5.6); a detailed proof is given in [39, p. 75]. Relation (3.5.7) completes the characterization of imbedding operator (3.5.2). Relations (3.4.14) and (3.5.7) imply that set (3.3.3) is compact in the topological space (3.4.13); in other words, the space

$$(\mathbf{T}(\mathcal{I}), \tau_*(\mathcal{I})|_{\mathbf{T}(\mathcal{I})}) \qquad (3.5.8)$$

is compact. Thus, (3.5.2) actually defines a compactification of the initial space E; we know the role played by these procedures in the problems of asymptotic optimization (see Chap. 2). In the next section, we shall briefly discuss its utilization in solving the cone optimization problem. Recall that the issues related to the representation similar to (3.5.6) for a measurable space with an algebra of sets are considered in the theory of Boolean algebras [5, 31]. For some special issues, and, in particular, the problems of "extending" ultrafilters, see [39].

3.6. Extension of the Cone Optimization Problem in the Class of Two-Valued Measures

Let us now discuss one possibility of using space (3.5.8) for constructing a compactificator in the extremal problem of optimization with respect to the cone corresponding to the pointwise ordering of the space of estimates. The general extension scheme for such a problem was discussed in Sec. 2.6 under the assumption that there exists a compactificator (Y, τ, m, g) satisfying the conditions of Theorem 2.6.1. Now, using (3.5.7), we are going to construct it explicitly. We fix, as in Sec. 2.6, the arbitrary nonempty set Q. Conventions (2.6.2), (2.6.3), and (2.6.6) are observed without additional explanations. We shall need the operators $\mathcal{J}[\mathcal{I}]$, $\mathcal{I} \in \Pi[E]$, defined by (3.5.3), and therefore we shall interpret (3.5.7) as an assertion about $*$-weakly dense imbeddings. Let $\forall \mathcal{I} \in \Pi[E]$:

$$\tilde{\mathbf{B}}[E, \mathcal{I}, Q] \triangleq H^Q|_{H=B(E,\mathcal{I})}. \tag{3.6.1}$$

The elements of (3.6.1) are, in fact, parametrized families of $\mathcal{I}$-stratum functionals on E. Let, moreover,

$$\mathbf{B}[E, Q] \triangleq H^Q|_{H=\mathbf{B}(E)}. \tag{3.6.2}$$

Relation (3.6.2) provides a similar set of all parametrized (by Q) families of bounded functionals on E. We shall use the elements of (3.6.2) as the objective operator s from Sec. 2.6. With the aid of definitions from Sec. 3.4, we can show that

$$\mathbf{B}[E; Q] = \bigcup_{\mathcal{I} \in \Pi[E]} \tilde{\mathbf{B}}[E, \mathcal{I}, Q], \tag{3.6.3}$$

and, therefore, according to (3.6.3), it is sufficient to consider the case where s in Sec. 2.6 is an element of a set of the form (3.6.1). The further specification of a compactificator is based on the use of (3.5.2) and (3.5.7). Let us consider this scheme in detail.

Let there be given an arbitrary operator

$$s_0 \in \mathbf{B}[E, Q]. \tag{3.6.4}$$

With due account of (3.6.3) and (3.6.4), we choose a semialgebra $\Sigma \in \Pi[E]$ for which

$$s_0 \in \tilde{\mathbf{B}}[E, \Sigma, Q]. \tag{3.6.5}$$

We shall not discuss now the technique of selecting Σ (in particular, one can try to choose an algebra or a σ-algebra of the subsets of E satisfying condition (3.6.5); this can always be done). Let us only note that using the information concerning s_0 (3.6.4), we can make this choice fairly feasible. In any case, we can always identify Σ with $\mathcal{P}(E)$. From (3.6.5), it follows that $s_0(q) \in B(E, \Sigma)$ for $q \in Q$. Taking this into account, we introduce $g_0 \in \mathcal{Q}_{\mathbf{T}(\Sigma)}$ setting $\forall q \in Q$:

$$g_0(q) \triangleq \left(\int_E s_0(q)(x)\, \mu(dx) \right)_{\mu \in \mathbf{T}(\Sigma)}.$$

It is easy to verify that $g_0(q)$ is a continuous (on the topological space (3.5.8)) functional, so that, in reality, we have

$$g_0 \in \mathcal{Q}^0_{\mathbf{T}(\Sigma)}(\tau_*(\Sigma)|_{\mathbf{T}(\Sigma)}). \tag{3.6.6}$$

As one can see from (3.5.2), $\forall q \in Q$:

$$s_0(q) = g_0(q) \circ \mathcal{J}[\Sigma]. \tag{3.6.7}$$

Now, the 4-tuple

$$(\mathbf{T}(\Sigma), \tau_*(\Sigma)|_{\mathbf{T}(\Sigma)}, \mathcal{J}[\Sigma], g_0) \tag{3.6.8}$$

satisfies the conditions of pointwise compatibility of Sec. 2.6; they are realized by conditions (3.6.7). In this case, (3.5.8) is a compact space, so that the 4-tuple (3.6.8) is quite suitable for using as an individual compactificator (Y, τ, m, g) of Sec. 2.6. Note that

$$S_0 \triangleq (\text{term}) [s_0] \tag{3.6.9}$$

plays the role of the objective operator from Sec. 2.6, denoted there by S (see (2.6.8)), $S_0 \colon E \to \mathbf{R}^Q$. As one can see from (2.6.7),

$$G_0 \triangleq (\text{term}) [g_0] \in C(\mathbf{T}(\Sigma), \tau_*(\Sigma)|_{\mathbf{T}(\Sigma)}, \mathbf{R}^Q, Q^0(\tau_{\mathbf{R}})).$$

Now, replacing, in Theorem 2.6.1, s by s_0, (Y, τ, m, g) by the 4-tuple (3.6.8), S by S_0, and G by G_0, we get a natural generalized representation of the set of asymptotically attainable estimates. A similar specification can be realized with respect to other assertions of Sec. 2.6, but we shall not do it here for lack of space. Note that this version of compactification makes it possible, in certain problems, to simplify, compared to Sec. 2.6, the construction of asymptotically efficient approximate solutions (see [39]). This ends the discussion of the construction of compactification in the class of two-valued measures; it can hardly serve as a working tool. In our exposition, this construction was, in fact, used for proving assertions of the type of existence theorems, the only condition in which is (3.6.4), which, as a rule, either occurs in "pure form," or is valid for a substantial part of the solution space (we shall come across a situation of the latter type in Chaps. 7 and 8). If condition (3.6.4) is satisfied for $s = s_0$, then all assertions of Sec. 2.6, proved there under the assumption of the existence of a compactificator, become unconditional. Thus, Theorem 2.6.2 can be reformulated as a theorem of existence of asymptotically efficient estimates: if s_0 satisfies (3.6.4), $\mathcal{X}$ is a filter-base in E, and $S = S_0$ (3.6.9), then $(\leqq - \mathrm{MIN})[Z^0] \neq \varnothing$, where Z^0

is the intersection of all sets $\operatorname{cl}(S^1(U), \otimes^Q(\tau_{\mathbf{R}}))$, $U \in \mathcal{X}$. Similarly, specifying Theorem 2.6.3 under the assumption $s = s_0$ (3.6.4), we get (in the case of (3.6.4)) the assertion of the existence of an asymptotically efficient approximate solution.

Chapter 4

Finitely Additive Measures and Indefinite Integrals

4.1. Introduction

In this chapter, we shall continue the summary of notions and relevant (rather simple) assertions connected with the properties of f.a. measures. However, the issues considered here have a more specific character and concern the approximation of f.a. measures by indefinite integrals. These structures will be needed in constructing extensions in a class of extremal problems, whose systematic study we begin in the next chapter. Note that the conditions allowing us to represent a measure by an indefinite integral have, in its classical version, the meaning of absolute continuity [9, Chap. 3]; they relate, however, to countably additive measures on a standard measurable space. The possibility of representing a measure by an indefinite integral is equivalent to the Radon–Nicodym property. This property was studied in many publications; suffice it to mention only the monograph [56], where one can find the corresponding bibliography. For the f.a. measures, this property in its pure form is, generally speaking, no longer valid [59]. Nevertheless, there are various approximate analogs of the Radon–Nicodym property for f.a. measures, the most natural of which is the assertion of Bochner [9, p. 343] concerning the conditions of approximation in the strong sense, i.e., in variation (see norm (3.3.9)). There arise, however, needs for approximate analogs of another type, concerning the possibility of approximating measures by integrals in the ∗-weak sense, and also approximating with respect to some nonstandard topologies, which is due, in particular, to applications to the extension of extremal problems. If we bear in mind the ∗-weak approximation (we speak here about the topological structure on the basis of the natural duality of the space of stratum functionals and the space of f.a. measures of bounded variation), then, fundamentally, a relevant need arises in connection with the nice compactness properties defined by the Alaoglu theorem [9, p. 459], so that this approximate analog is directly linked with the compactification of the space of "conventional" solutions; the latter is essential in settings of extremal problems (see Chap. 2). Note that the ∗-weak topology of strongly bounded sets in the space of f.a. measures of bounded variation (see Chap. 3) is not, generally speaking, metrizable, in contrast to the more traditional case of the duality $(\mathbf{C}(K,\tau), \mathbf{C}^*(K,\tau))$, where (K,τ) is a nonempty metric compact space. This does not hinder our constructions in deriving concrete inferences (see Chap. 1); we saw some applications of this kind in Chap. 3. We are now going to pass to a more specific procedure related to approximate analogs of the Radon–Nicodym property. We shall be concerned with using representation (3.4.11). According to this representation, the procedure of imbedding the stratum, and, in particular, step functionals into the space of f.a. measures of bounded variation can be defined by means of integration with respect to a given f.a. measure. This approach will clearly require investigation of the closure of images, i.e., the closure of sets whose elements are indefinite integrals. We are going to realize this program in this chapter, which, in turn, is a basis for constructing an extension of the space of "conventional" solutions, each of which is identified with a step-functional on the corresponding measurable space. We consider separately the case of extending the space of nonnegative step-functionals and the case concerning the extension of the integrally bounded part of this space, so that in this chapter these cases are distinguished as well. At the same time, some issues concerning the approximation of f.a. measures by indefinite integrals are given a broader consideration than is required for direct use in extension constructions. Thus, we employ the notion of component mentioned in the previous chapter, with the aim of a fuller characterization of the class of f.a. measures used in the sequel.

4.2. Weakly Absolutely Continuous Finitely Additive Measures

Throughout the remainder of the book, we fix the semialgebra $\mathcal{L} \in \Pi[E]$ and the f.a. measure $\eta \in (\mathrm{add})_+ [\mathcal{L}]$. The 3-tuple

$$(E, \mathcal{L}, \eta) \tag{4.2.1}$$

provides us with an analog of a measurable space with a measure. In the subsequent chapters, we shall formulate, in terms of space (4.2.1), extremal problems with restraints on the "finite" or "infinite" integrand of control $f \in B_0(E, \mathcal{L})$. Now we shall present one special construction of generalized objects, namely, measures replacing the "conventional" solutions f and allowing us to reduce the corresponding asymptotic settings to the standard form. We shall consider transformations that, generally speaking, do not reduce to compactifications but have the meaning of closures and set-theoretic limits in the appropriate subspace $\mathbf{A}(\mathcal{L})$. In this connection, we shall establish the relevant facts from the f.a. measure theory in a more general form than is required, at first glance, by the procedures of Chap. 2. Let $\mathbf{O} \triangleq \mathcal{O}_{\mathcal{L}}$, and then the relations

$$(\mathrm{add})^+ [\mathcal{L}, \eta] \triangleq \{\mu \in (\mathrm{add})_+ [\mathcal{L}]|\ \forall L \in \mathcal{L}\colon\ (\eta(L) = 0) \Rightarrow (\mu(L) = 0)\}, \tag{4.2.2}$$

$$\mathbf{A}_\eta[\mathcal{L}] \triangleq \{\mu \in \mathbf{A}(\mathcal{L})|\ v_\mu \in (\mathrm{add})^+ [\mathcal{L}, \eta]\}, \tag{4.2.3}$$

$$\mathbf{A}_\eta^d[\mathcal{L}] \triangleq \{\mu \in \mathbf{A}(\mathcal{L})|\ \forall \nu \in \mathbf{A}_\eta[\mathcal{L}]\colon\ v_\mu \wedge v_\nu = \mathbf{O}\}, \tag{4.2.4}$$

$$(\mathrm{add})_D^+ [\mathcal{L}, \eta] \triangleq \{\mu \in (\mathrm{add})_+ [\mathcal{L}]|\ \forall \nu \in (\mathrm{add})^+ [\mathcal{L}, \eta]\colon$$

$$(\nu \leqslant \mu) \Rightarrow (\nu = \mathbf{O})\} = \mathbf{A}_\eta^d[\mathcal{L}] \cap (\mathrm{add})_+ [\mathcal{L}] \tag{4.2.5}$$

define the decomposition of $\mathbf{A}(\mathcal{L})$ into an ordered direct sum, which is similar to the Hewitt–Yosida decomposition (but which is, generally speaking, not congruent with it), and also its "positive part." The fact is that (4.2.3) is a band (component) of $\mathbf{A}(\mathcal{L})$, for which (4.2.2) is a generating cone, so that $\mathbf{A}_\eta[\mathcal{L}]$ coincides with the set of all measures

$$\mu - \nu, \quad (\mu, \nu) \in (\mathrm{add})^+ [\mathcal{L}, \eta] \times (\mathrm{add})^+ [\mathcal{L}, \eta].$$

Relation (4.2.4) defines a disjunctive band (see (3.3.20)) generated by cone (4.2.5). We may apply to the duality

$$(\mathbf{A}_\eta[\mathcal{L}], \mathbf{A}_\eta^d[\mathcal{L}]),$$

understood in the sense of the vector lattice theory, the arguments of Sec. 3.3 and also the general propositions of this theory, which we shall not develop here for lack of space. Let us go over to the topological structures which are important for our purposes.

Together with the main topological space

$$(\mathbf{A}(\mathcal{L}), \tau_*(\mathcal{L})), \tag{4.2.6}$$

which is a realization of (3.4.13), we shall need to consider $\mathbf{A}(\mathcal{L})$ as a subspace of $\mathbf{R}^{\mathcal{L}}$ with its suitable topological structure; more precisely, we shall need Tychonoff powers of the spaces $(\mathbf{R}, \tau_{\mathbf{R}})$ and $(\mathbf{R}, \tau_{\partial})$ understood as (Tychonoff) products of the corresponding number of similar copies of the first or second space. The powers mentioned above are the spaces

$$(\mathbf{R}^{\mathcal{L}}, \otimes^{\mathcal{L}}(\tau_{\mathbf{R}})), \qquad (\mathbf{R}^{\mathcal{L}}, \otimes^{\mathcal{L}}(\tau_{\partial})) \tag{4.2.7}$$

respectively; $\mathbf{A}(\mathcal{L})$ can be regarded as a subspace of each of the topological spaces (4.2.7). This yields the following two topologies on $\mathbf{A}(\mathcal{L})$:

$$\tau_{\otimes}(\mathcal{L}) \triangleq \otimes^{\mathcal{L}}(\tau_{\mathbf{R}})|_{\mathbf{A}(\mathcal{L})}, \tag{4.2.8}$$

$$\tau_0(\mathcal{L}) \triangleq \otimes^{\mathcal{L}}(\tau_{\partial})|_{\mathbf{A}(\mathcal{L})}. \tag{4.2.9}$$

Along with the topology $\tau_*(\mathcal{L})$, used in (4.2.6), we get the following (topological) triad:

$$\mathfrak{M} \triangleq \{\tau_*(\mathcal{L}); \tau_0(\mathcal{L}); \tau_{\otimes}(\mathcal{L})\}; \tag{4.2.10}$$

it is a three-element set consisting of the topologies on $\mathbf{A}(\mathcal{L})$ defined by (4.2.6), (4.2.8), and (4.2.9) respectively. The topologies $\tau_*(\mathcal{L})$ and $\tau_0(\mathcal{L})$ are, generally speaking, unrelated (we shall give a congruent example); topology (4.2.8) is weaker than the other two topologies of the triad $\mathfrak{M}$ and does not coincide with them. The last assertion is obvious: the first of the topological spaces (4.2.7) is weaker than the second, and this is inherited by subspaces (4.2.8) and (4.2.9) respectively. However, $\{\chi_L \colon L \in \mathcal{L}\} \subset B(E, \mathcal{L})$ and this yields $\tau_{\otimes}(\mathcal{L}) \subset \tau_*(\mathcal{L})$ by the property of elementary integral (3.4.6). One can also easily verify the following property: if the set H, $H \subset \mathbf{A}(\mathcal{L})$, is strongly bounded in $\mathbf{A}(\mathcal{L})$, i.e.,

$$\{c \in [0, \infty) | \; \forall \mu \in H \colon v_{\mu}(E) \leqslant c\} \neq \varnothing,$$

then

$$\otimes^{\mathcal{L}}(\tau_{\mathbf{R}})|_H = \tau_{\otimes}(\mathcal{L})|_H = \tau_*(\mathcal{L})|_H. \tag{4.2.11}$$

Relation (4.2.11) provides, as a consequence, one simple property, similar to (4.2.11) but concerning the set H that is not strongly bounded, namely, we can replace H by the positive cone of $\mathbf{A}(\mathcal{L})$. Let

$$\tau_*^+(\mathcal{L}) \triangleq \tau_*(\mathcal{L})|_{(\mathrm{add})_+[\mathcal{L}]}, \quad \tau_0^+(\mathcal{L}) \triangleq \tau_0(\mathcal{L})|_{(\mathrm{add})_+[\mathcal{L}]}, \quad \tau_\otimes^+(\mathcal{L}) \triangleq \tau_\otimes(\mathcal{L})|_{(\mathrm{add})_+[\mathcal{L}]}.$$

Then the property mentioned above has the form

$$\tau_*^+(\mathcal{L}) = \tau_\otimes^+(\mathcal{L}) \subset \tau_0^+(\mathcal{L}). \tag{4.2.12}$$

The only thing that needs to be proved in (4.2.12) is the coincidence of the two first topologies. However, this easily follows from (4.2.11) and the definition of $\tau_*(\mathcal{L})$; in turn, relation (4.2.11) follows from the definition of the integral in Chap. 3. With the aid of (4.2.12), we can prove, in particular, that

$$\forall\, H \in \mathcal{P}((\mathrm{add})_+[\mathcal{L}]):\ \mathrm{cl}\,(H, \tau_0(\mathcal{L})) \subset \mathrm{cl}\,(H, \tau_*(\mathcal{L})). \tag{4.2.13}$$

The imbedding (4.2.13) will be used for constructing special limit representations for the relaxation of admissible sets in Chap. 5. The proof of (4.2.13) is obvious; in order to obtain closures in (4.2.13), we must use closures in the sense of topologies $\tau_0^+(\mathcal{L})$ and $\tau_*^+(\mathcal{L})$, respectively, for which we can use estimate (4.2.12). Let us observe a simple, but rather useful, representation of subspaces of $(\mathbf{A}(\mathcal{L}), \tau_0(\mathcal{L}))$: given a nonempty subset H of $\mathbf{A}(\mathcal{L})$, the (relative) topology

$$\tau_0(\mathcal{L})|_H = \otimes^{\mathcal{L}}(\tau_\partial)|_H \tag{4.2.14}$$

on H is the family of all sets G, $G \subset H$, such that $\forall\, \mu \in G\ \exists\, \mathcal{K} \in \mathrm{Fin}\,(\mathcal{L})$:

$$\{\nu \in H\,|\,\forall\, L \in \mathcal{K}:\ \mu(L) = \nu(L)\} \subset G.$$

Representation (4.2.14) of the topology can, in particular, be used to describe $\tau_0(\mathcal{L})$, $\tau_0^+(\mathcal{L})$.

We shall need to consider, along with the topological spaces corresponding to the triad $\mathfrak{M}$, some of their subspaces; in a number of cases we shall not state the passage to them explicitly. For some subspaces, we shall introduce special notation. For example, we set

$$\tau_\eta^*(\mathcal{L}) \triangleq \tau_*(\mathcal{L})|_{(\mathrm{add})^+[\mathcal{L},\eta]}, \tag{4.2.15}$$

$$\tau_*^{(\eta)}[\mathcal{L}] \triangleq \tau_*(\mathcal{L})|_{\mathbf{A}_\eta[\mathcal{L}]}. \tag{4.2.16}$$

Topologies (4.2.15) and (4.2.16) will be needed to formulate the main conditions imposed on the choice of a criterion in the extremal problems considered in the sequel.

Taking into account the fact that space (4.2.1) is fixed in all subsequent theoretical constructions, we shall often omit some of the letters E, $\mathcal{L}$, η in the subsequent notation; otherwise, the notation would be rather unwieldy and its identification with substantive concepts would be too complicated. In particular, this remark should be borne in mind in connection with the notation relating to some subspaces of the topological spaces from the triad $\mathfrak{M}$ (4.2.10). We set $\forall\, b \in [0, \infty)$:

$$M_b^+ \triangleq \left\{ f \in B_0^+(E, \mathcal{L}) \,\Big|\, \int_E f\, d\eta \leqslant b \right\}, \tag{4.2.17}$$

$$\widetilde{M}_b^+ \triangleq \{ f * \eta\colon\ f \in M_b^+ \}, \tag{4.2.18}$$

$$M_b \triangleq \left\{ f \in B_0(E, \mathcal{L}) \,\Big|\, \int_E |f|\, d\eta \leqslant b \right\}, \tag{4.2.19}$$

$$\widetilde{M}_b \triangleq \{ f * \eta\colon\ f \in M_b \}, \tag{4.2.20}$$

$$\Xi_+[b] \triangleq \{ \mu \in (\mathrm{add})^+[\mathcal{L}, \eta] \,|\, \mu(E) \leqslant b \}, \tag{4.2.21}$$

$$\Xi[b] \triangleq \{ \mu \in \mathbf{A}_\eta[\mathcal{L}] \,|\, v_\mu(E) \leqslant b \}, \tag{4.2.22}$$

$$\tilde{\tau}_b^*(\mathcal{L}) \triangleq \tau_*(\mathcal{L})|_{\Xi_+[b]} = \tau_\eta^*(\mathcal{L})|_{\Xi_+[b]}, \tag{4.2.23}$$

$$\tilde{\tau}_*^{(b)}(\mathcal{L}) \triangleq \tau_*(\mathcal{L})|_{\Xi[b]} = \tau_*^{(\eta)}[\mathcal{L}]|_{\Xi[b]}. \tag{4.2.24}$$

The sets (4.2.17)–(4.2.22) are characterized by an appropriate boundedness property; for our immediate purposes, the most important are sets (4.2.21) and (4.2.22), which are strongly bounded subsets of sets (4.2.2) and (4.2.3) respectively. One can observe a certain analogy between (4.2.17) and (4.2.21), (4.2.19), and (4.2.22), respectively; in the next section, we shall establish the relationship between these definitions in terms of the density of sets (4.2.18) and (4.2.20), which are images of sets (4.2.17) and (4.2.19), respectively, by virtue of the integration operator. Note that $\forall\, b \in [0, \infty]$

$$(\Xi_+[b], \tilde{\tau}_b^*(\mathcal{L})), \qquad (\Xi[b], \tilde{\tau}_*^{(b)}[\mathcal{L}]) \tag{4.2.25}$$

are compact topological spaces. The verification of this assertion is obvious: we must take into account (3.4.14) and the properties listed in Sec. 2.2. Note that the compact spaces (4.2.25) will play the main part in constructing extensions. Furthermore, it is easier to establish the assertions concerning the dense imbedding into the space of f.a. measures first for the first topological space in (4.2.25): the question is of imbedding set (4.2.17) into the compact space (4.2.25). The above-mentioned assertions concerning the density of the images in the sense of the triad $\mathfrak{M}$ will be considered in the next section.

4.3. Density Properties

For a number of reasons, this section plays an important part for all subsequent constructions because here we establish the density properties of $\mathcal{L}$-step-functionals on E in band (4.2.3) and also similar assertions for subsets of this band. In particular, let us consider explicitly the compactification in the class of weakly absolutely continuous measures and other propositions. The assertions concerning the dense imbedding into the cone $(\mathrm{add})^+[\mathcal{L}, \eta]$ can be established somewhat more easily, and we shall consider them first of all. At the same time, among the latter assertions, it is expedient to begin by proving the propositions concerning strongly bounded sets. In the sequel, we shall adhere precisely to this scheme of reasoning. However, let us first consider a repeatedly used net construction in the space of nonnegative step-functionals.

We set $\mathcal{D} \triangleq \mathbf{D}(E, \mathcal{L})$ and get (see Sec. 3.2) a nonempty set of (unordered) finite partitions of the "unity" E by the elements of $\mathcal{L}$. This set $\mathcal{D}$ is endowed with a natural direction characterized by the property of refining a partition by another one. Namely, we set $\forall \mathcal{A} \in \mathcal{D}, \mathcal{B} \in \mathcal{D}$ by definition:

$$(\mathcal{A} \prec \mathcal{B}) \Leftrightarrow (\forall B \in \mathcal{B} \, \exists A \in \mathcal{A}: B \subset A). \qquad (4.3.1)$$

The relationship (4.3.1) defines the binary relation $\prec$, $\prec \subset \mathcal{D} \times \mathcal{D}$, which is, as one can easily check, a direction, so that $(\mathcal{D}, \prec)$ is a (nonempty) directed set.

Now we shall consider a construction of an operator acting from $\mathcal{D}$ into $B_0^+(E, \mathcal{L})$. For our purposes, we initially introduce a special function of sets by setting $\mathfrak{N} \triangleq \{L \in \mathcal{L} | \, \eta(L) = 0\}$. Given $\mu \in (\mathrm{add})^+[\mathcal{L}, \eta]$, we assume, by definition, the functional

$$\theta_+[\mu]: \mathcal{L} \to [0, \infty) \qquad (4.3.2)$$

to be such that

$$(\forall L \in \mathfrak{N}: \theta_+[\mu](L) \triangleq 0) \& \left(\forall \tilde{L} \in \mathcal{L} \setminus \mathfrak{N}: \theta_+[\mu](\tilde{L}) \triangleq \frac{\mu(\tilde{L})}{\eta(\tilde{L})}\right). \qquad (4.3.3)$$

The following is the main property of functional (4.3.2), (4.3.3), which we use for integration: $\forall \mu \in (\mathrm{add})^+[\mathcal{L}, \eta], L \in \mathcal{L}$,

$$\mu(L) = \eta(L)\theta_+[\mu](L). \qquad (4.3.4)$$

Now, given $\mu \in (\mathrm{add})^+ (\mathcal{L}, \eta]$, $\mathcal{K} \in \mathcal{D}$, we assume, by definition, the functional

$$\Theta^+_\mu[\mathcal{K}] \colon E \to \mathbf{R} \qquad\qquad (4.3.5)$$

to be such that $\forall K \in \mathcal{K}$, $x \in K$:

$$\Theta^+_\mu[\mathcal{K}](x) \triangleq \theta_+[\mu](K). \qquad\qquad (4.3.6)$$

It is clear that (4.3.5) and (4.3.6) are well defined because $\forall\, \tilde{x} \in E \; \exists!\, \widetilde{K} \in \mathcal{K}$: $\tilde{x} \in \widetilde{K}$. Of course, $\forall\, \tilde{\mu} \in (\mathrm{add})^+ [\mathcal{L}, \eta]$ we have the operator

$$\mathcal{K} \mapsto \Theta^+_\mu[\mathcal{K}] \colon \mathcal{D} \to B^+_0(E, \mathcal{L}) \qquad\qquad (4.3.7)$$

denoted, as usual, by $\Theta^+_\mu[\cdot]$. With due account of (4.3.4) and (4.3.6), we get

$$\int\limits_E \Theta^+_\mu[\widetilde{\mathcal{K}}] \, d\eta = \mu(E) \qquad (\widetilde{\mathcal{K}} \in \mathcal{D}).$$

Then, as one can easily see, $\forall\, b \in [0, \infty)$, $\mu \in \Xi_+[b]$, $\mathcal{K} \in \mathcal{D}$:

$$\Theta^+_\mu[\mathcal{K}] \in M^+_b. \qquad\qquad (4.3.8)$$

Let us also stipulate the following notation for the natural duplicate of operator (4.3.7), setting (for $\mu \in (\mathrm{add})^+ [\mathcal{L}, \eta]$)

$$\Theta^+_\mu[\cdot] * \eta \triangleq (\Theta^+_\mu[\mathcal{K}] * \eta)_{\mathcal{K} \in \mathcal{D}} \qquad\qquad (4.3.9)$$

and thus obtaining, for $\mu \in \Xi_+[b]$ $(b \geqslant 0)$, an operator acting from $\mathcal{D}$ into $\widetilde{M}^+_b$; here, of course, we make allowance for (4.2.18) and (4.3.8). Let us now note one equivalent, easily verifiable representation of $\prec$, namely, $\forall\, \mathcal{A} \in \mathcal{D}$, $\mathcal{B} \in \mathcal{D}$:

$$(\mathcal{A} \prec \mathcal{B}) \Leftrightarrow (\forall\, U \in \mathcal{A} \setminus \{\varnothing\} \; \exists\, \mathcal{U} \in \mathbf{D}(U, \mathcal{L}) \colon \mathcal{U} \subset \mathcal{B}). \qquad (4.3.10)$$

Here we have replaced (4.3.1) by a description in the language of partitions: according to (4.3.10), the partition $\mathcal{B}$ follows $\mathcal{A}$ ($\mathcal{B}$ is "greater than or equal to" $\mathcal{A}$ in the sense of $\prec$) if every nonempty set from $\mathcal{A}$ can be decomposed into a sum of subsets from $\mathcal{B}$. With due regard for (4.2.12), (4.3.4), (4.3.10), and the Birkhoff theorem (see Sec. 2.2), we can easily prove the following lemma.

LEMMA 4.3.1. *Let* $\mu \in (\mathrm{add})^+ [\mathcal{L}, \eta]$. *Then* $\forall\, \tau \in \mathfrak{M}$ *the net*

$$(\mathcal{D}, \prec, \Theta^+_\mu[\cdot] * \eta) \qquad\qquad (4.3.11)$$

converges to μ in the topological space $(\mathbf{A}(\mathcal{L}), \tau)$.

Here we restrict ourselves to the following remarks (for details, see, for example, [40]). The essential part of the reasoning consists in proving the convergence in the sense of $\tau_0(\mathcal{L})$ which uses the earlier mentioned representation of topology (4.2.14) and is, in fact, an extension of relation (4.3.4) "by additivity" with the use of the definition of the elementary integral (3.4.6); it is here that we must take into account (4.3.10). Since (4.3.9) assumes values in $(\mathrm{add})_+[\mathcal{L}]$, in proving the $*$-weak convergence on the basis of the earlier established convergence in the sense of $\tau_0(\mathcal{L})$, one has to make use of estimate (4.2.12), thus obtaining the convergence of (4.3.11) in the sense of the topologies $\tau_*^+(\mathcal{L})$ and $\tau_\otimes^+(\mathcal{L})$. This implies the convergence with respect to $\tau_*(\mathcal{L})$ and $\tau_\otimes(\mathcal{L})$, which ends the proof.

Here (and in the sequel), we use, without additional explanations, the obvious relationships of the filters of neighborhoods of a point in the topological space and in its subspace. Namely, if (T, τ) is a (nonempty) topological space, S is a subset of T, $s \in S$, $\mathcal{U}$ and $\mathcal{V}$ are families of all neighborhoods of s in the sense of the topologies τ and $\tau|_S$, respectively, then

$$\mathcal{V} = \{S \cap U \colon U \in \mathcal{U}\}.$$

Lemma 4.3.1 plays an important part not only in assertions concerning the dense imbedding presented in this section and supplemented when there is a need in the sequel, but also directly in the extension constructions of extremal problems, where it will be used for constructing asymptotically optimal approximate solutions.

LEMMA 4.3.2. *Let $b \in [0, \infty)$. Then the operator of integration transforms the set M_b^+ (4.2.17) into a set everywhere dense in $\Xi_+[b]$ with respect to any topology of the triad $\mathfrak{M}$:* $\forall \tau \in \mathfrak{M}$

$$\Xi_+[b] = \mathrm{cl}\,(\widetilde{M_b^+}, \tau).$$

The proof uses (4.2.11), (4.2.12), Lemma 4.3.1, and Birkhoff's theorem. We also make allowance for the property of $*$-weak closeness of $\Xi_+[b]$ mentioned in Sec. 4.2 and the fact that $\widetilde{M_b^+}$ is a subset of the positive cone of $\mathbf{A}(\mathcal{L})$; the latter implies, by virtue of (4.2.12), the equality $\mathrm{cl}\,(\widetilde{M_b^+}, \tau_*(\mathcal{L})) = \mathrm{cl}\,(\widetilde{M_b^+}, \tau_\otimes(\mathcal{L}))$. The other obvious details of the proof are left for the reader.

As an immediate corollary of Lemma 4.3.2, we obtain a theorem on a dense imbedding of $B_0^+(E, \mathcal{L})$ and $B^+(E, \mathcal{L})$ into cone (4.2.2); we only have to bear in mind that

$$B_0^+(E, \mathcal{L}) = \bigcup_{b \geqslant 0} M_b^+, \qquad (\mathrm{add})^+[\mathcal{L}, \eta] = \bigcup_{b \geqslant 0} \Xi_+[b].$$

THEOREM 4.3.1. *By virtue of the integration operator, the image of every one of the sets $B_0^+(E,\mathcal{L})$ and $B^+(E,\mathcal{L})$ is an everywhere dense subset of* $(\mathrm{add})^+[\mathcal{L},\eta]$ *with respect to any topology from* $\mathfrak{M}$: $\forall\,\tau\in\mathfrak{M}$ *we have*

$$(\mathrm{add})^+[\mathcal{L},\eta] = \mathrm{cl}\,(\{f*\eta\colon f\in B_0^+(E,\mathcal{L}]\},\tau) = \mathrm{cl}\,(\{f*\eta\colon f\in B^+(E,\mathcal{L})\},\tau).$$

In turn, Theorem 4.3.1, which is significant by itself, allows us to make the following important assertion.

THEOREM 4.3.2. *The integration operator transforms $B_0(E,\mathcal{L})$ and $B(E,\mathcal{L})$ into sets everywhere dense in the band $\mathbf{A}_\eta[\mathcal{L}]$ with respect to any topology from the triad* $\mathfrak{M}$, *so that* $\forall\,\tau\in\mathfrak{M}$:

$$\mathbf{A}_\eta[\mathcal{L}] = \mathrm{cl}\,(\{f*\eta\colon f\in B_0(E,\mathcal{L})\},\tau) = \mathrm{cl}\,(\{f*\eta\colon f\in B(E,\mathcal{L})\},\tau).$$

SCHEME OF THE PROOF. With due account of the definitions from Secs. 3.3 and 4.2, it is easy to prove that

$$\mathbf{A}_\eta[\mathcal{L}] \in \mathcal{F}_{\tau_\otimes(\mathcal{L})}. \tag{4.3.12}$$

Indeed, $\forall\,\mu\in\mathbf{A}_\eta[\mathcal{L}]$, $L\in\mathfrak{N}$: $\mu(L)=0$ (a simple consequence of the definition of v_μ). To this we must add that $\{\tilde{L}\in\mathcal{L}\mid \tilde{L}\subset L\}\subset\mathfrak{N}$ for $L\in\mathfrak{N}$. The next part of the proof of (4.3.12) makes use of the definition of pointwise topology on $\mathbf{R}^{\mathcal{L}}$. Indeed, if

$$\omega\in\mathbf{A}(\mathcal{L})\setminus\mathbf{A}_\eta[\mathcal{L}],$$

then one can find $D\in\mathfrak{N}$ such that $v_\omega(D)$ does not vanish. But in this case, for some set $\Lambda\in\mathfrak{N}$, $\Lambda\subset D$, we find that $|\omega(\Lambda)|>0$, whereas $\mu(\Lambda)=0$ for $\mu\in\mathbf{A}_\eta[\mathcal{L}]$. This fact was already noted above. But then the set

$$M^0 \triangleq \{\,\nu\in\mathbf{A}(\mathcal{L})\mid |\omega(\Lambda)-\nu(\Lambda)|<|\omega(\Lambda)|\,\}$$

is an open neighborhood of ω in the sense of $\tau_\otimes(\mathcal{L})$ and, at the same time, satisfies the obvious condition $M^0\subset\mathbf{A}(\mathcal{L})\setminus\mathbf{A}_\eta[\mathcal{L}]$. This proves (4.3.12). At the same time, $f*\eta\in\mathbf{A}_\eta[\mathcal{L}]$ for $f\in B(E,\mathcal{L})$ by virtue of the obvious properties of the integral from Chap. 3. Taking account of (4.3.12), we now have

$$\mathrm{cl}\,(\{f*\eta\colon f\in B_0(E,\mathcal{L})\},\tau_\otimes(\mathcal{L}))\subset$$

$$\mathrm{cl}\,(\{f*\eta\colon f\in B(E,\mathcal{L})\},\tau_\otimes(\mathcal{L}))\subset\mathbf{A}_\eta[\mathcal{L}]. \tag{4.3.13}$$

Since $\forall\,H\in\mathcal{P}(\mathbf{A}(\mathcal{L}))$ we have the inclusion $\mathrm{cl}\,(H,\tau_*(\mathcal{L}))\subset\mathrm{cl}\,(H,\tau_\otimes(\mathcal{L}))$ (see Sec. 4.2), (4.3.13) can also realize an estimate for the $*$-weak closures. For this it suffices to take into account that

$$\mathrm{cl}\,(\{f*\eta\colon f\in B_0(E,\mathcal{L})\},\tau_*(\mathcal{L}))\subset\mathrm{cl}\,(\{f*\eta\colon f\in B(E,\mathcal{L})\},\tau_*(\mathcal{L}))\subset$$

$$\mathrm{cl}\left(\{f * \eta \colon f \in B(E, \mathcal{L})\}, \tau_{\otimes}(\mathcal{L})\right). \tag{4.3.14}$$

Actually, the chain of inclusions (4.3.13), (4.3.14) is a system of equalities. This can be easily checked with due account of Theorem 4.3.1 and the representation of band (4.2.3) as a linear space generated by cone (4.2.2). In view of the obvious estimate $\tau_{\otimes}(\mathcal{L}) \subset \tau_0(\mathcal{L})$ (see (4.2.8) and (4.2.9)), we obtain a chain of inclusions similar to (4.3.14) with the only difference that the topology $\tau_*(\mathcal{L})$ is replaced in it by $\tau_0(\mathcal{L})$. Finally, using Theorem 4.3.1 and the representation from Sec. 4.2 for the band $\mathbf{A}_\eta[\mathcal{L}]$ in terms of the generating cone (4.2.2), we complete the proof by converting the above chain of inclusions into equalities.

To conclude this section, we return to the assertions concerning the integrally bounded subsets of $B(E, \mathcal{L})$. The question will be of imbedding sets (4.2.19) into the space of f.a. measures of bounded variation and of their representation as subsets of (4.2.22). For the sake of completeness of the exposition, we introduce $\forall b \in [0, \infty)$ the sets

$$\mathcal{M}_b \triangleq \left\{f \in B(E, \mathcal{L}) \,\Big|\, \int_E |f|\, d\eta \leqslant b\right\}, \tag{4.3.15}$$

$$\widetilde{\mathcal{M}}_b \triangleq \{f * \eta \colon f \in \mathcal{M}_b\}. \tag{4.3.16}$$

Note that sets similar to (4.3.15) and (4.3.16) could also be considered in the case of nonnegative functionals and measures (we mean the analogs of (4.2.17) and (4.2.18) respectively). However, in the sequel we shall not use them in constructing extensions, confining ourselves to "physically realizable" $\mathcal{L}$-step-functionals on E. Now, for a change, we shall consider (4.3.15) and (4.3.16) along with (4.2.19) and (4.2.20), respectively, although the required assertions for these new sets can be formulated without particular difficulties.

THEOREM 4.3.3. *Let* $b \in [0, \infty)$. *Then* $\forall \tau \in \mathfrak{M} \colon \Xi[b] = \mathrm{cl}\,(\widetilde{\mathcal{M}}_b, \tau) = \mathrm{cl}\,(\widetilde{\mathcal{M}}_b, \tau)$.

The proof is quite simple and we shall note only some points of it. First of all, $\widetilde{\mathcal{M}}_b \subset \Xi[b]$ because $\forall f \in B(E, \mathcal{L})$

$$v_\eta(E)\big|_{\mu=f*\eta} = \int_E |f|\, d\eta; \tag{4.3.17}$$

in turn, (4.3.17) can be easily checked with due regard for the definitions from Chap. 3. Owing to the fact that set (4.2.22) is $*$-weakly closed, we now have, in the form of this set, an upper bound for the closure of $\widetilde{\mathcal{M}}_b$ in the sense of $\tau_*(\mathcal{L})$, so that

$$\mathrm{cl}\,(\widetilde{M}_b, \tau_*(\mathcal{L})) \subset \mathrm{cl}\,(\widetilde{\mathcal{M}}_b, \tau_*(\mathcal{L})) \subset \Xi[b].$$

Taking account of the strong boundedness of (4.2.20) and (4.3.16), we find that the $*$-weak closures of $\widetilde{M}_b$ and $\widetilde{\mathcal{M}}_b$ coincide with their closures in the sense of $\tau_{\otimes}(\mathcal{L})$. As a consequence, we obtain the estimates

$$\mathrm{cl}\,(\widetilde{M}_b, \tau_0(\mathcal{L})) \subset \mathrm{cl}\,(\widetilde{M}_b, \tau_*(\mathcal{L})), \qquad \mathrm{cl}\,(\widetilde{\mathcal{M}}_b, \tau_0(\mathcal{L})) \subset \mathrm{cl}\,(\widetilde{\mathcal{M}}_b, \tau_*(\mathcal{L})).$$

Finally, using the Jordan decomposition and Lemma 4.3.2, we get the estimate $\Xi[b] \subset \mathrm{cl}\,(\widetilde{M}_b, \tau_0(\mathcal{L}))$. The other arguments are quite clear, and therefore we omit them. Let us consider some other assertions concerning a dense imbedding by means of the integration operator, which will not be used in the sequel. We therefore give them in a very brief form, omitting proofs (the proofs are similar to those given above and can be easily reconstructed by the reader). Actually, we shall consider a certain supplement to the assertion of Lemma 4.3.2. Let $\forall\, b \in [0, \infty)$:

$$M_*^+[b] \triangleq \left\{ f \in B_0^+(E, \mathcal{L}) \Big| \int_E f \, d\eta = b \right\}, \tag{4.3.18}$$

$$\mathcal{M}_*^+[b] \triangleq \left\{ f \in B^+(E, \mathcal{L}) \Big| \int_E f \, d\eta = b \right\}, \tag{4.3.19}$$

$$\widetilde{M}_*^+[b] \triangleq \{ f * \eta \colon f \in M_*^+[b] \},$$
$$\widetilde{\mathcal{M}}_*^+[b] \triangleq \{ f * \eta \colon f \in \mathcal{M}_*^+[b] \},$$
$$\Xi_+^*[b] \triangleq \{ \mu \in (\mathrm{add})^+ [\mathcal{L}, \eta] | \, \mu(E) = b \}.$$

We have found that (4.3.18) and (4.3.19) are nonempty sets in the space of stratum functionals on $(E, \mathcal{L})$. Of course, $\widetilde{\mathcal{M}}_*^+[b] \subset \Xi_+^*[b]$ for $b \geqslant 0$.

THEOREM 4.3.4. *Let* $b \in [0, \infty)$. *Then* $\forall\, \tau \in \mathfrak{M}$: $\Xi_+^*[b] = \mathrm{cl}\,(\widetilde{M}_*^+[b], \tau) = \mathrm{cl}\,(\widetilde{\mathcal{M}}_*^+[b], \tau)$.

The proof uses arguments connected with construction (4.3.6)–(4.3.8), and also the relationships for the topologies of the triad $\mathfrak{M}$ repeatedly mentioned earlier.

Lemma 4.3.2 and Theorems 4.3.1–4.3.4 provide a series of assertions concerning the necessary and sufficient conditions of the $\mathfrak{M}$-approximation of f.a. measures by indefinite integrals in the class of generalized sequences (nets); we speak about the approximation in the sense of any topology from the triad $\mathfrak{M}$. These topologies (for example, $\tau_*(\mathcal{L})$ and $\tau_0(\mathcal{L})$) differ substantially; nevertheless, in this concrete case of constructing approximate analogs of the Radon–Nicodym property, they behave alike.

4.4. Examples

Let us consider some effects connected with the construction of dense imbeddings of conventional solutions (stratum functionals) into the space of f.a. measures of bounded variation considered in this chapter. Our constructions were based on the property of weak absolute continuity with respect to the fixed f.a. measure η (see (4.2.1)). We shall demonstrate by an example that this property may differ essentially from the familiar property of absolute continuity [9, Chap. 3]. Simultaneously, we shall clarify some circumstances connected, in particular, with components (4.2.3) and (4.2.4).

EXAMPLE 4.4.1. Let $E = [0, 1)$. Suppose that $\mathcal{L}$ is defined as the family of all half-open intervals $[a, b)$, $0 \leqslant a \leqslant b \leqslant 1$, including the empty interval of this type; then $\mathcal{L} \in \Pi[E]$. Let us define the measure η used in (4.2.1) as the length, so that $\eta \in (\mathrm{add})_+[\mathcal{L}]$ is a (unique) measure such that

$$\eta([a, b)) = b - a \qquad (0 \leqslant a < b \leqslant 1).$$

Then the obtained realization of space (4.2.1) has the following property: $\forall L \in \mathcal{L}$

$$(\eta(L) = 0) \Leftrightarrow (L = \varnothing);$$

hence, $\mathfrak{N} = \{\varnothing\}$. But in this case, as one can easily check, with due account of (4.2.2) and (4.2.5), the following equalities hold:

$$(\mathrm{add})^+[\mathcal{L}, \eta] = (\mathrm{add})_+[\mathcal{L}], \qquad (\mathrm{add})_D^+[\mathcal{L}, \eta] = \{\mathbf{O}\}. \qquad (4.4.1)$$

This yields $\mathbf{A}(\mathcal{L}) = \mathbf{A}_\eta[\mathcal{L}]$ (see (3.3.8)) and $\mathbf{A}_\eta^d[\mathcal{L}] = \{\mathbf{O}\}$. We can see that in this example every nonnegative f.a. measure on $\mathcal{L}$ possesses the property of weak absolute continuity with respect to η; however, one can easily find a $\mu \in (\mathrm{add})_+[\mathcal{L}]$, which is not absolutely continuous with respect to η (for μ we can take, for example, the Dirac measure $(\delta_x|\mathcal{L})$, where $x \in E$). Note that the obtained decomposition of $\mathbf{A}(\mathcal{L})$ into the ordered direct sum of components (4.2.3) and (4.2.4) does not agree, in our example, with the well-known decomposition of Hewitt–Yosida because (4.2.2) contains both c.a. (countably additive) and purely f.a. measures. This follows from (4.4.1).

Let us now consider an effect arising in extending the most simple (Dirac) measure to a larger measurable space. Namely, let us fix a point $q \in E$ and then consider a semialgebra $\mathcal{I}$ of subsets of $E = [0,1)$ for which $\mathcal{L} \subset \mathcal{I}$ and $\{q\} \in \mathcal{I}$ (such a semialgebra $\mathcal{I}$ can be chosen, if desired, rather feasibly; for example, we can define $\mathcal{I}$ as an algebra generated by the family $\mathcal{L} \cup \{\{q\}\}$ and obtained from it by means of a finite number of set-theoretical operations [25, Chap. 1]). It is easy to see that $(\delta_q|\mathcal{I})$ is a (countably additive) extension of $(\delta_q|\mathcal{L})$ to a measure from $(\mathrm{add})_+[\mathcal{I}]$. Let us show that $(\delta_q|\mathcal{L})$ also has a purely f.a. extension $\gamma \in (\mathrm{add})_+[\mathcal{I}]$, which is consequently a measure for which, for any choice of a c.a. measure $\mu \colon \mathcal{I} \to [0,\infty)$, we have

$$(\mu \leq \gamma) \Rightarrow (\mu = \mathcal{O}_{\mathcal{I}}).$$

First of all, let us observe that $\exists \lambda \in (\mathrm{add})_+[\mathcal{I}] \colon \eta(L) = \lambda(L)$ for $L \in \mathcal{L}$. It is a simple consequence of the Hahn–Banach theorem (of course, here we have to take into account that $\mathbf{A}(\widetilde{\mathcal{L}})$ and $B^*(E, \widetilde{\mathcal{L}})$ are isometrically isomorphic for $\widetilde{\mathcal{L}} \in \Pi[E]$; this was already discussed in Chap. 3). Then $\forall k \in \mathcal{N}$

$$a_k \triangleq \lambda\left(\left[q, q + \frac{1-q}{k}\right)\right) = \eta\left(\left[q, q + \frac{1-q}{k}\right)\right);$$

hence, the sequence $(a_i)_{i \in \mathcal{N}}$ converges to 0. However, we have

$$0 \leqslant \lambda(\{q\}) \leqslant a_k \qquad (k \in \mathcal{N}),$$

so that $\lambda(\{q\}) = 0$. Let us introduce the set $Q_0 \triangleq (q, 1]$, which we shall endow with the ordering $\angle$, dual with respect to the natural one; in other words, for $t_1 \in Q_0$ and $t_2 \in Q_0$ the statements $t_1 \angle t_2$ and $t_2 \leqslant t_1$ are equivalent. Finally, let us associate every number $t \in Q_0$ with the function

$$F[t] \triangleq \frac{1}{t - q} \chi_{[q,t[} \in B_0^+(E, \mathcal{L}),$$

whose η-integral on the set E is equal to 1; but, of course, $F[t] * \lambda \in (\mathrm{add})_+[\mathcal{I}]$ and the λ-integral $F[t]$ on the set E coincides with the similar η-integral and, consequently, is also equal to 1. Thus,

$$\widetilde{F} \triangleq F[\cdot] * \lambda \triangleq (F[t] * \lambda)_{t \in Q_0} \tag{4.4.2}$$

is an operator mapping into a set similar to (4.2.18) under the condition that $\mathcal{L}$ in (4.2.18) is replaced by $\mathcal{I}$ and η by λ. Then (4.4.2) can also be regarded as an operator mapping into a set of (4.2.21)-type on the assumption that space (4.2.1) is replaced by $(E, \mathcal{I}, \lambda)$. Then, taking account of the compactness of the topological space (4.2.25) under the conditions of the above replacement, we can indicate: (1) a measure $\gamma_* \in (\mathrm{add})_+[\mathcal{I}]$, $\gamma_*(E) \leqslant 1$, for which (for $\Lambda \in \mathcal{I}$) we have the implication

$$(\lambda(\Lambda) = 0) \Rightarrow (\gamma_*(\Lambda) = 0);$$

(2) the directed set $(\Omega, \leqslant)$, $\Omega \neq \varnothing$; (3) an operator $\rho \colon \Omega \to Q_0$, isotonically cofinal with respect to $(\Omega, \leqslant)$ and $(Q_0, \angle)$, for which the net $(\Omega, \leqslant, \widetilde{F} \circ \rho)$ converges to γ_* in the sense of $\tau_*(\mathcal{I})$. It is easy to check, with due regard for the definition of $\tau_*(\mathcal{I})$, that $(\gamma_* | \mathcal{L}) = (\delta_q | \mathcal{L})$ and γ_* is a purely f.a. measure, i.e., a measure without nonnegative countably additive identically nonvanishing minorants. This assertion can be easily verified with regard for the continuity property at $\varnothing$ (on monotonically convergent set sequences), which every c.a. positive measure must possess. Thus, the f.a. Dirac measure $(\delta_q | \mathcal{L})$ has, along with the natural countably additive extension of $(\delta_q | \mathcal{I})$ to the semialgebra $\mathcal{I}$, $\mathcal{L} \subset \mathcal{I}$, a "pathological" extension. We could also consider some intermediate cases, combining, for example, the measures $(\delta_q | \mathcal{I})$ and γ_* with weight coefficients. If $\mathcal{I}$ consists only of Borel-measurable sets, then the countably additive extension of the measure $(\delta_q | \mathcal{L})$ is, of course, unique; in particular, such a situation occurs when $\mathcal{I}$ is defined as an algebra generated by the family $\mathcal{L} \cup \{\{q\}\}$, but the measure $(\delta_q | \mathcal{L})$ has many f.a. extensions to $\mathcal{I}$ in this simple case of the space $(E, \mathcal{I})$, and among them there necessarily exists a "pathological" (purely f.a.) extension.

    EXAMPLE 4.4.2. Let $E = \mathcal{N} = \{1, 2, \ldots\}$. We set, as usual, $\forall\, k \in \mathcal{N}$, $l \in \mathcal{N}$,

$$\overline{k, l} \triangleq \{i \in \mathcal{N} \mid (k \leqslant i)\,\&\,(i \leqslant l)\}. \tag{4.4.3}$$

We denote by $\mathcal{L}_1$ the family of all intervals (4.4.3) for k and l ranging over $\mathcal{N}$. Furthermore, $\forall\, m \in \mathcal{N} \colon \overrightarrow{m, \infty} \triangleq \{i \in \mathcal{N} \mid m \leqslant i\}$. We suppose that $\mathcal{L}_2$ is, by definition, the family of all (semiinfinite) intervals $\overrightarrow{m, \infty}$, where m ranges over $\mathcal{N}$. Throughout this example, we assume that $\mathcal{L} \triangleq \mathcal{L}_1 \cup \mathcal{L}_2$, obtaining, as a result, a semialgebra of subsets of $E \colon \mathcal{L} \in \Pi[E]$. The measurable space under consideration $(E, \mathcal{L})$ has the following features: (1) $B_0(E, \mathcal{L})$ is the set of all stationary sequences of real numbers; (2) $B(E, \mathcal{L})$ is the set of all convergent sequences of real numbers; for more details, see [36, p. 117]. Suppose that the function

$$\omega_\infty \colon \mathcal{L} \to [0, \infty)$$

is, by definition, such that $(\forall\, L \in \mathcal{L}_1 \colon \omega_\infty(L) \triangleq 0)\,\&\,(\forall\, \widetilde{L} \in \mathcal{L}_2 \colon \omega_\infty(\widetilde{L}) \triangleq 1)$. Then $\omega_\infty \in (\mathrm{add})_+[\mathcal{L}]$: moreover, ω_∞ is purely f.a. Now we shall consider some specific sequences whose elements are f.a. measures on $\mathcal{L} = \mathcal{L}_1 \cup \mathcal{L}_2$. These sequences will have the property of convergency in the sense of some topologies from the triad $\mathfrak{M}$ and not have this property in the sense of other topologies from $\mathfrak{M}$. Let us first illustrate the unique reason for the difference of the topologies $\tau_*(\mathcal{L})$ and $\tau_{\otimes}(\mathcal{L})$ (we shall give an example of this sort); here it is useful to take into account the equality on the left-hand side of (4.2.12). Thus, let us construct a $*$-weakly convergent (and strongly bounded) sequence in $\mathbf{A}(\mathcal{L})$ for which the corresponding sequences of Jordan's decomposition components will not be convergent to the similar components of the limit measure. Thus, let $\forall\, s \in \mathcal{N}$:

$$\mu_s \triangleq (\delta_s|\mathcal{L}) - (\delta_{s+1}|\mathcal{L}). \tag{4.4.4}$$

By means of (4.4.4) we have introduced a strongly bounded sequence, every element of which is the difference of two Dirac measures restricted to $\mathcal{L}$. This implies, as one can easily verify with due regard for the representation of $B(E, \mathcal{L})$ in terms of convergent sequences, that $\forall\, g \in B(E, \mathcal{L})$ the sequence

$$\int_E g\, d\mu_s \to 0 \quad \text{as} \quad s \to \infty.$$

This means that the sequence $(\mu_s)_{s \in \mathcal{N}}$ converges to $\mathbf{O}$ in space (4.2.6). Let us observe in passing that the convergence of $(\mu_s)_{s \in \mathcal{N}}$ to $\mathbf{O}$ obviously takes place both in the sense of the topology $\tau_0(\mathcal{L})$ and in the sense of the topology $\tau_{\otimes}(\mathcal{L})$. Indeed, $\forall\, L \in \mathcal{L}_1$ we have $\mu_s(L) = 0$ for almost all s because $\delta_s(L) = \delta_{s+1}(L) = 0$ from some moment. But if $L \in \mathcal{L}_2$, then, on the contrary, $\delta_s(L) = \delta_{s+1}(L) = 1$ from some moment, so that again $\mu_s(L) = 0$ for almost all s. This shows the convergence of $(\mu_1, \mu_2, \ldots)$ to $\mathbf{O}$ in the sense of $\tau_0(\mathcal{L})$, and this implies the convergence in $\tau_{\otimes}(\mathcal{L})$. Thus, the sequence of f.a. measures (4.4.4) converges in the triad $\mathfrak{M}$ to $\mathbf{O}$. Of course, we have $\mathbf{O}^+ = \mathbf{O}^- = \mathbf{O}$. However, the sequences

$$s \mapsto (\mu_s)^+ \colon \mathcal{N} \to (\mathrm{add})_+ [\mathcal{L}], \tag{4.4.5}$$

$$s \mapsto (\mu_s)^- \colon \mathcal{N} \to (\mathrm{add})_+ [\mathcal{L}] \tag{4.4.6}$$

do not converge to $\mathbf{O}$ but to the measure ω_∞. Indeed, $\mu_s^+ \triangleq (\mu_s)^+ = (\delta_s|\mathcal{L})$ since $(\delta_s|\mathcal{L}) \in \mathbf{A}(\mathcal{L})$ majorizes $\{\mu_s, \mathbf{O}\}$. If now $\gamma \in \mathbf{A}(\mathcal{L})$ majorizes $\{\mu_s, \mathbf{O}\}$, then $\gamma(L) \geqslant 0$, and we have $\mu_s(L) \leqslant \gamma(L)$ on $\mathcal{L}$. If $L \in \mathcal{L}$ and $s \in L$, then $\{s\} \subset L$ and, hence,

$$1 = \mu_s(\{s\}) \leqslant \gamma(\{s\}) \leqslant \gamma(L).$$

This means that $(\delta_s|\mathcal{L}) \le \gamma$. Thus, $(\delta_s|\mathcal{L})$ is the least majorant of $\{\mu_s, \mathbf{O}\}$ in $\mathbf{A}(\mathcal{L})$, so that $\mu_s^+ = (\delta_s|\mathcal{L})$. But the sequence

$$s \mapsto (\delta_s|\mathcal{L})\colon \mathcal{N} \to (\mathrm{add})_+ [\mathcal{L}]$$

converges to ω_∞ even in the sense of $\tau_0(\mathcal{L})$, and, consequently, sequence (4.4.5) converges to ω_∞ in $\tau_0^+(\mathcal{L})$, and, hence, in the sense of $\tau_*^+(\mathcal{L}) = \tau_\otimes^+(\mathcal{L})$. We have established the convergence of sequence (4.4.5) to ω_∞ in the sense of any topology of the triad $\mathfrak{M}$. The convergence of sequence (4.4.6) to ω_∞ can be verified by analogy. Here we take into account that $\forall s \in \mathcal{N}\colon \mu_s^- \triangleq (\mu_s)^- = (\delta_{s+1}|\mathcal{L})$. Indeed, $(\delta_{s+1}|\mathcal{L})$ majorizes $\{-\mu_s, \mathbf{O}\}$. But if $\zeta \in \mathbf{A}(\mathcal{L})$ is an arbitrary majorant of this two-element set, i.e., $(-\mu_s \le \zeta) \& (\mathbf{O} \le \zeta)$, then $\zeta \in (\mathrm{add})_+ [\mathcal{L}]$, and, in particular, ζ is monotonic on $\mathcal{L}$ [36, p. 97]. Then we have $\forall L \in \mathcal{L}$:

$$\delta_{s+1}(L) - \delta_s(L) \le \zeta(L);$$

if $s + 1 \in L$ in this case, then the last inequality yields

$$1 = \delta_{s+1}(\{s+1\}) \le \zeta(\{s+1\}) \le \zeta(L)$$

and, consequently, $\delta_{s+1}(L) \le \zeta(L)$. This means that ζ majorizes $(\delta_{s+1}|\mathcal{L})$, so that this measure is the least majorant of $\{-\mu_s, \mathbf{O}\}$, i.e.,

$$(\delta_{s+1}|\mathcal{L}) = (-\mu_s)^+ = \mu_s^-.$$

This representation also guarantees the convergence of (4.4.6) to ω_∞ with respect to any topology of the triad $\mathfrak{M}$. Indeed, the convergence in the sense of $\tau_0^+(\mathcal{L})$ being obvious (we must use the representation of topology (4.2.14) for $H = (\mathrm{add})_+ [\mathcal{L}]$ in terms of coincidence of values of f.a. measures on finite subfamilies of $\mathcal{L}$), all other types of convergence can be verified with the aid of (4.2.12). We have verified that the sequences consisting of components of the Jordan decomposition of measures (4.4.4) converge with respect to the triad $\mathfrak{M}$, but not where it is required, i.e., not to $\mathbf{O}^+$ and $\mathbf{O}^-$ respectively. Thus, the projection in the sense of Jordan is not sequentially continuous (and, hence, is not continuous at all) with respect to the topologies of the triad $\mathfrak{M}$. It is this circumstance which, as we shall see now, makes it impossible to establish, with the aid of the left-hand equality in (4.2.12), the coincidence of $\tau_*(\mathcal{L})$ and $\tau_\otimes(\mathcal{L})$. Let us consider the sequence

$$s \mapsto s\mu_s\colon \mathcal{N} \to \mathbf{A}(\mathcal{L}), \tag{4.4.7}$$

where μ_s satisfies (4.4.4). This sequence of f.a. measures is no longer strongly bounded. Nevertheless, sequence (4.4.7) converges to $\mathbf{O}$ in the sense of $\tau_0(\mathcal{L})$. Indeed, if $L \in \mathcal{L}_1$, then $\delta_s(L) = \delta_{s+1}(L) = 0$ for almost all s. Therefore, $(s\mu_s)(L) = 0$ (for almost all $s \in \mathcal{N}$) according to (4.4.4). But if $\widetilde{L} \in \mathcal{L}_2$, then $\delta_s(\widetilde{L}) = \delta_{s+1}(\widetilde{L}) = 1$ for almost all $s \in \mathcal{N}$; therefore, we again have $(s\mu_s)(\widetilde{L}) = 0$ from some moment. In all cases, the values of $s\mu_s$ on any set from $\mathcal{L}$ begin to coincide with similar values of the measure $\mathbf{O}$ for sufficiently large s. Of course, this "stationary state" is not uniform in $L \in \mathcal{L}$, but this is not required for proving the convergence in the sense of $\tau_0(\mathcal{L})$. The established property implies, of course, the convergence of sequence (4.4.7) in the sense of the topology $\tau_\otimes(\mathcal{L})$. We shall show, however, that sequence (4.4.7) does not converge to $\mathbf{O}$ in the sense of $\tau_*(\mathcal{L})$. Indeed, let us introduce a sequence $\varphi \colon E \to \mathbf{R}$, assuming for $\mathcal{M} \triangleq \{2k \colon k \in \mathcal{N}\}$ that

$$\Big(\forall m \in \mathcal{M}\colon \varphi(m) \triangleq \frac{1}{m}\Big) \& \Big(\forall i \in \mathcal{N} \setminus \mathcal{M}\colon \varphi(i) \triangleq -\frac{1}{i}\Big).$$

In this case, φ is a convergent sequence in $\mathbf{R}$, so that $\varphi \in B(E, \mathcal{L})$. If $s \in \mathcal{M}$, then

$$\int_E \varphi \, d(s\mu_s) = \Big(s \int_E \varphi \, d(\delta_s | \mathcal{L})\Big) - \Big(s \int_E \varphi \, d(\delta_{s+1} | \mathcal{L})\Big)$$

$$= s(\varphi(s) - \varphi(s+1)) = s\Big(\frac{1}{s} + \frac{1}{s+1}\Big) = 1 + \frac{s}{s+1} \geqslant 1.$$

This means, in particular, that the sequence

$$s \mapsto \int_E \varphi \, d(s\mu_s) \colon \mathcal{N} \to \mathbf{R}$$

does not converge to 0, which would necessarily occur in the case of a $*$-weak convergence of sequence (4.4.7) to $\mathbf{O}$. We have established that in space (4.2.6) sequence (4.4.7) does not converge to $\mathbf{O}$; this shows that $\tau_*(\mathcal{L}) \neq \tau_\otimes(\mathcal{L})$. It stands to reason that the topologies $\tau_*(\mathcal{L})$ and $\tau_0(\mathcal{L})$ do not coincide here either. Moreover, they are incongruent. Indeed, we have already seen a sequence converging to $\mathbf{O}$ in the sense of $\tau_0(\mathcal{L})$ but not converging in the sense of $\tau_*(\mathcal{L})$. It is easy to construct a sequence with the converse property. In fact, if we fix $l \in \mathcal{N}$, then the sequence

$$k \mapsto \frac{1}{k}(\delta_l | \mathcal{L}) \colon \mathcal{N} \to (\mathrm{add})_+[\mathcal{L}] \tag{4.4.8}$$

converges to $\mathbf{O}$ strongly and, in particular, $*$-weakly (i.e., in the sense of $\tau_*(\mathcal{L})$); however, it does not converge to $\mathbf{O}$ in the sense of $\tau_0(\mathcal{L})$ since $\forall\, k \in \mathcal{N}$:

$$\left(\frac{1}{k}(\delta_l|\mathcal{L})\right)(E) = \frac{1}{k} \neq 0 = \mathbf{O}(E).$$

Note, in passing, that $\tau_0(\mathcal{L})$ and the topology generated by the (strong) variation-norm (3.3.9) are also incongruent. Indeed, we have already seen in examples (see sequence (4.4.8)) that convergence in the strong topology does not imply, generally speaking, convergence in the sense of $\tau_0(\mathcal{L})$. On the other hand, as we already know, the sequence of Dirac measures $(\delta_s|\mathcal{L})$ converges to ω_∞ in the sense of $\tau_0(\mathcal{L})$. However, $\forall\, s \in \mathcal{N}$:

$$1 = |(\delta_s|\mathcal{L})(\overrightarrow{s+1,\infty}) - \omega_\infty(\overrightarrow{s+1,\infty})|$$

$$= |((\delta_s|\mathcal{L}) - \omega_\infty)(\overrightarrow{s+1,\infty})| \leqslant v_\mu(\overrightarrow{s+1,\infty})|_{\mu=(\delta_s|\mathcal{L})-\omega_\infty},$$

so that (see (3.3.8)) the sequence of Dirac measures given above does not strongly converge to ω_∞.

Chapter 5

Admissible Sets and Their Relaxations

5.1. Introduction

Making use of the propositions of Chap. 4 (on the imbedding of conventional solutions (step-functionals) into the space of f.a. measures), we consider a key problem for the subsequent extension constructions for extremal problems, namely, that of the behavior of admissible sets (sets of admissible elements) under the perturbation of certain parameters of a given complex of constraints. The problem of dependence of admissible sets requires regularization, which was shown in the examples of Chap. 1; as before, by a perturbation of a complex of constraints we mean their weakening. Such a progressive weakening of constraints results in a corresponding asymptotics of admissible sets, and different constructions of perturbations are associated with different asymptotics. It is of interest, not only from the theoretical but also from the practical point of view, to compare these asymptotics with a view toward making some inferences concerning the degree of (asymptotic) sensitivity of a problem to perturbation of its different parameters.

In what follows, we consider the constraints on the vector integrand that are conceptually equivalent to condition (1.3.3). By selecting a conventional solution (a step-functional f), we must realize the value of the vector integrand as an element of a given set (in (1.3.3) this set is denoted by Y). However, in a number of cases we can encounter situations more general than that of (1.3.3), in particular, a need to consider infinite (more precisely, infinite-dimensional) integrands. By the way, in a control problem similar to that discussed in Sec. 1.3, the system of periods may often not be specified and, at the same time, the requirement of alternation of "impulses" and "pauses" from Sec. 1.3 may, in essence, be present but admit arbitrary shifts of the normal picture of alternation in time. To represent such a constraint, we should, however, give up the requirement that the set $\{s_1, \ldots, s_n\}$ be finite. Other situations of this kind can occur; for example, a problem can be encountered of a constraint of the type of an infinite system of inequalities imposed on the integrands of the "control" f and perturbations of such a constraint on finite subsystems. In any case, it is of interest to consider constraints on an infinite system of integrands. Thus, we shall suppose that we are given a nonempty set Γ and an operator

$$(S_\gamma)_{\gamma \in \Gamma} \colon \Gamma \to B(E, \mathcal{L}), \tag{5.1.1}$$

putting every element (index) $\gamma \in \Gamma$ into correspondence with a stratum functional S_γ on the (fixed) measurable space $(E, \mathcal{L})$, $E \neq \varnothing$, with a semialgebra of sets. If we select the conventional solution $f \in B_0(E, \mathcal{L})$, i.e., $\mathcal{L}$ is a step-functional on E, then we have the functional

$$\gamma \mapsto \int_E S_\gamma f \, d\eta \colon \Gamma \to \mathbf{R}, \tag{5.1.2}$$

on which we shall impose our constraints. For their formulation we introduce a set $\mathbf{Y} \in \mathcal{P}(\mathbf{R}^\Gamma)$, i.e.,

$$\mathbf{Y}, \quad \mathbf{Y} \subset \mathbf{R}^\Gamma, \tag{5.1.3}$$

assumed to be closed in the topology $\otimes \Gamma(\tau_{\mathbf{R}})$ of the pointwise convergence in $\mathbf{R}^\Gamma$. Then our main condition on the choice of $f \in B_0(E, \mathcal{L})$ has the form

$$\left(\int_E S_\gamma f \, d\eta \right)_{\gamma \in \Gamma} \in \mathbf{Y}. \tag{5.1.4}$$

In other words, the realization of functional (5.1.2) obeys the requirement of coincidence with an element of set (5.1.3). In a number of cases, we will also impose the condition of nonnegativity on the choice of f, i.e., $f \in B_0^+(E, \mathcal{L})$, which is typical for problems of practical interest. Now, however, we shall focus our attention on discussing inclusion (5.1.4). In speaking about it, however, we should bear in mind that the dependence of the set of admissible elements on the parameters defining $\mathbf{Y}$ is rather irregular. The corresponding examples from Chap. 1 confirm it well enough and indicate natural methods of relaxation. First of all, we must obviously give up the idea of a complete coincidence of the integrand as a functional on Γ and an element of $\mathbf{Y}$; instead, we should, apparently, tend to an approximate coincidence. Following this path, we shall come across some natural means of weakening this condition. Perhaps, the most preferable variant would be the postulation of an approximate coincidence with an estimation of the proximity in the uniform metrics. If Γ is finite, as it was in Chap. 1, then this version is realized almost automatically as the most natural one and all our subsequent conclusions are completely applicable for the analysis of this situation. If the set Γ is infinite, then serious difficulties arise. For a general method of relaxation, we select, in this connection, a different construction based on perturbation over "finite subsystems." A simple example of this construction consists in replacing (5.1.4) by the requirement of a choice of f such that the f-integrand (the functional on the left-hand side of (5.1.4)) would coincide with the values of an appropriate functional $y \in \mathbf{Y}$ at the points γ of a (finite) set $K \in \mathrm{Fin}\,(\Gamma)$; the latter plays the role of a parameter. One could require the realization (on K) of the ε-coincidence with $y(\gamma)$, $\gamma \in K$, where $\varepsilon > 0$. In that case, we would get another, still less severe version of constructing weakened constraints. We shall try, however, to get such a construction for which the two above-mentioned versions of weakening the entire system of constraints would be special cases. The following is a simple way of realizing these relaxations. We introduce a set Γ_0, $\Gamma_0 \subset \Gamma$, of singular indices γ. For them we shall try to achieve a precise coincidence of the values of the f-integrand with $y(\gamma)$, and outside Γ_0, merely an ε-coincidence. In this case, we again use, as parameters, $K \in \mathrm{Fin}\,(\Gamma)$ and $\varepsilon > 0$, taking into account the singular indices within the set $K \cap \Gamma_0$. Other indices, for which we postulate the ε-coincidence, we consider within the set $K \setminus \Gamma_0$. Of course (for given K and ε) the functional $y = y(f, K, \varepsilon)$, with respect to which we realize the approximate coincidence of the f-integrand, is constrained by the unique condition $y \in \mathbf{Y}$.

Note that Γ_0 is a parameter defining the concrete version of asymptotics of admissible sets. It is of great interest to compare different asymptotics, i.e., different sets Γ_0 in the sense of regularizations obtained. It turns out that over wide limits there occurs an asymptotic equivalence of different regularizations. The case $\Gamma_0 = \varnothing$ is not excluded. It plays an important role; the relaxations arising in this case can be conventionally called Tychonoff relaxations, since here we replace condition (5.1.4) by the requirement of ε-coincidence of the values of the f-integrand and a certain functional $y \in \mathbf{Y}$ on the set $K \in \mathrm{Fin}\,(\Gamma)$. This extreme case will be distinguished in the sequel with the use of a special notation and separate justifications, although we could do without its separate consideration. However, we shall consider it with methodical aims, since this construction admits natural analogies with the classical notion of the Tychonoff product and is, to some extent, canonical. As for the other cases of the choice of Γ_0, $\Gamma_0 \subset \Gamma$, we shall establish a general condition connected with the requirement that

$$S_\gamma \in B_0(E, \mathcal{L}) \qquad (\gamma \in \Gamma_0). \tag{5.1.5}$$

Condition (5.1.5) means that operator (5.1.1) is step-valued on Γ_0; it will specify the class of equivalent asymptotics for admissible sets. It is of interest to consider the case where operator (5.1.1) is fully step-valued, so that in (5.1.5) one can set $\Gamma_0 = \Gamma$ (we have already discussed a concrete method of perturbing condition (5.1.4) in terms of precise matching of the f-integrand with a "suitable" functional $y \in \mathbf{Y}$ on the finite subsets of Γ); in Chap. 1 we saw concrete examples of this kind (see (1.3.6)). Some general issues of computational stability will be discussed in Chap. 9, where, by analogy with Chap. 1, we shall study the case of the finite set Γ and the step-valued operator (5.1.1). Now we shall consider the general case. The results obtained will be further developed in two directions: (1) in the next chapter we shall consider perturbations on the so-called attainable (in the strength of an operator) sets, (2) the limit representations of admissible sets will define generalized extremal problems which, in turn, will play, in our study, the role of the standard problems of Chap. 2. Within the framework of these generalized representations, we shall admit different versions of the set Γ_0, which will allow us to make rather substantive inferences concerning the relaxations of the corresponding concrete extremal problem. We will separately consider the setting of the problem in which the choice of a "conventional" solution f will be, along with (5.1.4), constrained by the additional nonnegativity condition. In this case, we succeed in moving further ahead in studying relaxations of the general kind and derive additional propositions whose extension to the "alternating" case is difficult for a number of reasons, among which we should note the effects observed in the last section of the preceding chapter.

5.2. Asymptotics of Perturbed Constraints and Its Limit Representation (Case of Positive Measures)

Let us consider a basic construction of relaxations of the general condition (5.1.4); an appropriate discussion of this construction was carried out in Sec. 5.1. We shall now introduce the admissible sets of the perturbed constraints mentioned above and consider a generalized representation of the corresponding limit set. Let us assume the following notation: if $\mathcal{H}$ is a nonempty family of subsets of $B(E,\mathcal{L})$ and τ is a topology in $\mathbf{A}(\mathcal{L})$, then

$$(\eta - \mathrm{Lim})[\mathcal{H},\tau] \triangleq \bigcap_{H \in \mathcal{H}} \mathrm{cl}\left(\{f * \eta : f \in H\}, \tau\right). \tag{5.2.1}$$

Sets of this type will be used in the sequel for writing down the corresponding set-theoretic limits (of course, we endow $\mathcal{H}$ with the properties that make the given definition of this limit natural). The simplest version of perturbation of condition (5.1.4) has the form

$$f \in B_0(E,\mathcal{L}), \qquad \left\{ y \in \mathbf{Y} : \sup_{\gamma \in K}\left| \int_E S_\gamma f \, d\eta - y(\gamma) \right| \leqslant \varepsilon \right\} \neq \varnothing, \tag{5.2.2}$$

where $K \in \mathrm{Fin}\,(\Gamma)$ and $\varepsilon > 0$. In essence, (5.2.2) imposes a constraint on the choice of f, namely, the choice is restrained by the condition of the ε-proximity of the values of the f-integrand and an element of $\mathbf{Y}$ at the points of $\gamma \in K$. Condition (5.2.2) naturally generates an admissible set or a set of all elements f admissible in the sense of (5.2.2). In reality, condition (5.2.2) can be considered as a special case of a more general perturbation construction that we discussed in the preceding section; this construction uses the special set $\Gamma_0 \in \mathcal{P}(\Gamma)$ as a parameter by forming the condition

$$f \in B_0(E,\mathcal{L}), \qquad \left\{ y \in \mathbf{Y} \left| \left(\forall \gamma \in K \cap \Gamma_0 : \int_E S_\gamma f \, d\eta = y(\gamma) \right) \& \right. \right.$$

$$\left. \left(\forall \gamma \in K \setminus \Gamma_0 : \left| \int_E S_\gamma f \, d\eta - y(\gamma) \right| \leqslant \varepsilon \right) \right\} \neq \varnothing,$$

where K and ε are the same as in (5.2.2). Namely, the reduction of this condition to the form (5.2.2) is realized by imposing the requirement $\Gamma_0 = \varnothing$. Nevertheless, we shall distinguish (5.2.2) among the versions with various sets Γ_0. We shall call relaxations of this type, which have the most natural character from the topological standpoint, Tychonoff relaxations, thus attracting attention to their natural relations with the constructions of Tychonoff's product. We set $\forall K \in \mathrm{Fin}\,(\Gamma)$, $\varepsilon \in [0,\infty)$:

$$\Omega(K,\varepsilon) \triangleq \left\{ f \in B_0(E,\mathcal{L}) \left| \exists y \in \mathbf{Y} : \sup_{\gamma \in K}\left| \int_E S_\gamma f \, d\eta - y(\gamma) \right| \leqslant \varepsilon \right. \right\}; \tag{5.2.3}$$

$$\Omega^+(K,\varepsilon) \triangleq \Omega(K,\varepsilon) \cap B_0^+(E,\mathcal{L})$$

$$= \left\{ f \in B_0^+(E,\mathcal{L}) \,\middle|\, \exists\, y \in \mathbf{Y}: \sup_{\gamma \in K} \left| \int_E S_\gamma f \, d\eta - y(\gamma) \right| \leqslant \varepsilon \right\}. \qquad (5.2.4)$$

In the case of $\varepsilon > 0$, we get as (5.2.3) the admissible set with respect to condition (5.2.2); the set (5.2.4) is admissible from the point of view of a condition similar to (5.2.2) but restrained by the additional requirement of nonnegativity of f. We also introduce similar definitions for admissible sets of the general form, by setting $\forall\, \Gamma_* \in \mathcal{P}(\Gamma)$, $K \in \operatorname{Fin}(\Gamma)$, $\varepsilon \in [0,\infty)$:

$$\Omega_0(\Gamma_*,K,\varepsilon) \triangleq \left\{ f \in B_0(E,\mathcal{L}) \,\middle|\, \exists\, y \in \mathbf{Y}: \left(\forall\, \gamma \in K \cap \Gamma_*: \int_E S_\gamma f \, d\eta = y(\gamma) \right) \& \right.$$

$$\left. \left(\forall\, \gamma \in K \setminus \Gamma_*: \left| \int_E S_\gamma f \, d\eta - y(\gamma) \right| \leqslant \varepsilon \right) \right\}; \qquad (5.2.5)$$

$$\Omega_0^+(\Gamma_*,K,\varepsilon) \triangleq \Omega_0(\Gamma_*,K,\varepsilon) \cap B_0^+(E,\mathcal{L}). \qquad (5.2.6)$$

For $\varepsilon > 0$, condition (5.2.5) defines the admissible set with respect to the general condition if $\Gamma_0 = \Gamma_*$. Of course, here we have

$$\forall\, K \in \operatorname{Fin}(\Gamma), \quad \varepsilon \in [0,\infty): \ \Omega(K,\varepsilon) = \Omega_0(\varnothing,K,\varepsilon), \qquad (5.2.7)$$

$$\Omega^+(K,\varepsilon) = \Omega_0^+(\varnothing,K,\varepsilon). \qquad (5.2.8)$$

As we have already noticed, relations (5.2.7) and (5.2.8) allow us to interpret construction (5.2.3), (5.2.4) as a special case of (5.2.5), (5.2.6).

We now set

$$\mathcal{T}_+ \triangleq \{\Omega^+(K,\varepsilon): (K,\varepsilon) \in \operatorname{Fin}(\Gamma) \times (0,\infty)\}, \qquad (5.2.9)$$

$$\mathcal{T} \triangleq \{\Omega(K,\varepsilon): (K,\varepsilon) \in \operatorname{Fin}(\Gamma) \times (0,\infty)\}. \qquad (5.2.10)$$

In (5.2.9), (5.2.10), we defined the families of admissible sets corresponding to Tychonoff's relaxations. Moreover, $\forall\, \Gamma_* \in \mathcal{P}(\Gamma)$:

$$\mathcal{T}_0^+[\Gamma_*] \triangleq \{\Omega_0^+(\Gamma_*,K,\varepsilon): (K,\varepsilon) \in \operatorname{Fin}(\Gamma) \times (0,\infty)\}, \qquad (5.2.11)$$

$$\mathcal{T}_0[\Gamma_*] \triangleq \{\Omega_0(\Gamma_*,K,\varepsilon): (K,\varepsilon) \in \operatorname{Fin}(\Gamma) \times (0,\infty)\} \qquad (5.2.12)$$

are families of admissible sets corresponding to relaxations of special type; in this case we obviously have, by virtue of (5.2.7), (5.2.8), the equalities $\mathcal{T}_0^+[\varnothing] = \mathcal{T}_+$, $\mathcal{T}_0[\varnothing] = \mathcal{T}$. By relations (5.2.9)–(5.2.12) we have introduced nonempty families of subsets of $B_0(E,\mathcal{L})$ so that operation (5.2.1) can be applied to any of these families. The study of the values of (5.2.1) at the points of $\mathcal{T}_+$, $\mathcal{T}$, $\mathcal{T}_0[\Gamma_*]$, $\mathcal{T}_0^+[\Gamma_*]$ will be the subject of our investigation in this chapter. This section is devoted to the case where we take for $\mathcal{H}$ (in relation (5.2.1)) the families $\mathcal{T}_+$ and $\mathcal{T}_0^+[\Gamma_*]$ for some $\Gamma_* \in \mathcal{P}(\Gamma)$. It is advisable to distinguish this case for some reasons. The choice of the parameter Γ_* will be restrained by one additional requirement that is essential for the assertions obtained in the sequel. We set

$$(\text{Step})\,[\Gamma] \triangleq \{\Gamma_0 \in \mathcal{P}(\Gamma)\,|\, \forall\, \gamma \in \Gamma_0\colon S_\gamma \in B_0(E, \mathcal{L})\}; \tag{5.2.13}$$

this introduces the family of all sets of step-valuedness for operator (5.1.1). In the subsequent constructions, we shall take for Γ_* the elements of (5.2.13), so that the "directions" of the exact coincidence with the elements of $\mathbf{Y}$ will be subjected to the requirement that (5.1.1) must be step-valued; with the examples given in Sec. 1.2, it is already easy to see that this nontraditional condition is essential. Let

$$\tilde{\Omega}_+ \triangleq \left\{\mu \in (\text{add})^+\,[\mathcal{L}, \eta]\,\Big|\, \left(\int_E S_\gamma\, d\mu\right)_{\gamma \in \Gamma} \in \mathbf{Y}\right\}. \tag{5.2.14}$$

We have defined an admissible set corresponding to the generalized conditions imposed on the "full" integrand, bearing in mind the work in the positive cone of the space of weakly absolutely continuous (with respect to η) f.a. measures.

THEOREM 5.2.1. *Let* $\tau \in \mathfrak{M}$. *Then*

$$(\tilde{\Omega}_+ = (\eta - \text{Lim})[\mathcal{T}_+, \tau])\,\&\,(\forall\, \Gamma_* \in (\text{Step})\,[\Gamma]\colon \tilde{\Omega}_+ = (\eta - \text{Lim})[\mathcal{T}_0^+[\Gamma_*], \tau]).$$

The proof uses obvious estimates for the topologies of the triad $\mathfrak{M}$, and, in particular, essential use is made of (4.2.12). We first establish the inclusion

$$(\eta - \text{Lim})[\mathcal{T}_+, \tau_*(\mathcal{L})] \subset \tilde{\Omega}_+;$$

for this we make use of the simplest properties of the Tychonoff product $(\mathbf{R}^\Gamma, \otimes^\Gamma(\tau_\mathbf{R}))$, with respect to which the set $\mathbf{Y}$ is closed, and of the definition of $*$-weak topology. The inclusion

$$\tilde{\Omega}_+ \subset (\eta - \text{Lim})[\mathcal{T}_0^+[\Gamma_*], \tau_0(\mathcal{L})]$$

is established for $\Gamma_* \in (\text{Step})\,[\Gamma]$ with due regard for Theorem 4.3.1; we also use Birkhoff's theorem (see Sec. 2.2) and relation (4.2.12). The remaining constructions take into account the comparative estimates for the triad topologies and are clear. We restrict ourselves to this brief scheme, omitting the unwieldy details.

Theorem 5.2.1 claims the existence for (5.2.11) of a common asymptotics for the indicated family (invariant with respect to the choice of the parameter Γ_* within the set (5.2.13)); if the condition that Γ_* belongs to the family (5.2.13) is violated, then, as the example of relaxations of the form (1.2.9) for the "severe" condition (1.2.8) shows, there may occur, generally speaking, some other asymptotics. In this statement, we restrict ourselves to meaningful reasoning (the corresponding rigorous computation can be easily done by the reader with the use of the assertions from Chap. 4). Let us note, as a simple corollary of Theorem 5.2.1, a statement concerning the sufficient conditions of asymptotic consistency. Namely, we have

$$(\widetilde{\Omega}_+ \neq \varnothing) \Rightarrow ((\varnothing \notin \mathcal{T}_+)\&(\forall\,\Gamma_* \in (\text{Step})\,[\Gamma]: \varnothing \notin \mathcal{T}_0^+[\Gamma_*])). \qquad (5.2.15)$$

Relation (5.2.15) implies, with due account of (5.2.9) and (5.2.11), that if the set (5.2.14) is nonempty, each of the families $\mathcal{T}_+$ and $\mathcal{T}_0^+[\Gamma_*]$, $\Gamma_* \in (\text{Step})\,[\Gamma]$, is the base for the respective filter. It should be emphasized that until now we did not assume any boundedness conditions for the admissible sets; the set (5.2.14) is not, generally speaking, strongly bounded. We shall return to this issue in Sec. 5.4.

5.3.   Asymptotics of Perturbed Constraints and Its Limit Representation (Case of Alternating Measures)

Here we consider assertions similar to Theorem 5.2.1 but relating to the asymptotics of families (5.2.10) and (5.2.12). The situation is now, generally speaking, more complicated, which is, to some extent, due to the effects illustrated in Sec. 4.4 by concrete examples. We also introduce the generalized admissible set

$$\widetilde{\Omega} \triangleq \left\{ \mu \in \mathbf{A}_\mu[\mathcal{L}] \,\bigg|\, \left(\int_E S_\gamma\, d\mu \right)_{\gamma \in \Gamma} \in \mathbf{Y} \right\}, \qquad (5.3.1)$$

which is conceptually similar to the set (5.2.14).

THEOREM 5.3.1. *The generalized admissible set (5.3.1) defines the ∗-weak set-theoretic limit of asymptotics (5.2.10) and (5.2.12):*

$$((\eta - \text{Lim})[\mathcal{T}, \tau_*(\mathcal{L})] = \widetilde{\Omega})\&$$

$$(\forall\,\Gamma_* \in (\text{Step})\,[\Gamma]: (\eta - \text{Lim})[\mathcal{T}_0[\Gamma_*]; \tau_*(\mathcal{L})] = \widetilde{\Omega});$$

furthermore, $\forall\, \Gamma^* \in (\mathrm{Step})\,[\Gamma]$:

$$\tilde{\Omega} \subset (\eta - \mathrm{Lim}\,)[\mathcal{T}_0[\Gamma^*], \tau_0(\mathcal{L})].$$

PROOF. From the definitions given in Sec. 5.2 it easily follows that $\forall\, T \in \mathcal{T}$

$$\mathrm{cl}\,(\{f * \eta\colon f \in T\}, \tau_*(\mathcal{L})) \subset \mathbf{A}_\eta[\mathcal{L}], \tag{5.3.2}$$

$$(\eta - \mathrm{Lim}\,)[\mathcal{T}_0[\Gamma_*], \tau_*(\mathcal{L})] \subset (\eta - \mathrm{Lim}\,)[\mathcal{T}, \tau_*(\mathcal{L})], \tag{5.3.3}$$

$\Gamma_* \in (\mathrm{Step})\,[\Gamma]$. Let us show that

$$(\eta - \mathrm{Lim}\,)[\mathcal{T}, \tau_*(\mathcal{L})] \subset \tilde{\Omega}. \tag{5.3.4}$$

We select an arbitrary element λ of the set on the left-hand side of (5.3.4). Taking account of (5.3.2), we have $\lambda \in \mathbf{A}_\eta[\mathcal{L}]$. Let

$$u \triangleq \left(\int_E S_\gamma\, d\lambda \right)_{\gamma \in \Gamma}.$$

Then $u \in \mathbf{Y}$. Indeed, let us assume the contrary, i.e., $u \notin \mathbf{Y}$. With due account of the closedness of the set (5.1.3) in the sense of the topology $\otimes^\Gamma(\tau_\mathbf{R})$, we can find $n \in \mathcal{N}$, $\varkappa \in (0, \infty)$, and a collection

$$(p_i)_{i \in \overline{1,n}}\colon \overline{1, n} \to \Gamma,$$

for which the set $\Lambda \triangleq \{s \in \mathbf{R}^\Gamma \mid \forall\, i \in \overline{1,n}\colon |s(p_i) - u(p_i)| < \varkappa\}$ is contained in $\mathbf{R}^\Gamma \setminus \mathbf{Y}$. We find that $Q \triangleq \{p_i\colon i \in \overline{1,n}\} \in \mathrm{Fin}\,(\Gamma)$ so that $\Omega(Q, \varkappa/4) \in \mathcal{T}$; the choice of λ immediately implies that

$$\lambda \in \mathrm{cl}\,(\{f * \eta\colon f \in \Omega(Q, \varkappa/4)\}, \tau_*(\mathcal{L})).$$

This implies, in turn, that the set

$$V \triangleq \left\{ \nu \in \mathbf{A}(\mathcal{L}) \,\middle|\, \forall\, i \in \overline{1,n}\colon \left| \int_E S_{p_i}\, d\lambda - \int_E S_{p_i}\, d\nu \right| < \frac{\varkappa}{4} \right\}$$

cuts the set $\{f * \eta\colon f \in \Omega(Q, \varkappa/4)\}$; this fact follows directly from the definition of the $*$-weak topology. Let us choose $\varphi \in \Omega(Q, \varkappa/4)$ so that $\varphi * \eta \in V$. Then, for some choice of $\rho \in \mathbf{Y}$, we have $\forall\, \gamma \in Q$:

$$\left| \int_E S_\gamma \varphi\, d\eta - \rho(\gamma) \right| \leqslant \frac{\varkappa}{4}. \tag{5.3.5}$$

Now we have to take account of the definition of V, which implies that $\forall \gamma \in Q$:

$$\left| \int_E S_\gamma \, d\lambda - \int_E S_\gamma \varphi \, d\eta \right| \leqslant \frac{\varkappa}{4}.$$

Taking account of (5.3.5) and the definition of the function u, we infer that $\forall\, i \in \overline{1, n}$:

$$|u(p_i) - \rho(p_i)| \leq \frac{\varkappa}{2} < \varkappa,$$

so that $\rho \in \Lambda$, and, as a consequence, $\rho \in \mathbf{R}^\Gamma \setminus \mathbf{Y}$, and this is impossible. This contradiction shows that $u \in \mathbf{Y}$. But, by virtue of (5.3.1), this implies that $\lambda \in \widetilde{\Omega}$, and this establishes inclusion (5.3.4). Note that from the definition of $\tau_0(\mathcal{L})$ it easily follows that for any choice of the net $(D, \angle, d)$ in $\mathbf{A}(\mathcal{L})$ and the measure $\mu \in \mathbf{A}(\mathcal{L})$ the convergence of $(D, \angle, d)$ to μ takes place if and only if $\forall\, L \in \mathcal{L}$ we have $d(\delta)(L) = \mu(L)$ from some moment, i.e., for $\delta_0 \angle \delta$, where $\delta_0 = \delta_0(L) \in D$. This property can be easily extended to the elementary integrals of Chap. 3 because they can be represented by finite sums of the values of the measure with the corresponding coefficients. Let now $\Gamma_* \in (\mathrm{Step})\,[\Gamma]$. Then $S_\gamma \in B_0(E, \mathcal{L})$ for $\gamma \in \Gamma_*$ so that we can apply the stationarity property that we have just mentioned to the integrals of the indicated functions. Let us prove the inclusion

$$\widetilde{\Omega} \subset (\eta - \mathrm{Lim}\,)[\mathcal{T}_0[\Gamma_*], \tau_*(\mathcal{L})]. \tag{5.3.6}$$

We take an arbitrary $\omega \in \widetilde{\Omega}$ so that $\omega \in \mathbf{A}_\eta[\mathcal{L}]$ and

$$v \triangleq \left(\int_E S_\gamma \, d\omega \right)_{\gamma \in \Gamma} \in \mathbf{Y}. \tag{5.3.7}$$

Then $\omega^+ \in (\mathrm{add})^+\,[\mathcal{L}, \eta]$ and $\omega^- \in (\mathrm{add})^+\,[\mathcal{L}, \eta]$ by the properties of the band $\mathbf{A}_\eta[\mathcal{L}]$. As a result, with respect to ω^+ and ω^- we can apply the construction of the net of Lemma 4.3.1 by identifying μ in its conditions either with ω^+ or with ω^-. Then, as one can easily check, the net

$$(\mathcal{D}, \prec, ((\Theta_\alpha^+[\mathcal{K}] - \Theta_\beta^+[\mathcal{K}]) * \eta)_{\mathcal{K} \in \mathcal{D}}), \tag{5.3.8}$$

where $\alpha \triangleq \omega^+$ and $\beta \triangleq \omega^-$, converges to ω in the sense of $\tau_*(\mathcal{L})$ and $\tau_0(\mathcal{L})$. The further reasoning used to substantiate the inclusion

$$\omega \in (\eta - \mathrm{Lim}\,)[\mathcal{T}_0[\Gamma_*], \tau_*(\mathcal{L})]$$

is, in fact, an immediate consequence of (5.2.5), (5.3.7), and the above-mentioned stationarity property of the net of elementary integrals corresponding to the convergent net (5.3.8). Note that with the aid of similar arguments, we can establish the inclusion which is similar to (5.3.6) but which concerns closures in the sense of $\tau_0(\mathcal{L})$. To complete the proof, it now suffices to compare (5.3.4) and (5.3.6), taking into account the relations between the sets (5.2.3) and (5.2.5), and definitions (5.2.10), (5.2.12). We leave these elementary arguments to the reader.

THEOREM 5.3.2. *Let* $\forall\, \gamma \in \Gamma\colon S_\gamma \in B_0(E,\mathcal{L})$. *Moreover, let* $\tau \in \mathfrak{M}$. *Then*

$$((\eta - \mathrm{Lim})[\mathcal{T},\tau] = \tilde{\Omega})\,\&\,(\forall\, \Gamma_* \in (\mathrm{Step})[\Gamma]\colon \tilde{\Omega} = (\eta - \mathrm{Lim})[\mathcal{T}_0[\Gamma_*],\tau]).$$

As for the proof, we only note the obvious fact that under the conditions of the theorem for any choice of the net $(D, \angle, h)$ in $\mathbf{A}(\mathcal{L})$ and the measure $\mu \in \mathbf{A}(\mathcal{L})$ the convergence of $(D, \angle, h)$ to μ in the sense of $\tau_\otimes(\mathcal{L})$ implies the convergence

$$\int\limits_E S_\gamma\, dh(\delta) \to \int\limits_E S_\gamma\, d\mu \qquad (\gamma \in \Gamma).$$

This property allows us to replace, to some extent, in arguments similar to the previous substantiation the topology $\tau_*(\mathcal{L})$ by the topology $\tau_\otimes(\mathcal{L})$, which, as we mentioned earlier, is weaker than $\tau_0(\mathcal{L})$. In other respects, the scheme of the reasoning remains the same and, therefore, we omit the details.

We shall not analyze in the general part the case where the main condition of Theorem 5.3.2 is satisfied (the case where operator (5.1.1) is step-valued). However, we shall touch upon it in Chap. 9 when we shall discuss the conditions of computational stability of the corresponding extremal problem (in that chapter we shall assume, in addition, that the set Γ is finite; this condition is not imposed now). To conclude the section, let us observe one important circumstance concerning the limit representations given in Secs. 5.2 and 5.3. Namely, these representations should not be regarded as compactifications, because both the set (5.2.14) and the set (5.3.1) are not, generally speaking, strongly bounded, and, therefore, they are not ∗-weakly compact. The question is of the "closure" of the asymptotic setting, the description of the limits of the convergent generalized sequences that are in agreement with the asymptotics of perturbed conditions under study. Hidden "chaoses" are, however, possible (a relevant example will be considered in Chap. 7, see also [41]); they are characterized by the condition that they respect constraints of asymptotic character and diverge as nets in $\mathbf{A}(\mathcal{L})$. In this connection, in the next section we shall touch upon some rather important cases for the compactification procedures in which these pathologies are absent.

As for Theorem 5.3.1 itself, note that in the case where $\widetilde{\Omega} \neq \varnothing$, the sets from $\mathcal{T}$ are nonempty, and the same is true for the sets from $\mathcal{T}_0[\Gamma_*]$ for $\Gamma_* \in (\mathrm{Step})\,[\Gamma]$.

5.4. Strong Boundedness Conditions and Relaxations of Admissible Sets. I

We now consider the problem of strong boundedness of admissible sets as the latter are imbedded into the space of f.a. measures. However, we should first clarify the setting of this problem since we are concerned with the asymptotics of perturbed conditions. We shall be guided by one natural condition concerning relaxations (5.2.4) in order to embrace the entire spectrum of perturbed conditions (5.2.6) when the parameter Γ_* runs over the set (5.2.13), although we could also study other cases corresponding to some other value of this parameter (actually, in the imposed general condition we use the convention concerning the case $\Gamma_* = \varnothing$; this circumstance is clarified in relation (5.2.8)). Thus, we shall be interested in the admissible sets (5.2.4); the corresponding conditions of ε-proximity restrict the choice of f. However, the η-integrals of $f \in \Omega^+(K,\varepsilon)$ may not be bounded in totality; a relevant example will be considered in Chap. 7. In this connection, we introduce

CONDITION 5.4.1. $\exists T \in \mathcal{T}_+,\ c \in [0,\infty):$

$$\left\{ \int\limits_E f\,d\eta\colon f \in T \right\} \subset [0, c].$$

A simpler version of Condition 5.4.1 was utilized in [42, 43]; in that version of the global condition a resource bound was present in pure form: the η-integral of $f \in B_0^+(E,\mathcal{L})$ must not be greater than a given constant. From the formal point of view, this circumstance is taken into account by a requirement placed on the operator (5.1.1) and on the set $\mathbf{Y}$. Namely, in this case it is required that among the functions S_γ, $\gamma \in \Gamma$, there should be present an identity function, and the set $\mathbf{Y}$ should have the structure of the product $(-\infty, c] \times \mathbf{Y}_1$, where the half-infinite interval imposes a condition on the η-integral $S_{\gamma_*} f$, $S_{\gamma_*}(x) \equiv 1$, and $\mathbf{Y}_1$ is a set in the space of functionals on $\Gamma \setminus \{\gamma_*\}$. A similar representation was considered in Sec. 1.3 when constraint

(1.3.4) was split into a conjunction of conditions (1.3.5), (1.3.6), and also in (1.3.11), and therefore we shall not discuss this simple case any more. Let us consider the following somewhat more general condition.

CONDITION 5.4.2. $\exists\, m \in \mathcal{N}$, $(\gamma_i)_{i \in \overline{1,m}} \in \Gamma^m$, $(a_i)_{i \in \overline{1,m}} \in [0,\infty)^m$, $(b_i)_{i \in \overline{1,m}} \in (0,\infty)^m$, $(L_i)_{i \in \overline{1,m}} \in \mathcal{L}^m$:

$$(\forall\, k \in \overline{1,m}\colon (S_{\gamma_k} \in B^+(E,\mathcal{L}))) \& (\forall\, x \in L_k\colon b_k \leqslant S_{\gamma_k}(x)) \&$$

$$(\mathbf{Y} \subset \{h \in \mathbf{R}^\Gamma \mid h(\gamma_k) \leqslant a_k\})) \& \left(E = \bigcup_{i=1}^{m} L_i\right).$$

This condition, perhaps formulated in a somewhat formal way, nevertheless, is rather visual. Namely, in fact, it shows (in Condition 5.4.2) that the full condition (5.1.4) includes a finite number of inequality-type constraints concerning the integrals of the nonnegative functions $S_{\gamma_1}, \ldots, S_{\gamma_k}$; moreover, the latter are subject to the conditions of strict positive minorization on the sets $L_1 \in \mathcal{L}, \ldots, L_m \in \mathcal{L}$ that form, in their totality, a covering of E. In Condition 5.4.2, we do not tend to ultimately general conditions; our main purpose here is to illustrate Condition 5.4.1. Actually, the following theorem is true.

THEOREM 5.4.1. *Condition 5.4.2 is sufficient for satisfaction of Condition 5.4.1.*

PROOF. We shall follow the notation adopted in Condition 5.4.2, which we assume to be fulfilled. Then $\forall\, k \in \overline{1,m}$, $f \in B_0^+(E,\mathcal{L})$

$$\mu_k^{(f)} \triangleq (S_{\gamma_k} f) * \eta = \left(\int_L S_{\gamma_k} f\, d\eta\right)_{L \in \mathcal{L}} \in (\mathrm{add})_+[\mathcal{L}]$$

and we have, as a consequence,

$$b_k(f * \eta)(L_k) \leqslant \int_{L_k} S_{\gamma_k} f\, d\eta = \mu_k^{(f)}(L_k) \leqslant \mu_k^{(f)}(E) = \int_E S_{\gamma_k} f\, d\eta,$$

which implies the obvious estimate

$$(f * \eta)(L_k) \leqslant \frac{1}{b_k} \int_E S_{\gamma_k} f\, d\eta. \tag{5.4.1}$$

Let $K_* \triangleq \{\gamma_i\colon i \in \overline{1,m}\}$, $K_* \in \mathrm{Fin}\,(\Gamma)$. Then

$$T_* \triangleq \Omega^+(K_*, 1) \in \mathcal{T}_+, \qquad c_* \triangleq \sum_{i=1}^{m} \frac{a_i + 1}{b_i} \in (0,\infty). \tag{5.4.2}$$

Take an arbitrary $f \in T_*$. Then estimate (5.4.1) holds for the values of the f.a. measure $\omega \triangleq f * \eta \in (\mathrm{add})_+ [\mathcal{L}]$,

$$\omega(E) = \int_E f \, d\eta \leqslant \sum_{k=1}^{m} (f * \eta)(L_k) = \sum_{k=1}^{m} \omega(L_k). \qquad (5.4.3)$$

The proof of (5.4.3) is quite obvious. Nevertheless, we note that in this case it is convenient to introduce an additive extension λ of the measure ω to the f.a. measure on the algebra $\mathcal{A}$ generated by $\mathcal{L}$; in this case, $\lambda \in (\mathrm{add})_+ [\mathcal{A}]$ and $\omega = (\lambda | \mathcal{L})$. But then $\forall A_1 \in \mathcal{A}, A_2 \in \mathcal{A}$ the following estimate holds:

$$\lambda(A_1 \cup A_2) \leqslant \lambda(A_1 \cup A_2) + \lambda(A_1 \cap A_2) = \lambda(A_1) + \lambda(A_2).$$

This estimate can be extended by induction to the case of the union of any finite number of sets from $\mathcal{A}$ and, in particular, from $\mathcal{L}$. On $\mathcal{L}$ the values of the measures ω and λ coincide, and this completes the verification of (5.4.3). We now observe that, according to (5.2.4), for some choice of $y \in \mathbf{Y}$,

$$\int_E S_\gamma f \, d\eta \leqslant y(\gamma) + 1 \qquad (\gamma \in K_*). \qquad (5.4.4)$$

With due regard to Condition 5.4.2, $y(\gamma_i) \leqslant a_i \, \forall i \in \overline{1, m}$. By the choice of the set K_* we now have from (5.4.4) that $\forall i \in \overline{1, m}$:

$$\int_E S_{\gamma_i} f \, d\eta \leqslant a_i + 1.$$

This implies that (5.4.1)–(5.4.3) yield the inequality

$$\int_E f \, d\mu \leqslant c_*.$$

Since f was arbitrary, this establishes the integral boundedness of T_*. Thus, if we define T and c as in (5.4.2), i.e., $T = T_*$ and $c = c_*$, then we clearly get Condition 5.4.1.

Throughout the remainder of this section, we shall assume that Condition 5.4.1 is fulfilled. With due regard for (5.2.9), we select and fix a triple

$$K_0 \in \mathrm{Fin}\,(\Gamma), \qquad \varepsilon_0 \in (0, \infty), \qquad c_0 \in [0, \infty), \qquad (5.4.5)$$

for which the following inclusion holds:

$$\left\{ \int_E f \, d\eta \colon f \in \Omega^+(K_0, \varepsilon_0) \right\} \subset [0, c_0] \qquad (5.4.6)$$

(of course, the choice of the triple (5.4.5) is not unique; we make it arbitrarily with (5.4.6) being the only constraint). Comparing (4.2.17) and (5.4.6), we get the imbedding

$$\Omega^+(K_0, \varepsilon_0) \subset M_{c_0}^+. \tag{5.4.7}$$

THEOREM 5.4.2. *Under (5.4.5) and (5.4.6), the following inclusion occurs:*

$$\widetilde{\Omega}_+ \subset \Sigma_+[c_0].$$

The proof is practically an obvious corollary of Lemma 4.3.2 and Theorem 5.2.1. Indeed, according to (5.2.9) and (5.4.5), we have $\Omega^+(K_0, \varepsilon_0) \in \mathcal{T}_+$ so that, by virtue of (4.2.18), (5.2.1), and (5.4.7), we immediately get the inclusion

$$(\eta - \mathrm{Lim}\,)[\mathcal{T}_+, \tau_*(\mathcal{L})] \subset \mathrm{cl}\,(\{f * \eta\colon\, f \in \Omega^+(K_0, \varepsilon_0)\}, \tau_*(\mathcal{L})) \subset \mathrm{cl}\,(\widetilde{M_{c_0}^+}, \tau_*(\mathcal{L})). \tag{5.4.8}$$

A straightforward combination of Lemma 4.3.2, Theorem 5.2.1, and relation (5.4.8) completes the proof.

Taking account of the compactness of spaces (4.2.25), we derive from Theorem 5.4.2 (under Condition 5.4.1) a new important fact. Namely, now the question may be of compactification of a substantial part of the solution space and regularization, on this basis, of the family of admissible sets corresponding to the perturbation of the system of constraints. This inference is substantiated by the following obvious property. Namely (see (5.2.4) and (5.4.7)), $\forall K \in \mathrm{Fin}\,(\Gamma)$

$$(K_0 \subset K) \Rightarrow (\forall \varepsilon \in (0, \varepsilon_0]\colon\, \Omega^+(K, \varepsilon) \subset M_{c_0}^+). \tag{5.4.9}$$

By virtue of (5.4.9), "almost all" sets of the family $\mathcal{T}_+$ are contained in $M_{c_0}^+$, and this is quite sufficient for restricting, in the constructions of asymptotic optimization, the choice of the solution space by the set $M_{c_0}^+$. For the sake of a more rigorous realization of this idea, we introduce

$$\widetilde{\mathcal{T}}_+ \triangleq \{\Omega^+(K, \varepsilon)\colon\, (K, \varepsilon) \in \mathrm{Fin}\,(\Gamma) \times (0, \varepsilon_0],\ K_0 \subset K\}, \tag{5.4.10}$$

thus clearly obtaining a nonempty family of subsets of $M_{c_0}^+$. If, however, $\Gamma_* \in \mathcal{P}(\Gamma)$, then, by virtue of (5.2.4)–(5.4.6), and (5.4.9), we also have $\forall K \in \mathrm{Fin}\,(\Gamma)$

$$(K_0 \subset K) \Rightarrow (\forall \varepsilon \in (0, \varepsilon_0]\colon\, \Omega_0^+(\Gamma_*, K, \varepsilon) \subset M_{c_0}^+).$$

Taking this into account, we introduce $\forall \Gamma_* \in \mathcal{P}(\Gamma)$

$$\widetilde{\mathcal{T}}_0^+[\Gamma_*] \triangleq \{\Omega_0^+(\Gamma_*, K, \varepsilon)\colon\, (K, \varepsilon) \in \mathrm{Fin}\,(\Gamma) \times (0, \varepsilon_0],\ K_0 \subset K\}, \tag{5.4.11}$$

thus obtaining a nonempty family of subsets of $M_{c_0}^+$. In this section, we deal with families (5.4.10), (5.4.11) instead of (5.2.9), (5.2.11) respectively, and we clearly have

$$(\tilde{\mathcal{T}}_+ \subset \mathcal{T}_+)\&(\forall\, \Gamma_* \in \mathcal{P}(\Gamma)\colon \tilde{\mathcal{T}}_0^+[\Gamma_*] \subset \mathcal{T}_0^+[\Gamma_*]).$$

However, the most important case for us is where we have $\Gamma_* \in (\mathrm{Step})\,[\Gamma]$ in (5.4.11). An elementary verification shows that

$$\tilde{\Omega}_+ = (\eta - \mathrm{Lim}\,)[\mathcal{T}_+, \tau_*(\mathcal{L})] = (\eta - \mathrm{Lim}\,)[\tilde{\mathcal{T}}_+, \tau_*(\mathcal{L})]$$

$$= \bigcap_{T \in \tilde{\mathcal{T}}_+} \mathrm{cl}\,(\{f * \eta\colon f \in T\}, \tilde{\tau}_{c_0}^*(\mathcal{L})); \qquad (5.4.12)$$

here we have taken into account (4.2.18) and Lemma 4.3.2. By analogy with relation (5.4.12) we have $\forall\, \Gamma_* \in (\mathrm{Step})\,[\Gamma]$:

$$\tilde{\Omega}_+ = (\eta - \mathrm{Lim}\,)[\mathcal{T}_0^+[\Gamma_*], \tau_*(\mathcal{L})] = (\eta - \mathrm{Lim}\,)[\tilde{\mathcal{T}}_0^+[\Gamma_*], \tau_*(\mathcal{L})]$$

$$= \bigcap_{T \in \tilde{\mathcal{T}}_0^+[\Gamma_*]} \mathrm{cl}\,(\{f * \eta\colon f \in T\}, \tilde{\tau}_{c_0}^*(\mathcal{L})). \qquad (5.4.13)$$

The role of (5.4.12) and (5.4.13) is determined in the sequel by the extension constructions of Chap. 2 because (see Sec. 4.2) the topological space

$$(\Sigma_+[c_0], \tilde{\tau}_{c_0}^*(\mathcal{L})) \qquad (5.4.14)$$

is compact; this space is the first one from (4.2.25).

THEOREM 5.4.3. *The following conditions are equivalent:*
(1) $\tilde{\Omega}_+ \neq \varnothing$,
(2) $\varnothing \notin \mathcal{T}_+$,
(3) $\varnothing \notin \tilde{\mathcal{T}}_+$,
(4) $\{\Gamma_* \in (\mathrm{Step})\,[\Gamma]\,|\ \varnothing \notin \mathcal{T}_0^+[\Gamma_*]\} \neq \varnothing$,
(5) $\{\Gamma_* \in (\mathrm{Step})\,[\Gamma]\,|\ \varnothing \notin \tilde{\mathcal{T}}_0^+[\Gamma_*]\} \neq \varnothing$,
(6) $\forall\, \Gamma^* \in (\mathrm{Step})\,[\Gamma]$ *we have*

$$\varnothing \notin \mathcal{T}_0^+[\Gamma^*],$$

(7) $\forall\, \Gamma^* \in (\mathrm{Step})\,[\Gamma]$ *we have*

$$\varnothing \notin \tilde{\mathcal{T}}_0^+[\Gamma^*].$$

The proof is an obvious consequence of the compactness of (5.4.14) and of the well-known assertions concerning the properties of systems of closed sets with finite intersection property in compact spaces. Nevertheless, we shall give a short scheme of the proof. We already know from Sec. 5.2 (see (5.2.15)) that (1) $\Rightarrow$ (2) and (1) $\Rightarrow$ (6). However, (2) $\Rightarrow$ (3), (4) $\Rightarrow$ (5), and (6) $\Rightarrow$ (7) by the construction of families (5.4.10) and (5.4.11). Since $\varnothing$ is an element of (5.2.13), we get the implication (3) $\Rightarrow$ (4) as a consequence of (5.2.8). The implication (5) $\Rightarrow$ (1) follows from (5.4.13) by the compactness of (5.4.14) and the properties of systems of closed sets with the finite intersection property. We have obtained the equivalence of conditions (1)–(5) so that (5) $\Rightarrow$ (6). But, since the family (Step) $[\Gamma]$ is nonempty, we have (7) $\Rightarrow$ (5). Hence, properties (5)–(7) are also equivalent, which completes the proof.

5.5. Strong Boundedness Conditions and Relaxations of Admissible Sets. II

Now we examine a problem similar to the one considered in Sec. 5.4, but concerning alternating controls f and the conditions on the integrands of these controls. In this case, we suppose, in addition to the conditions imposed earlier, that the choice of controls is restricted by "energy" constraints. Thus, we are concerned with conditions of the form

$$f \in B_0(E, \mathcal{L}), \quad \int_E |f| \, d\eta \leqslant c, \quad \left(\int_E S_\gamma f \, d\eta \right)_{\gamma \in \Gamma} \in \mathbf{Y}, \qquad (5.5.1)$$

where we assume (and fix throughout the remainder of the section) that $c \in [0, \infty)$; the other elements are the same as in Sec. 5.1. As an example of such a situation, we can again mention the problems of optimal control of linear systems of the form (1.3.1), where (in contrast to Sec. 1) we assume the possibility of using alternating control functions f with a bounded total impulse. This situation is more natural for control problems; it can serve as a motivation for the further study of general conditions of the type (5.5.1), and then of the extremal problems including conditions of this kind (we shall not reproduce the substantive settings of control problems in full generality, referring the interested reader to Sec. 1.3, where similar conditions were investigated in the case of nonnegative controls f). At the same time, this section is a natural continuation of Sec. 5.3. We note here the natural analogies with [30, 40, 42] and a number of other works which we shall not consider in detail for lack of space. In the sequel, we shall make extensive use of the compactification procedure for the solution space on the basis of Theorem 4.3.3.

As before, we shall consider the relaxations of condition (5.5.1) rather than condition (5.5.1) itself, and we shall again use several ways of introducing perturbed constraints. One of the versions of these perturbed constraints has the form

$$
f \in B_0(E, \mathcal{L}), \qquad \int_E |f|\, d\eta \leqslant c + \varepsilon,
$$

$$
\left\{ y \in \mathbf{Y} \,\middle|\, \sup_{\gamma \in K} \left| \int_E S_\gamma f\, d\eta - y(\gamma) \right| \leqslant \varepsilon \right\} \neq \varnothing,
\tag{5.5.2}
$$

where $K \in \mathrm{Fin}\,(\Gamma)$, $\varepsilon > 0$. The meaning of the new constraint on the choice of f is the same as in Sec. 5.2: see (5.2.2). In addition, we shall consider analogs of condition (5.2.3), namely, conditions of the form

$$
f \in B_0(E, \mathcal{L}), \quad \int_E |f|\, d\eta \leqslant c, \quad \left\{ y \in \mathbf{Y} \,\middle|\, \left(\forall t \in K \cap \Gamma_0 \colon \int_E S_t f\, d\eta = y(t) \right) \& \right.
$$

$$
\left. \left(\forall \gamma \in K \setminus \Gamma_0 \colon \left| \int_E S_\gamma f\, d\eta - y(\gamma) \right| \leqslant \varepsilon \right) \right\} \neq \varnothing,
\tag{5.5.3}
$$

where $\Gamma_0 \in \mathcal{P}(\Gamma)$, $K \in \mathrm{Fin}\,(\Gamma)$, $\varepsilon > 0$. In this case, Γ_0 specifies the type of the corresponding perturbation construction, and K and ε vary within the respective sets. However, if K and ε are specified, then condition (5.5.3) is, generally speaking, more severe. As before, we shall impose a condition on the choice of Γ_0, which provides a certain asymptotic equivalence of the constructions on the basis of (5.5.2) and (5.5.3). In particular, this technique will allow us in the sequel to construct a certain regularization of the dependence of the value of an extremal problem with constraints (5.5.1) under an increasing perturbation of the resource parameter. Just as in Sec. 5.2, we introduce admissible sets (the sets of all admissible elements) corresponding to conditions (5.5.2) and (5.5.3). Namely, we assume that $\forall K \in \mathrm{Fin}\,(\Gamma)$, $\varepsilon \in [0, \infty)$

$$
\widehat{\Omega}[K, \varepsilon | c] \triangleq \left\{ f \in B_0(E, \mathcal{L}) \,\middle|\, \left(\int_E |f|\, d\eta \leqslant c + \varepsilon \right) \& \right.
$$

$$
\left. \left(\exists\, y \in \mathbf{Y} \colon \sup_{\gamma \in K} \left| \int_E S_\gamma f\, d\eta - y(\gamma) \right| \leqslant \varepsilon \right) \right\} = \Omega(K, \varepsilon) \cap M_{c+\varepsilon}
\tag{5.5.4}
$$

(see (4.2.19), (5.2.3)); we associate (5.5.4) with condition (5.5.2) in the form of an admissible set whose parameters K and ε characterize (for $\varepsilon > 0$) the extent of perturbation of the rigid condition (5.5.1). We do the same for condition (5.5.3), setting

$$\forall\, \Gamma_* \in \mathcal{P}(\Gamma), \quad K \in \mathrm{Fin}\,(\Gamma), \quad \varepsilon \in [0, \infty):$$

$$\widehat{\Omega}_0[\Gamma_*, K, \varepsilon | c] \triangleq \Omega_0(\Gamma_*, K, \varepsilon) \cap M_c. \tag{5.5.5}$$

According to (4.2.19) and (5.2.5), we have, in the form of (5.5.5), an admissible set corresponding to condition (5.5.3), where we replace Γ_0 by Γ_*. Note that, as before in Sec. 5.2, the parameter Γ_* can be given various values; in particular, we can assume that $\Gamma_* = \varnothing$ and $\Gamma_* = \Gamma$. In this case, as one can easily see, we have $\forall\, K \in \mathrm{Fin}\,(\Gamma)$, $\varepsilon \in [0, \infty)$

$$\widehat{\Omega}_0[\varnothing, K, \varepsilon | c] = \Omega(K, \varepsilon) \cap M_c = \widehat{\Omega}[K, \varepsilon | c] \cap M_c, \tag{5.5.6}$$

$$\widehat{\Omega}_0[\Gamma, K, \varepsilon | c] = \left\{ f \in M_c \,\middle|\, \exists\, y \in \mathbf{Y} \,\forall\, \gamma \in K : \int_E S_\gamma f\, d\eta = y(\gamma) \right\}; \tag{5.5.7}$$

relations (5.5.6) and (5.5.7) define the two extreme cases. At the same time, some intermediate cases may be of interest in specific problems. In the sequel, we shall not consider separate versions of conditions (5.5.2) and (5.5.3), but their asymptotics. This is formalized by introducing the families of the corresponding admissible sets ((5.5.4) or (5.5.5)) by exhausting all $K \in \mathrm{Fin}\,(\Gamma)$ and $\varepsilon > 0$. Let

$$\widehat{\mathcal{T}}_c \triangleq \{ \widehat{\Omega}[K, \varepsilon | c] : (K, \varepsilon) \in \mathrm{Fin}\,(\Gamma) \times (0, \infty) \}. \tag{5.5.8}$$

We regard (5.5.8) as the family of all perturbations of condition (5.5.1) up to (5.5.2). From the formal standpoint, (5.5.8) is a nonempty family of subsets of $B_0(E, \mathcal{L})$, clearly satisfying the semimultiplicativity property, namely, $\forall\, T_1 \in \widehat{\mathcal{T}}_c,\ T_2 \in \widehat{\mathcal{T}}_c\ \exists\, T_3 \in \widehat{\mathcal{T}}_c : T_3 \subset T_1 \cap T_2$. We further set $\forall\, \Gamma_* \in \mathcal{P}(\Gamma)$

$$\widehat{\mathcal{T}}_c^0[\Gamma^*] \triangleq \{ \widehat{\Omega}_0[\Gamma_*, K, \varepsilon | c] : (K, \varepsilon) \in \mathrm{Fin}\,(\Gamma) \times (0, \infty) \}; \tag{5.5.9}$$

family (5.5.9) is nonempty and possesses a property similar to that mentioned above for $\widehat{\mathcal{T}}_c$. Observe that all elements of the family (5.5.9) are subsets of M_c. Note one special case. As one can easily see,

$$\widehat{\mathcal{T}}_c^0[\varnothing] = \{ M_c \cap T : T \in \widehat{\mathcal{T}}_c \}, \tag{5.5.10}$$

so that for $\Gamma_* = \varnothing$ relation (5.5.9) defines (in the form of (5.5.10)) the trace of the family (5.5.8) on

$$M_c = \left\{ f \in B_0(E, \mathcal{L}) \Big| \int_E |f| \, d\eta \leqslant c \right\}.$$

In the sequel, we shall consider an important special case, where $\Gamma_* \in$ (Step) $[\Gamma]$ in (5.5.9). We shall obtain, for asymptotics (5.5.8) and (5.5.9), a general limit representation in terms of the generalized admissible set

$$\tilde{\Omega}_*^{(c)} \triangleq \left\{ \mu \in \Xi[c] \Big| \left(\int_E S_\gamma \, d\mu \right)_{\gamma \in \Gamma} \in \mathbf{Y} \right\}, \qquad (5.5.11)$$

which, obviously, coincides with the intersection of the set (5.3.1) with the ball $\{\mu \in \mathbf{A}(\mathcal{L})| \, v_\mu(E) \leqslant c\}$. Namely, the following theorem holds.

THEOREM 5.5.1. *Let* $\tau \in \mathfrak{M}$. *Then*

$$((\eta - \mathrm{Lim})[\widehat{\mathcal{T}}_c, \tau] = \tilde{\Omega}_*^{(c)}) \& (\forall \, \Gamma_* \in (\text{Step})[\Gamma]: (\eta - \mathrm{Lim})[\widehat{\mathcal{T}}_c^0[\Gamma_*], \tau] = \tilde{\Omega}_*^{(c)}).$$

In great measure, the proof is similar to the proof of Theorem 5.3.1, and we shall not give it in full, confining ourselves to discussing fragments of the scheme by which the reader can easily restore the entire reasoning. The inclusion

$$(\eta - \mathrm{Lim})[\widehat{\mathcal{T}}_c, \tau_*(\mathcal{L})] \subset \tilde{\Omega}_*^{(c)}$$

is established with due account of Theorem 5.3.1 and representation of the set (5.5.11) in terms of the intersection of $\tilde{\Omega}$ and a ball corresponding to the strong norm. Let $\Gamma_* \in$ (Step) $[\Gamma]$; the inclusion

$$\tilde{\Omega}_*^{(c)} \subset (\eta - \mathrm{Lim})[\widehat{\mathcal{T}}_c^0[\Gamma_*], \tau_0(\mathcal{L})]$$

is established with the use of Theorem 4.3.3 by analogy with the proof of (5.3.6) (see the proof of Theorem 5.3.1), but is much simpler. The further reasoning uses (4.2.11) and the obvious relationships between spaces (4.2.7) and their subspaces.

We see that Theorem 5.5.1 provides a more detailed characteristic of the limit representation, complete in the sense of the triad $\mathfrak{M}$. This is due to the fact that relaxations (5.5.2) and (5.5.3) are a priori subject to the condition of integral boundedness, just as the severe condition (5.5.1). Here we have a sort of analogy with the case of positive measures, and also some regularizing action of the "automatically" arising conditions of strong boundedness, in fact, the construction of compactification without any supplementary constraints. However, it turns out to be expedient to isolate a bounded part in asymptotics (5.5.8); here we must bear in mind that the admissible sets (5.5.4) are not, generally speaking, integrally bounded in their totality. Hence we introduce the following auxiliary family:

$$\widehat{\mathcal{T}}_c^{(1)} \triangleq \{\widehat{\Omega}[K,\varepsilon|c]\colon (K,\varepsilon) \in \mathrm{Fin}\,(\Gamma) \times (0,1]\,\}, \qquad (5.5.12)$$

thus obtaining a nonempty family of subsets M_{c+1}, which contain, together with every pair of its sets, a subset of their intersection. The family $\widehat{\mathcal{T}}_c^{(1)}$ is a substantial part of (5.5.8). Indeed,

$$\widehat{\mathcal{T}}_c^{(1)} \subset \widehat{\mathcal{T}}_c \qquad (5.5.13)$$

as one can see from (5.5.8) and (5.5.12); furthermore,

$$\forall T \in \widehat{\mathcal{T}}_c \quad \exists \widetilde{T} \in \widehat{\mathcal{T}}_c^{(1)}\colon \widetilde{T} \subset T \qquad (5.5.14)$$

(verification is based on (5.5.4): this dependence is monotonic in $\varepsilon > 0$). Combining (5.5.13) and (5.5.14), and taking into account the fact that $\forall T \in \widehat{\mathcal{T}}_c^{(1)}$

$$\{f * \eta\colon f \in T\} \in 2^{\widetilde{M}_{c+1}}, \qquad (5.5.15)$$

we obtain a "precompact" (see (4.2.25) and Theorem 4.3.3) part of $\widehat{\mathcal{T}}_c$ which is sufficient for the subsequent asymptotic constructions. Note that we select this part under the condition of imbedding into the precompact set $\widetilde{M}_{c+1}$ (see (5.5.15)), but we could try to get imbedding into an arbitrary set $\widetilde{M}_{c+\varkappa}$, $\varkappa > 0$. We use family (5.5.12) only to substantiate the main assertions concerning asymptotics (5.5.8), (5.5.10); from the standpoint of the proof the only important thing is the possibility of imbedding into a compact space. For definiteness, we shall be oriented, in the sequel, toward using the precompact space $\widetilde{M}_{c+1}$. Taking account of (5.5.15), we have $\forall T \in \widehat{\mathcal{T}}_c^{(1)}$

$$\mathrm{cl}\,(\{f * \eta\colon f \in T\}, \widetilde{\tau}_*^{(c+1)}[\mathcal{L}]) = \mathrm{cl}\,(\{f * \eta\colon f \in T\}, \tau_*(\mathcal{L})) \in \mathcal{P}(\Xi[c+1]);$$

here, of course, we take account of (4.2.20), (4.2.24), and Theorem 4.3.3. As a result, we find, in accordance with (5.2.1), that

$$(\eta - \mathrm{Lim}\,)[\widehat{\mathcal{T}}_c^{(1)}, \tau_*(\mathcal{L})] = \bigcap_{T \in \widehat{\mathcal{T}}_c^{(1)}} \mathrm{cl}\,(\{f * \eta\colon f \in T\}, \widetilde{\tau}_*^{(c+1)}[\mathcal{L}]). \qquad (5.5.16)$$

Taking account of (5.5.14), (5.5.16), and Theorem 5.5.1, we get a relationship between the generalized admissible set (5.5.11) and the "truncated" asymptotics (5.5.12), namely,

$$\tilde{\Omega}_*^{(c)} = \bigcap_{T \in \widehat{\mathcal{T}}_c^{(1)}} \mathrm{cl}\left(\{f * \eta \colon f \in T\}, \tilde{\tau}_*^{(c+1)}[\mathcal{L}]\right). \tag{5.5.17}$$

In relation (5.5.17), we use closures in a compact space (see (4.2.25)) because this provides some convenience from the standpoint of the subsequent assertions concerning the conditions of asymptotic consistency. Expressly, the following theorem holds.

THEOREM 5.5.2. *The following assertions are equivalent:*

(1) $\tilde{\Omega}_*^{(c)} \neq \varnothing$,

(2) $\varnothing \notin \widehat{\mathcal{T}}_c$,

(3) $\varnothing \notin \widehat{\mathcal{T}}_c^{(1)}$,

(4) $\forall \, \Gamma_* \in (\mathrm{Step})\,[\Gamma] \colon \varnothing \notin \widehat{\mathcal{T}}_c^0[\Gamma_*]$,

(5) $\{\Gamma_* \in (\mathrm{Step})\,[\Gamma] \,|\, \varnothing \notin \widehat{\mathcal{T}}_c^{(0)}[\Gamma_*]\} \neq \varnothing$.

PROOF. Relation (5.2.1) and Theorem 5.5.1 imply that $(1) \Rightarrow (2)$ and $(1) \Rightarrow (4)$; these implications follow directly from the closure operator axioms. By virtue of (5.5.13), we have the implication $(2) \Rightarrow (3)$. Finally, $(4) \Rightarrow (5)$ because $(\mathrm{Step})\,[\Gamma]$ is a nonempty set. To verify $(3) \Rightarrow (1)$, it suffices to use (5.5.17) and the properties of systems of closed sets with the finite intersection property in the compact space

$$(\Xi[c + 1], \tilde{\tau}_*^{(c+1)}[\mathcal{L}]). \tag{5.5.18}$$

This kind of reasoning was already used above and we shall not repeat it. It should be pointed out that

$$\forall \, \Gamma_* \in (\mathrm{Step})\,[\Gamma], \quad T \in \widehat{\mathcal{T}}_c^0[\Gamma_*] \colon$$

$$\mathrm{cl}\left(\{f * \eta \colon f \in T\}, \tau_*(\mathcal{L})\right) = \mathrm{cl}\left(\{f * \eta \colon f \in T\}, \tilde{\tau}_*^{(c)}[\mathcal{L}]\right). \tag{5.5.19}$$

Relation (5.5.19) follows directly from (4.2.24) and (5.5.9). Again making use of (5.2.1) and Theorem 5.5.1, we get

$$\tilde{\Omega}_*^{(c)} = \bigcap_{T \in \widehat{\mathcal{T}}_c^0[\Gamma_*]} \mathrm{cl}\left(\{f * \eta \colon f \in T\}, \tilde{\tau}_*^{(c)}[\mathcal{L}]\right).$$

The implication $(5) \Rightarrow (1)$ can now be established by the scheme mentioned above (for proving the implication $(3) \Rightarrow (1)$), in which, instead of (5.5.18), we must use a similar compact set from (4.2.25) for the case $b = c$.

Theorem 5.5.2 obviously yields the sufficiency of the consistency property of the generalized problem for the asymptotic consistency; this will be utilized in Chap. 7 for optimization.

Chapter 6

Relaxations of Attainable Sets

6.1. Introduction

In this chapter, we shall slightly deviate from the main focus of our exposition and consider an issue that may not seem to be immediately related to extensions of extremal problems; the question is of the attainability of elements of a given topological space on the values of a continuous operator under conditions where the choice of the argument is constrained by restrictions of an asymptotic character. The latter, in turn, may arise in concrete settings when we consider real perturbations of the complex of constraints of a problem. As an example, we can mention the problem of investigating the attainability domains of controllable systems. Confining ourselves, for simplicity, to system (1.3.1) of Chap. 1, we shall discuss this case in greater detail at the substantive level, because practically all effects considered further, which occur in general settings, can be illustrated by this example. Thus, let us consider system (1.3.1) under the conditions of constraints (1.3.3) on the choice of the control program f (we could discuss even more general conditions). The set G of all (finite) states $x_f(\vartheta_0)$ attainable from the initial position (t_0, x_0) with a suitable choice of the admissible control f is usually called [22, 28] the attainability domain (we now ignore some conceptually inessential differences in terminology). In a number of cases, it is of interest to consider the following question: how will the attainability domain change relative to G under "small" perturbations of the complex of constraints; will the corresponding modifications of the attainability domain also be "small"? In the latter case, we can speak of stability. However, in many cases, the problem of Sec. 1.3 may not have such a stability, but there occurs a "structural stability" with respect to a part of the parameters. Thus, under the conditions of constraints (1.3.11), the problem is always "structurally stable" relative to a possible weakening of the conditions on the choice of the resource parameter. To be more precise, if we consider for $\varepsilon > 0$ the attainability domain G_ε corresponding to the perturbation of the set Y_1 up to its ε-neighborhood and the attainability domain $G^{(\varepsilon)}$ corresponding to the weakening by ε of both constituent constraints in (1.3.11), then the sets $G_\varepsilon, G^{(\varepsilon)}, G_\varepsilon \subset G^{(\varepsilon)}$, are close for small ε (the elements of G_ε are the vectors $x_f(\vartheta_0)$ generated by the programs f which strictly observe the c-constraint, and the constraint in the form of an inclusion into Y_1 up to the ε; the definition of $G^{(\varepsilon)}$ is similar but concerns a fuller perturbation). Thus, we have a certain asymptotic nonsensitivity of the attainability domain relative to the perturbation of the resource parameter. At the same time, the attainability domain (we mean the domain defined by the constraints of problem (1.3.12)) may not have the ordinary continuity with respect to the parameter $c \geqslant 0$ (when it is increasing). Relevant examples were given in Sec. 1.2, where it was shown, in particular, that by a small variation of the resource parameter the empty attainability domain is transformed by a jump into a nonempty set. Thus, we are concerned with situations which are unstable in the traditional meaning of this word and which, at the same time, have a sort of stability in some directions. It is this last property that we shall consider in the general case, giving up the specific features of a control problem.

Some constructions from Sec. 2.5 which are important for our purposes should be pointed out, in particular, assertion (2.5.1) and Theorem 2.5.2, where, actually, the question is of representing the asymptotic attainability set.

6.2.   Limit Representation of Attainable Sets

We shall now consider some general relationships characterizing asymptotic accessibility in terms of the generalized representations (involving the operator of integration) considered in Chap. 4. We shall not, however, require any boundedness conditions for the admissible sets; in other words, in this section we do not assume the fulfillment of Condition 5.4.1. We shall focus our attention on the case where, together with Condition (5.1.4), the choice of the control f is restrained by an additional condition of nonnegativeness, so that the question will be of deriving some rather elementary corollaries of Theorem 5.2.1 in a, generally speaking, "unbounded" case.

Thus, we assume in this and the following sections that we are given a nonempty set $\mathbf{X}$ and the operator

$$\varphi\colon (\mathrm{add})^+\,[\mathcal{L}, \eta] \to \mathbf{X}. \tag{6.2.1}$$

By virtue of the operator φ, we consider the subsets of cone (4.2.2) and their images as some generalized objects (generalized attainable sets). For the conventional solutions f and their transformations into elements of $\mathbf{X}$ by means of (6.2.1), we shall introduce a "conventional" operator mapping $B_0^+(E,\mathcal{L})$ into $\mathbf{X}$. Namely, we shall assume as such the operator

$$\Phi \triangleq \big(\varphi(f * \eta)\big)_{f \in B_0^+(E,\mathcal{L})} \in \mathbf{X}^{B_0^+(E,\mathcal{L})}, \tag{6.2.2}$$

obtained from (6.2.1); in fact, we can consider (6.2.2) as a constriction of (6.2.1). This representation will lead to deriving various substantive conclusions if φ is subjected to some additional condition. For the latter we shall impose a condition of continuity in the corresponding $*$-weak sense. Indeed,

$$((\mathrm{add})^+\,[\mathcal{L}, \eta], \tau_\eta^*(\mathcal{L})) \tag{6.2.3}$$

(see (4.2.15)) is a subspace of (4.2.6). We shall endow the set $\mathbf{X}$ with a topology τ so that $(\mathbf{X}, \tau)$ is a specified topological space and the operator φ is assumed to be continuous with respect to the topological spaces (6.2.3) and $(\mathbf{X}, \tau)$ throughout this and the next section.

Note that in concrete problems we are frequently initially given precisely the operator (6.2.2), acting on the space of conventional solutions (in our case, the mapping

$$\Phi\colon B_0^+(E, \mathcal{L}) \to \mathbf{X}$$

plays the role of such an operator), whereas we have to invent somehow (6.2.1), observing, in this case, the condition $\Phi(f) = \varphi(f * \eta)$, $f \in B_0^+(B, \mathcal{L})$. In a number of cases, we succeed in doing this. We shall check such a possibility later in the example of the problem of controlling system (1.3.1); for the time being we shall suppose that operator (6.2.1) is given a priori and possesses all the enumerated properties. Now we shall consider the action of operator (6.2.2) on the elements of families (5.2.9) and (5.2.11). In other words, we shall speak about the Φ-images of sets (5.2.4) and (5.2.6) with a subsequent limiting process, as a result of which we shall get the sets

$$\left(\Phi \triangleq \bigcap_{T \in \mathcal{T}_+} \mathrm{cl}\,(\Phi^1(T), \tau)\right) \& \left(\forall\, \Gamma_* \in \mathcal{P}(\Gamma)\colon \Phi_0(\Gamma_*) \triangleq \bigcap_{T \in \mathcal{T}_0^+[\Gamma_*]} \mathrm{cl}\,(\Phi^1(T), \tau)\right). \tag{6.2.4}$$

The elements of sets (6.2.4) can be treated as asymptotically attainable points of $\mathbf{X}$. Indeed, if $\bar{x} \in \Phi$, then, for any $K \in \mathrm{Fin}\,(\Gamma)$ and $\varepsilon > 0$, we satisfy the condition $\bar{x} \in \mathrm{cl}\,(\{\Phi(f)\colon f \in \Omega^+(K,\varepsilon)\}, \tau)$; this condition means the possibility of an arbitrarily accurate (in the sense of the topology τ) approximation of $\bar{x}$ by the elements $\Phi(f)$, $f \in \Omega^+(K,\varepsilon)$; thus $\bar{x}$ "touches" all images of admissible sets corresponding to the perturbed conditions. A similar reasoning is valid concerning the second limit set participating in (6.2.4). We could, carrying on substantive discussion, have singled out important special cases from the control theory, which we mentioned in the introduction to this chapter. Such a specification will be carried out at the end of this chapter; now we shall try to find a relationship between the set (6.2.4) and the set

$$\varphi^1(\widetilde{\Omega}_+) = \{\varphi(\mu)\colon \mu \in \widetilde{\Omega}_+\} \in \mathcal{P}(\mathbf{X}). \tag{6.2.5}$$

Here we introduced the image of a generalized admissible set; conceptually similar sets are frequently considered in problems of control theory and differential games [20, 21] as idealizations of possibilities of real controllable processes. By comparing Theorem 5.2.1 and (6.2.4), we easily find that $\forall\,\Gamma_* \in (\mathrm{Step})\,[\Gamma]$

$$\varphi^1(\widetilde{\Omega}_+) \subset \Phi_0(\Gamma_*) \subset \Phi. \tag{6.2.6}$$

In particular, $\varphi^1(\widetilde{\Omega}_+) \subset \Phi$. In proving (6.2.6), we take into account the "rigidity scale" of relaxations from Sec. 5.2, bearing in mind, in particular, that $\Omega_0^+(\Gamma_*, K, \varepsilon) \subset \Omega^+(K,\varepsilon)$. The last circumstance leads (by virtue of the monotonicity of the image and closure operations) to the inclusion $\Phi_0(\Gamma_*) \subset \Phi$. The first inclusion in (6.2.6) follows from the familiar [19, 54] properties of continuous operators

$$\varphi^1(\mathrm{cl}\,(H, \tau_\eta^*(\mathcal{L}))) \subset \mathrm{cl}\,(\varphi^1(H), \tau) \tag{6.2.7}$$

if H is a subset of (4.2.2); moreover, here we take into account Theorem 5.2.1 and the obvious property

$$\mathrm{cl}\,(H, \tau_*(\mathcal{L})) = \mathrm{cl}\,(H, \tau_\eta^*(\mathcal{L})),$$

where H is the same as in (6.2.7). We will refrain from a more detailed discussion. Let us only note the natural consequence

$$(\widetilde{\Omega}_+ \neq \varnothing) \Rightarrow ((\Phi \neq \varnothing)\,\&\,(\forall\,\Gamma_* \in (\mathrm{Step})\,[\Gamma]\colon \Phi_0(\Gamma_*) \neq \varnothing))$$

concerning the sufficient conditions of existence of asymptotically attainable elements of $\mathbf{X}$. Now, the next theorem is practically obvious.

THEOREM 6.2.1. *Let the operator φ be:* (1) *closed, i.e.,* $\forall H \in \mathcal{F}_{\tau_\eta^*(\mathcal{L})}$

$$\varphi^1(H) \in \mathcal{F}_\tau,$$

(2) *injective:* $\forall \mu \in (\mathrm{add})^+[\mathcal{L}, \eta],\ \nu \in (\mathrm{add})^+[\mathcal{L}, \eta]$

$$(\varphi(\mu) = \varphi(\nu)) \Rightarrow (\mu = \nu).$$

Then $\varphi^1(\widetilde{\Omega}_+) = \Phi$ *and, moreover,* $\forall \Gamma_* \in (\mathrm{Step})[\Gamma]$:

$$\varphi^1(\widetilde{\Omega}_+) = \Phi_0(\Gamma_*).$$

The proof is first of all connected with the fact that in the hypothesis of the theorem, inclusion (6.2.7) becomes an equality (because the operator φ is closed); this argument can, in particular, be applied to images on the strength of operator (6.2.2), so that $\forall T \in \mathcal{T}_+$:

$$\varphi^1(\mathrm{cl}\,(\{f * \eta\colon f \in T\}, \tau_*(\mathcal{L}))) = \mathrm{cl}\,(\varphi^1(\{f * \eta\colon f \in T\}), \tau)$$

$$= \mathrm{cl}\,(\{\varphi(f * \eta)\colon f \in T\}, \tau) = \mathrm{cl}\,(\{\Phi(f)\colon f \in T\}, \tau) = \mathrm{cl}\,(\Phi^1(T), \tau).$$

By virtue of (6.2.4), this equality means that

$$\Phi = \bigcap_{T \in \mathcal{T}_+} \varphi^1(\mathrm{cl}\,(\{f * \eta\colon f \in T\}, \tau_*(\mathcal{L})))$$

$$= \varphi^1\!\left(\bigcap_{T \in \mathcal{T}_+} \mathrm{cl}\,(\{f * \eta\colon f \in T\}, \tau_*(\mathcal{L}))\right) = \varphi^1(\widetilde{\Omega}_+);$$

in the last relation we took into account the fact that φ is injective. This proves the theorem because (6.2.6) allows us to extend the obtained generalized limit representation to the case of asymptotics (5.2.11) under conditions in which Γ_* is the set of step-valuedness of operator (5.1.1).

We did not try to obtain ultimate generality in Theorem 6.2.1; let us only note that this assertion concerns, generally speaking, the unbounded case of the problem.

6.3. Asymptotic Nonsensitivity and Relaxations of Attainable Sets (Case of Positive Measures)

In this section, we continue the investigation started in Sec. 6.2; throughout the section, we assume Condition 5.4.1 to be fulfilled. This essential circumstance will allow us to get a number of qualitative results concerning the property of asymptotic nonsensitivity, which was already discussed, at the substantive level, in Sec. 6.1 for a control problem. Now we shall present the corresponding assertions in general form, and we shall return to the "control" interpretation of these assertions at the end of this chapter. Thus, throughout this section we assume condition (5.4.1) to be fulfilled. As in Sec. 5.4, we select and fix a triple $(K_0, \varepsilon_0, c_0)$ (5.4.5) satisfying condition (5.4.6). Finally, for more convenience, we introduce the truncated asymptotics $\widetilde{\mathcal{T}}_+$ and $\widetilde{\mathcal{T}}_0^+[\Gamma_*]$ defined by (5.4.10) and (5.4.11) respectively (for the $K_0 \in \mathrm{Fin}\,(\Gamma)$ and $\varepsilon_0 > 0$ selected above); the families $\widetilde{\mathcal{T}}_+$ and $\widetilde{\mathcal{T}}_0^+[\Gamma_*]$, $\Gamma_* \in (\mathrm{Step})\,[\Gamma]$, are convenient to use at intermediate stages of reasoning in connection with the possibility of compactifying a substantial part of the space of conventional solutions guaranteed by condition (5.4.1). We shall use, without further explanation, the limit representations (5.4.12) and (5.4.13), bearing in mind their natural consequences connected with the compactness of space (5.4.14).

LEMMA 6.3.1. *Sets (6.2.4) are defined by the truncated asymptotics:*

$$\left(\Phi = \bigcap_{T \in \widetilde{\mathcal{T}}_+} \mathrm{cl}\,(\Phi^1(T), \tau) \right) \&$$

$$\left(\forall \, \Gamma_* \in (\mathrm{Step})\,[\Gamma]: \ \Phi_0(\Gamma_*) = \bigcap_{T \in \widetilde{\mathcal{T}}_0^+[\Gamma_*]} \mathrm{cl}\,(\Phi^1(T), \tau) \right).$$

We shall omit the obvious proof of the lemma. Let us note, however, that this assertion allows us, in fact, to reduce the problem under consideration to the study of the asymptotic behavior of the images by a continuous operator on a compact space with values in $\mathbf{X}$, and this is a much simpler problem. Indeed,

$$\varphi_0 \triangleq (\varphi | \Xi_+[c_0]) \in \mathbf{X}^{\Xi_+[c_0]} \tag{6.3.1}$$

is an operator which is continuous with respect to the pair of topological spaces

$$(\Xi_+[c_0], \tilde{\tau}_{c_0}^*(\mathcal{L})), \qquad (\mathbf{X}, \tau). \tag{6.3.2}$$

Of course, in a substantial part of the space of the generalized solutions, operator (6.3.1) coincides with φ, so that the introduction of this new operator is nothing but a convenient technical method. Note that in establishing the continuity of (6.3.1), we make use of (4.2.23) and of the transitivity property of the operation of transition to a subspace of a topological space. Throughout the remainder of this section, we assume the following.

CONDITION 6.3.1. The space $(\mathbf{X}, \tau)$ is Hausdorff.

Thus, we assume for the sequel that every pair of distinct points of $\mathbf{X}$ possesses a pair of their disjoint neighborhoods. Under these conditions, operator (6.3.1) is closed, so that the image of every subset of $\Xi_+[c_0]$ closed in space (5.4.14) (this is the first of the spaces used in (6.3.2)) is an element of $\mathcal{F}_\tau$. Of course, in this case we have the obvious property of preserving the closure operation by mapping (6.3.1): given a subset H of $\Xi_+[c_0]$, we have

$$\mathrm{cl}\,(\varphi_0^1(H), \tau) = \varphi_0^1(\mathrm{cl}\,(H, \tilde{\tau}_{c_0}^*(\mathcal{L}))), \tag{6.3.3}$$

where $\varphi_0^1(\cdot)$, as before, is used to denote the image operator in the sense of (6.3.1). In addition to (6.3.3) we observe that $\forall\, H \in \mathcal{P}(\Xi_+[c_0])$

$$\varphi_0^1(H) = \varphi^1(H). \tag{6.3.4}$$

Relation (6.3.4) allows us not to distinguish φ and φ_0 from the standpoint of the issues that are of interest to us.

THEOREM 6.3.1. *The set $\varphi^1(\tilde{\Omega}_+)$ defines the generalized representation of the sets of asymptotic accessibility*

$$(\varphi^1(\tilde{\Omega}_+) = \mathbf{\Phi})\,\&\,(\forall\,\Gamma_* \in (\mathrm{Step})\,[\Gamma]\colon \varphi^1(\tilde{\Omega}_+) = \mathbf{\Phi}_0(\Gamma_*)).$$

SCHEME OF PROOF. By virtue of (6.2.6), it suffices to establish the inclusion $\mathbf{\Phi} \subset \varphi^1(\tilde{\Omega}_+)$. Let $u \in \mathbf{\Phi}$ so that $u \in \mathrm{cl}\,(\mathbf{\Phi}^1(T), \tau)$ for $T \in \tilde{\mathcal{T}}_+$. At the same time, for such T we have $f * \eta \in \Xi_+[c_0]$ for $f \in T$ (see Lemma 4.3.2) and, hence, $\varphi_0(f * \eta) = \varphi(f * \eta) = \mathbf{\Phi}(f)$. Therefore, $\mathbf{\Phi}^1(T) = \varphi_0^1(\{f * \eta\colon f \in T\})$, and, as a consequence (see (6.3.3)),

$$\mathrm{cl}\,(\mathbf{\Phi}^1(T), \tau) = \varphi_0^1(\mathrm{cl}\,(\{f * \eta\colon f \in T\}, \tilde{\tau}_{c_0}^*(\mathcal{L})))$$

for $T \in \tilde{\mathcal{T}}_+$. But then, by virtue of the choice of the point u, we obviously have $\forall\, T \in \tilde{\mathcal{T}}_+$:

$$U_T \triangleq \varphi_0^{-1}(\{u\}) \cap \mathrm{cl}\,(\{f * \eta\colon f \in T\}, \tilde{\tau}_{c_0}^*(\mathcal{L})) \neq \varnothing, \tag{6.3.5}$$

where $\varphi_0^{-1}(\{u\}) = \varphi^{-1}(\{u\}) \cap \Xi_+[c_0]$; moreover, the set on the left-hand side of (6.3.5) is closed in space (5.4.14). Finally, note that the dependence of U_T on $T \in \tilde{\mathcal{T}}_+$ is monotonic. With due allowance for this circumstance and (6.3.5), it is easy to show that $\mathcal{U} \triangleq \{U_T\colon T \in \tilde{\mathcal{T}}_+\}$ is a system of closed subsets of compact space (5.4.14) with the finite intersection property. It has a nonempty intersection. Let us take an arbitrary f.a. measure $\lambda \in \Xi_+[c_0]$ from the intersection of all sets of the family $\mathcal{U}$. Then $\lambda \in U_T$ for $T \in \tilde{\mathcal{T}}_+$. The last assertion means that, on one hand, $\varphi(\lambda) = \varphi_0(\lambda) = u$, and, on the other hand, we have $\lambda \in \widetilde{\Omega}_+$ by virtue of (5.4.12) and the definition of U_T, $T \in \widetilde{\mathcal{T}}_+$. But in this case $u \in \varphi^1(\widetilde{\Omega}_+)$, so that the required inclusion is established, which completes the proof as a whole.

This theorem provides a number of useful corollaries which are quite obvious from the topological viewpoint. The question is of the properties of stability type; therefore, they will be formulated in terms of neighborhoods (see Sec. 2.2).

THEOREM 6.3.2. *Let H be a neighborhood of the set $\varphi^1(\widetilde{\Omega}_+)$ in the space $(\mathbf{X}, \tau)$. Then $\exists K_* \in \mathrm{Fin}\,(\Gamma)$, $\varepsilon_* \in (0, \infty)\; \forall K \in \mathrm{Fin}\,(\Gamma)$:*

$$(K_* \subset K) \Rightarrow (\forall \varepsilon \in (0, \varepsilon_*]\colon \mathrm{cl}\,(\{\Phi(f)\colon f \in \Omega^+(K, \varepsilon)\}, \tau) \subset H).$$

The meaning of the theorem obviously consists in approaching sets attainable under relaxed constraints to a generalized attainable set as the system of perturbed constraints becomes tougher. For the proof, let us select a set $G_0 \in \tau$ for which

$$\varphi^1(\widetilde{\Omega}_+) \subset G_0 \subset H.$$

We shall consider a (nonempty) set $\varphi^1(\Xi_+[c_0])$,

$$\varphi^1(\Xi_+[c_0]) \subset \mathbf{X},$$

which is compact in $(\mathbf{X}, \tau)$. This obviously means that the space

$$(\varphi^1(\Xi_+[c_0]), \tau^0), \tag{6.3.6}$$

where

$$\tau^0 \triangleq \tau\big|_{\varphi^1(\Xi_+[c_0])},$$

is compact. Let us now consider the family $\mathcal{F}_{\tau^0}$. Any system of sets with the finite intersection property in $\mathcal{F}_{\tau^0}$, has, by virtue of the compactness of (6.3.6), a nonempty intersection. Let us note the obvious relationship between $\mathcal{F}_\tau$ and $\mathcal{F}_{\tau^0}$ following from the representation of space (6.3.6) as a subspace of $(\mathbf{X}, \tau)$. Namely, $\mathcal{F}_{\tau^0}$ is a trace of $\mathcal{F}_\tau$ on $\varphi^1(\Xi_+[c_0])$. Note that if we select $T \in \widetilde{\mathcal{T}}_+$, the following property holds (by virtue of the observed relationship between $\mathcal{F}_\tau$ and $\mathcal{F}_{\tau^0}$):

$$\mathrm{cl}\,(\{\varphi(f * \eta)\colon f \in T\}, \tau) \in \mathcal{F}_{\tau^0}. \tag{6.3.7}$$

Suppose that the assertion of the theorem is not valid, i.e., $\forall K_* \in \mathrm{Fin}\,(\Gamma)$, $\varepsilon_* \in (0, \infty)\; \exists K \in \mathrm{Fin}\,(\Gamma)$:

$$(K_* \subset K)$$

$$\&(\{\varepsilon \in (0,\varepsilon_*]| \operatorname{cl}(\{\varphi(f*\eta): f \in \Omega^+(K,\varepsilon)\},\tau) \setminus H \neq \varnothing\} \neq \varnothing). \qquad (6.3.8)$$

Let us further note that

$$G^0 \triangleq \varphi^1(\Xi_+[c_0]) \cap G_0 \in \tau^0, \qquad F^0 \triangleq \varphi^1(\Xi_+[c_0]) \setminus G^0 \in \mathcal{F}_{\tau^0}.$$

Taking account of (6.3.7) and (6.3.8), we find, by the definition of G^0 and F^0, that $\forall T \in \widetilde{\mathcal{T}}_+$

$$V_T \triangleq \operatorname{cl}(\{\varphi(f*\eta): f \in T\},\tau) \setminus G_0$$

$$= \operatorname{cl}(\{\varphi(f*\eta): f \in T\},\tau) \cap F^0 \in \mathcal{F}_{\tau^0} \setminus \{\varnothing\}.$$

But then, by the definition of $\widetilde{\mathcal{T}}_+$ and the monotonicity property of the closure operator, we have that $\mathcal{V} \triangleq \{V_T: T \in \widetilde{\mathcal{T}}_+\}$ (this notation is assumed only within this proof) is a system of sets with the finite intersection property from $\mathcal{F}_{\tau^0}$, so that its intersection is nonempty; let us select an element λ from this intersection of sets from $\mathcal{V}$. Then, clearly,

$$\lambda \in \varphi^1(\Xi_+[c_0]) \setminus G_0. \qquad (6.3.9)$$

However, λ is an element of the intersection of all sets $\operatorname{cl}(\{\varphi(f*\eta): f \in T\},\tau)$, $T \in \widetilde{\mathcal{T}}_+$. But, according to (6.2.2) and (6.2.4), this implies $\lambda \in \Phi$, and, consequently, we have, with due allowance for Theorem 6.2.1, the inclusion $\lambda \in \varphi^1(\widetilde{\Omega}_+)$, which contradicts (6.3.9) by the choice of G_0. This completes the proof.

COROLLARY. *Let H be a neighborhood of $\varphi^1(\widetilde{\Omega}_+)$ in the space $(\mathbf{X},\tau)$. Then $\exists K_* \in \operatorname{Fin}(\Gamma)$, $\varepsilon_* \in (0,\infty] \,\forall K \in \operatorname{Fin}(\Gamma):$*

$$(K_* \subset K) \Rightarrow (\forall \varepsilon \in (0,\varepsilon_*], \, \Gamma^* \in (\text{Step})[\Gamma]:$$

$$\operatorname{cl}(\{\Phi(f): f \in \Omega_0^+(\Gamma^*,K,\varepsilon)\},\tau) \subset H).$$

The proof of the corollary is obvious; it suffices to compare (5.2.4) and (5.2.6) and take into account Theorem 6.3.2. This theorem and its corollary admit a very simple but useful, in a number of practically interesting problems, specification of the metric space $\mathbf{X}$. By their meaning, these two assertions have the form of regularizations of perturbed attainable sets, namely, the images by the operator Φ. We can, however, consider the question concerning the proximity of these attainable sets; the fact is that relaxations with the use of $\Gamma_* \in (\text{Step})[\Gamma]$ have a tougher character. This can be seen from the juxtaposition of (5.2.4) and (5.2.6). The proximity of the respective images — attainable (on the strength of Φ) sets — would mean a certain asymptotic nonsensitivity of an attainable set to perturbation of a part of the constraints. We are going to check it for the special case where $(\mathbf{X}, \tau)$ is metrizable [9]. Thus, let

$$\rho\colon \mathbf{X} \times \mathbf{X} \to [0, \infty)$$

be a metric [9, pp. 29, 30] generating the topology τ (we shall adhere to the agreement throughout the remainder of this section). Given a subset A of $\mathbf{X}$ and $\varepsilon \in (0, \infty)$, we set

$$\mathbf{U}^0_\rho(A, \varepsilon) \triangleq \{x \in \mathbf{X} \mid \exists y \in A\colon \rho(x, y) < \varepsilon\}. \qquad (6.3.10)$$

The case $A = \varnothing$ is not excluded; under this condition the set (6.3.10) is empty. In all cases, (6.3.10) defines an element of τ. Moreover, under the conditions defining (6.3.10), the set $\mathbf{U}^0_\rho(A, \varepsilon)$ is a τ-open neighborhood of A. Here, $\forall A \in 2^X$, $\varepsilon \in (0, \infty)$:

$$\mathbf{U}^0_\rho(A, \varepsilon) = \{x \in \mathbf{X} \mid \inf_{y \in A} \rho(x, y) < \varepsilon\}.$$

We can use as the sets A the Φ-images of the respective relaxed admissible sets.

THEOREM 6.3.3. *Let $\beta \in (0, \infty)$ be a specified accuracy parameter. Then $\exists K_* \in \mathrm{Fin}\,(\Gamma)$, $\varepsilon_* \in (0, \infty) \; \forall K \in \mathrm{Fin}\,(\Gamma)$:*

$$(K_* \subset K) \Rightarrow (\forall \varepsilon \in (0, \varepsilon_*], \; \Gamma_* \in (\mathrm{Step})\,[\Gamma]\colon$$

$$\Phi^1(\Omega^+_0(\Gamma^*, K, \varepsilon)) \subset \Phi^1(\Omega^+(K, \varepsilon)) \subset \mathbf{U}^0_\rho(\Phi^1(\Omega^+_0(\Gamma_*, K, \varepsilon)), \beta)).$$

The proof of the theorem easily follows from the assertion of Theorem 6.3.2 with due account of Theorem 5.2.1 and the obvious properties of metric spaces. Through its meaning, Theorem 6.3.3 characterizes the asymptotic nonsensitivity of the Φ-images of admissible sets with respect to the conditions of coincidence of the f-integrand and an appropriate element of $\mathbf{Y}$ on finite subsets of Γ^* (see the hypothesis of Theorem 6.3.3). We shall postpone different substantive interpretations of Theorem 6.3.3 until Sec. 6.5. In the next section, we consider alternating analogs of the assertions made in this section, which will be illustrated mainly by the examples of control problems; in these problems, the constraints on the total impulse of the type used in (5.5.1) are more typical.

6.4. Asymptotic Nonsensitivity and Relaxations of Attainable Sets (Case of Alternating Measures)

Here we consider applications of the assertions given in Secs. 5.3 and 5.5 to the problems connected with the study of attainable sets under perturbation of the corresponding complex of constraints. The expediency of such a separate consideration is motivated, on one hand, by the requirements of control theory, which was observed in Sec. 6.3 (see also Sec. 6.5), and, on the other hand, by some features of the alternating case that were discussed in Chap. 4, and, in particular, in the example of Sec. 4.4. Now, without going into details of a substantive character (see Sec. 6.5), we shall discuss, as we did in Secs. 6.2 and 6.3, this issue in the general form, as a problem of "regularization" of the image by a continuous operator. Thus, we shall assume, as before, that we are given a topological space $(\mathbf{X}, \tau)$, $\mathbf{X} \neq \varnothing$; furthermore, let

$$\psi \colon \mathbf{A}_\eta[\mathcal{L}] \to \mathbf{X} \tag{6.4.1}$$

be an operator which is continuous with respect to the topological spaces

$$(\mathbf{A}_\eta[\mathcal{L}], \tau_*^{(\eta)}[\mathcal{L}]), \quad (\mathbf{X}, \tau), \tag{6.4.2}$$

the first of which was defined by (4.2.16). Taking into account that, according to Theorem 4.3.2, $f * \eta \in \mathbf{A}_\eta[\mathcal{L}]$ for $f \in B_0(E, \mathcal{L})$, we introduce, along with the "generalized" operator (6.4.1), the respective operator on the space of conventional solutions. Namely, we assume that

$$\Psi \triangleq (\psi(f * \eta))_{f \in B_0(E, \mathcal{L})} \tag{6.4.3}$$

so that $\Psi \colon B_0(E, \mathcal{L}) \to \mathbf{X}$ and $\Psi(\tilde{f}) = \psi(\tilde{f} * \eta)$ if $\tilde{f} \in B_0(E, \mathcal{L})$. Of course, the restriction of ψ to the operator Ψ on the space of step-functionals on $(E, \mathcal{L})$, and not stratum functionals of the general form, is due to the use in the conventional setting of more realizable solutions. Generally, operator (6.4.3) can be regarded as the trace of (6.4.1) on $B_0(E, \mathcal{L})$. Let us introduce, by analogy with (6.2.4), the sets

$$\left(\Psi \triangleq \bigcap_{T \in \mathcal{T}} \mathrm{cl}\, (\Psi^1(T), \tau) \right) \&$$

$$\left(\forall \Gamma_* \in \mathcal{P}(\Gamma) \colon \Psi_0(\Gamma_*) \triangleq \bigcap_{T \in \mathcal{T}_0[\Gamma_*]} \mathrm{cl}\, (\Psi^1(T), \tau) \right). \tag{6.4.4}$$

The elements of sets (6.4.4) are characterized by the property of asymptotic attainability. Thus, every element $\tilde{x} \in \mathbf{\Psi}$ is located arbitrarily close to the set $\{\Psi(f): f \in \Omega(K, \varepsilon)\}$ for any choice of $K \in \mathrm{Fin}\,(\Gamma)$ and $\varepsilon > 0$. This means that for an arbitrarily small perturbation of condition (5.1.4), the element $\tilde{x}$ is realizable in the form $\Psi(f)$ for an appropriate choice of the control f admissible in the sense of the weakened condition. An elementary verification shows that $\forall\,\Gamma_* \in (\mathrm{Step})\,[\Gamma]$:

$$\psi^1(\tilde{\Omega}) \subset \mathbf{\Psi}_0(\Gamma_*) \subset \mathbf{\Psi}.$$

This assertion is an immediate corollary of Theorem 5.3.1 and the elementary properties of continuous functions. By analogy with Theorem 6.2.1, we establish

THEOREM   6.4.1. *Let the operator ψ be closed, i.e., $\forall\,H \in \mathcal{F}_{\tau_*^{(\eta)}[\mathcal{L}]}$*:

$$\psi^1(H) \in \mathcal{F}_\tau$$

and injective $\forall\,\mu \in \mathbf{A}_\eta[\mathcal{L}]$, $\nu \in \mathbf{A}_\eta[\mathcal{L}]$:

$$(\psi(\mu) = \psi(\nu)) \Rightarrow (\mu = \nu).$$

Then $\mathbf{\Psi} = \psi^1(\tilde{\Omega})$ and, moreover, $\forall\,\Gamma_* \in (\mathrm{Step})\,[\Gamma]$ we have $\mathbf{\Psi}_0(\Gamma_*) = \psi^1(\tilde{\Omega})$.

We shall not go into a detailed discussion of Theorem 6.4.1, which concerns, generally speaking, the unbounded case. Instead, we shall concentrate our attention on studying the case connected with relaxations considered in Sec. 5.5. In turn, for this purpose it is appropriate to note a number of auxiliary circumstances. We shall use the notation used in Sec. 5.5, fixing (as in Sec. 5.5) throughout the remainder of this section the value of the "resource" parameter $c \in [0, \infty]$; in particular, at intermediate stages we shall have to use family (5.5.12), taking into account the easily verifiable fact that

$$\bigcap_{T \in \widehat{\mathcal{T}}_c^{(1)}} \mathrm{cl}\,(\Psi^1(T), \tau) = \bigcap_{T \in \widehat{\mathcal{T}}_c} \mathrm{cl}\,(\Psi^1(T), \tau). \tag{6.4.5}$$

An important and simplifying feature (compared to Sec. 6.3) is the fact that the set

$$\psi^1(\Xi[c]) \in \mathcal{P}(\mathbf{X}) \tag{6.4.6}$$

is compact in $(\mathbf{X}, \tau)$. In general, we have to bear in mind that, whereas in Sec. 6.3 we had additionally to assume that Condition 5.4.1 is satisfied, in the case under consideration the requirement of integral boundedness is imposed a priori (see (5.5.1)). The above-mentioned compactness of set (6.4.6) readily follows from the elementary properties of continuous functions over compact spaces. In fact, every function obtained by the restriction of ψ to any set $\Xi[b]$, $b \geqslant 0$, is a function of this type if we regard this restriction as an operator defined on the compact space (4.2.24), (4.2.25). Note that (6.4.5) defines the set of asymptotic attainability in the strength of the operator Ψ if we consider relaxations of the strong condition (5.5.1). Here we omit the substantial argument, which is similar to the previous cases (see Sec. 6.2 and further on). Let us note a natural analog of (6.2.6) and also a similar property for the general case of an alternating construction whose verification is easy: $\forall \Gamma_* \in (\mathrm{Step})\,[\Gamma]$

$$\psi^1(\widetilde{\Omega}_*^{(c)}) \subset \bigcap_{T \in \widehat{T}_c^0[\Gamma_*]} \mathrm{cl}\,(\Psi^1(T), \tau) \subset \bigcap_{T \in \widehat{T}_c} \mathrm{cl}\,(\Psi^1(T), \tau). \tag{6.4.7}$$

Throughout the remainder of this section, we assume Condition 6.3.1 (that the space $(\mathbf{X}, \tau)$ is Hausdorff).

THEOREM 6.4.2. *Sets of asymptotic accessibility have the following generalized representation:*

$$\left(\psi^1(\widetilde{\Omega}_*^{(c)}) = \bigcap_{T \in \widehat{T}_c} \mathrm{cl}\,(\Psi^1(T), \tau)\right)\&$$

$$\left(\forall\, \Gamma_* \in (\mathrm{Step})\,[\Gamma]\colon \psi^1(\widetilde{\Omega}_*^{(c)}) = \bigcap_{T \in \widehat{T}_c^0[\Gamma_*]} \mathrm{cl}\,(\Psi^1(T), \tau)\right).$$

A significant part of the proof consists, with due allowance for (6.4.5) and (6.4.7), in substantiating the inclusion

$$\bigcap_{T \in \widehat{T}_c^{(1)}} \mathrm{cl}\,(\Psi^1(T), \tau) \subset \psi^1(\widetilde{\Omega}_*^{(c)}). \tag{6.4.8}$$

In turn, the proof of (6.4.8) is, in great measure, similar to the arguments used in the proof of Theorem 6.3.1. Here we make essential use of the compact topology (4.2.24) for $b = c+1$ and, in this connection, replace the family $\widetilde{T}_c$ by a replica of (5.5.12). We omit the rather obvious argument referring to the proof of Theorem 6.3.1. Let us note some rather elementary (but useful from the standpoint of applications) consequences, confining ourselves to statements and discussions at the substantial level.

THEOREM 6.4.3. *Let H be a neighborhood of the set $\psi^1(\widetilde{\Omega}_*^{(c)})$ in the topological space $(\mathbf{X}, \tau)$. Then $\exists\, K_* \in \mathrm{Fin}\,(\Gamma)$, $\varepsilon_* \in (0, \infty)\ \forall K \in \mathrm{Fin}\,(\Gamma)$:*

$$(K_* \subset K) \Rightarrow (\forall \varepsilon \in (0, \varepsilon_*]\colon \mathrm{cl}\,(\{\Psi(f)\colon f \in \widehat{\Omega}[K, \varepsilon|c]\}, \tau) \subset H).$$

We note that this assertion (and its proof) is conceptually similar to the statement of Theorem 6.3.2 but concerns a somewhat different setting of the problem. Let us observe, by the way, as a simple corollary of Theorem 6.4.2, that $\forall K \in \mathrm{Fin}\,(\Gamma)$, $\varepsilon \in (0,\infty)$:

$$\psi^1(\widetilde{\Omega}_*^{(c)}) \subset \mathrm{cl}\,(\Psi^1(\widehat{\Omega}[K,\varepsilon|c],\tau)).$$

The following is now quite obvious.

COROLLARY (of Theorem 6.4.3). *Let H be a neighborhood of the set* $\psi^1(\widetilde{\Omega}_*^{(c)})$ *in* $(\mathbf{X},\tau)$. *Then* $\exists K_* \in \mathrm{Fin}\,(\Gamma)$, $\varepsilon_* \in (0,\infty)$ $\forall K \in \mathrm{Fin}\,(\Gamma)$:

$$(K_* \subset K) \Rightarrow (\forall \varepsilon \in (0,\varepsilon_*], \ \Gamma^* \in (\mathrm{Step})\,[\Gamma]\colon \ \psi^1(\widetilde{\Omega}_*^{(c)}) \subset \mathrm{cl}\,(\{\Psi(f):$$

$$f \in \widehat{\Omega}_0[\Gamma^*,K,\varepsilon|c]\},\tau) \subset H).$$

The substance of Theorem 6.4.3 and of its corollary consists in establishing a certain general regularization of attainable sets corresponding to a perturbation of the strong condition (5.5.1); these attainable sets seem to be contracted from outside (independently of the choice of $\Gamma_* \in (\mathrm{Step})\,[\Gamma]$) toward the set of generalized accessibility $\psi^1(\widetilde{\Omega}_*^{(c)})$. Furthermore, in the case of the metric space of images, one can extract from these arguments consequences concerning the asymptotic nonsensitivity of a part of the constraints to perturbation. Here we shall use (6.3.10). Throughout the remainder of this section, we assume that $\mathbf{X}$ is provided with the metric ρ of Sec. 6.3 and the topology τ is generated by this metric ρ.

THEOREM 6.4.4. *Let $\beta \in (0,\infty)$. Then* $\exists \widetilde{K} \in \mathrm{Fin}\,(\Gamma)$, $\tilde{\varepsilon} \in (0,\infty)$ $\forall K \in \mathrm{Fin}\,(\Gamma):$

$$(\widetilde{K} \subset K) \Rightarrow (\forall \varepsilon \in (0,\tilde{\varepsilon}], \ \Gamma_* \in (\mathrm{Step})\,[\Gamma]\colon \ \Psi^1(\widehat{\Omega}[K,\varepsilon|c]) \subset$$

$$\mathbf{U}_\rho^0(\Psi^1(\widehat{\Omega}_0[\Gamma_*,K,\varepsilon|c]),\beta)).$$

The meaning of Theorem 6.4.4 is similar to that of Theorem 6.3.3, so that the arguments concerning the latter, given in Sec. 6.3, remain valid in our case, with an obvious reduction of the setting of the problem under study (replacement of the initial condition (5.1.4) by (5.5.1)). Both these assertions admit an immediate application to some problems connected with the study of the accessibility domains for controllable systems. In the next section we shall study this issue more thoroughly, confining ourselves to using Theorem 6.4.4 in a rather specific linear control problem.

6.5.   Attainability Domains of Controllable Systems and Their Relaxations

This section has an illustrative character although its role is rather significant because here we consider the problem of the behavior of attainability domains of dynamical systems under a modification of the complex of constraints which is important for applications. Note that the investigation of attainability domains in the modern control theory is a well established trend (see, e.g., [22, 28]). The results of these investigations have both theoretical and practical significance; they are important for the design and construction of decision systems, industrial controllable systems, etc. We shall now confine ourselves to illustrating the assertions of Sec. 6.4, although we can consider, in the same way, the case studied in Sec. 6.3 (see also the settings of the problems of multichannel control similar to that given in Sec. 1.4). Thus, we consider the linear control system (1.3.1), for which we retain all assumptions except for one, namely, we shall not assume the control f to be nonnegative. Thus we admit the realization in system (1.3.1), functioning on I_0, of any p.c. and r.c. control program f. Recall that $x_f(\cdot)$ is the f-solution of (1.3.1) satisfying the initial condition $x_f(t_0) = x_0$ and defined by Cauchy's relation (1.3.2). If we denote (in this section) by F the set of all p.c. and r.c. functions $f\colon I \to \mathbf{R}$, then we are thus given an operator

$$f \mapsto x_f(\vartheta_0)\colon \ F \to \mathbf{X}, \tag{6.5.1}$$

where $\mathbf{X}$ is identified (in this example) with a k-dimensional arithmetic space. We call (6.5.1) the system operator. When the choice of $f \in F$ is restrained by some additional conditions, we get the set of all possible or attainable, by means of admissible controls f, finite states $x_f(\vartheta_0)$; it is customary to call this set an attainability domain. Of course, an attainability domain depends on the system of constraints, and, as we saw in Sec. 1.2, this dependence is rather irregular. Thus, within this section, we shall adhere to the above agreement concerning the symbols F and $\mathbf{X}$ (in Chap. 1, the symbol F was used in a somewhat different sense). We assume that $\mathbf{X}$ is provided with the metric ρ (see Sec. 6.3), which associates every pair of points

$$x \triangleq (x_i)_{i \in \overline{1,k}} \in \mathbf{R}^k, \qquad \tilde{x} \triangleq (\tilde{x}_i)_{i \in \overline{1,k}} \in \mathbf{R}^k$$

with the number $\rho(x, \tilde{x})$ equal to the greatest of the values $|x_j - \tilde{x}_j|$, $j \in \overline{1,k}$. The system operator (6.5.1) admits of an explicit description in terms of Cauchy's relation (1.3.2), where we should set $t = \vartheta_0$.

REMARK. In a number of practically interesting control problems, there arises a necessity to study the dependence on the complex of conditions not only of the attainability domain at the last moment but of the entire bundle of f-solutions defined by (1.3.2). In this case, we must specify $\mathbf{X}$ in a different way, defining this set as a space of vector-functions with the topology of pointwise convergence. Then, for characterizing the operator Ψ, we shall have to associate every f-program with the entire vector-function $x_f(\cdot)$ described by (1.3.2). We do not consider this setting in detail (although it can be naturally imbedded into the general scheme of Sec. 6.4), confining ourselves, for brevity, to studying a more simple case. Namely, we shall confine ourselves to the above-mentioned specification in the interests of regularization of the dependence of the attainability domain under the perturbation of the complex of constraints.

Thus, we define the operator Ψ of Sec. 6.4 by (1.3.2) and (6.5.1). Now we have to introduce the generalized operator ψ. For this purpose, we recall that in (1.3.2) $\forall\, t \in I_0$ the mapping

$$\xi \mapsto \Phi(t,\xi)b(\xi)\colon I \to \mathbf{R}^k$$

has, as its components, functions from $B(E,\mathcal{L})$ under the conditions that $E = I$ and the semialgebra $\mathcal{L}$ of subsets of E is defined as the family of all half-intervals $[a,b)$, $t_0 \leqslant a \leqslant b \leqslant \vartheta_0$. We shall now omit the verification of this obvious circumstance, referring the reader to [29] for details concerning systems of linear differential equations. Throughout the remainder of this section, we shall adhere to the above-mentioned agreement concerning the choice of a specific realization of the space $(E,\mathcal{L})$; the measure η on the semi-algebra $\mathcal{L}$ will be defined as the length function as in Example 4.4.1: $\eta([a,b)) = b - a$ for $t_0 \leqslant a \leqslant b \leqslant \vartheta_0$. It stands to reason that for system (1.3.1) we follow the agreements made in Sec. 1.3. Taking account of these circumstances, we introduce $\forall\, \mu \in \mathbf{A}_\eta[\mathcal{L}]$, $t \in I_0$:

$$\tilde{x}_\mu(t) \triangleq \Phi(t,t_0)x_0 + \int\limits_{[t_0,t)} \Phi(t,\xi)b(\xi)\,\mu(d\xi), \qquad (6.5.2)$$

where the integral of the vector-function is defined componentwise. Notice that (6.5.2) defines a vector-function, which, under the conditions $\mu = f * \eta$, $f \in B_0(E,\mathcal{L})$, reduces to (1.3.2) so that $\forall\, t \in I_0\colon \tilde{x}_\mu(t) = x_f(t)$. The vector-functions (6.5.2) are generalized solutions of (1.3.1). In particular, (6.5.2) can be considered for $t = \vartheta_0$. The points $\tilde{x}_\mu(\vartheta_0)$, with μ being within the relevant constraints, form a generalized attainability domain. For our current purposes, related to constructing a required specification of ψ, we can construct this operator in the form

$$\mu \to \tilde{x}_\mu(\vartheta_0)\colon \mathbf{A}_\eta[\mathcal{L}] \to \mathbf{X}. \tag{6.5.3}$$

Thus, we shall use (6.5.1) as Ψ and operator (6.5.3) as ψ. In order to specify the relationship between these two operators, observe that in the case under consideration, the set F used in (6.5.1) is nothing other than the space $B_0(E, \mathcal{L})$ under the conditions of the above specification of space (4.2.1). The relation $\tilde{x}_\mu(\vartheta_0) = x_f(\vartheta_0)$, where $\mu = f * \eta$, leads to the coincidence $\Psi(f) = \psi(f * \eta)$, $f \in F$. It remains to check the continuity of ψ (6.5.3). This checking is, however, quite easy if we use representation (6.5.2) and take into account that the integral continuously depends on the measure (with respect to which the integration is performed) in the sense of the topological space (4.2.6). We shall not discuss this circumstance in detail. Notice that in the case under consideration, $(\mathbf{X}, \tau)$ is a metrizable space, and, therefore Theorem 6.4.4, with due specification of condition (5.5.1) and of the set Γ_*, is quite applicable to the problem of studying asymptotically relaxed attainability domains. In turn, this specification can be introduced, with the use of the prescriptions of Chap. 1, in the form of the condition

$$f \in F, \qquad \int_{t_0}^{\vartheta_0} |f(t)|\, dt \leqslant c,$$

$$\left(\int_{t_0}^{\vartheta_0} s_1(t) f(t)\, dt, \ldots, \int_{t_0}^{\vartheta_0} s_n(t) f(t)\, dt \right) \in Y. \tag{6.5.4}$$

Here, we must take the set $\overline{1, n}$ for Γ in (5.5.1), S_i must be identified with s_i for $i \in \overline{1, n}$, providing, of course, the proper conditions on $(s_1, \ldots, s_n)$ of the type mentioned in Chap. 1 (we mean the properties of piecewise continuity and right continuity); the set $\mathbf{Y}$ in (5.5.1) must be identified with Y. We must choose for Γ_* (see Theorem 6.4.4) the set of all $i \in \overline{1, n}$ such that s_i is a piecewise-constant and right-continuous function.

Chapter 7

Asymptotic Values and Their Generalized Representation

7.1. Introduction

This chapter continues the investigations begun in Chap. 5. We discuss the problem of constructing the asymptotic values of extremal problems with constraints on the vector integrand. The essence of the problem is sufficiently clarified in Chap. 2, where asymptotic values are introduced not only for the problems of scalar optimization but also for more general problems of cone optimization, which, in particular, include multicriteria settings. In this chapter, we consider their simplest specification and find some of their new properties. Here, we first of all bear in mind the asymptotic equivalence with respect to the results of various versions of perturbation of the complete system of constraints. As a consequence of this equivalence, we shall construct a peculiar regularization procedure for the value function of the corresponding extremal problem under perturbation of a part of the constraints. This possibility is guaranteed, however, by certain conditions; moreover, such a regularization takes place only in certain "directions," which was visually demonstrated by the examples of Chap. 1. The definition of these directions will be one of the main aims of this chapter. A special role will be played here by generalized extremal problems in the space of f.a. measures that are weakly absolutely continuous with respect to η. Under the condition of integral boundedness, these problems actually contain exhaustive information about the asymptotics of relaxed extremal problems in the class of conventional controls. We shall start the discussion of these problems with

an example that shows the importance of Condition 5.4.1 for the validity
of the generalized representations of the asymptotic value. This example
characterizes the bad, in some natural sense, properties of the "unbounded"
problems.

7.2.   Example

Let us discuss an extremal problem whose constraints have the form
(5.1.4) while the criterion is defined by a continuous dependence on the η-integral. Looking ahead, we note that precisely these objective functionals
will be the subject of the subsequent examination, though this is probably
clear in connection with the assumptions made in the preceding chapter (see,
e.g., (6.2.2)). A visual idea of these quality indices is given by the concrete
problems considered in Chap. 1. The example below has an illustrative
character and does not claim to have any practical significance. Thus, we
identify in this section the set E with the set of positive integers $\mathcal{N}$ and the
semialgebra $\mathcal{L}$ with the family of all (finite) intervals $\overline{k,l}$, $k \in \mathcal{N}$, $l \in \mathcal{N}$,
and all (infinite) intervals $\overrightarrow{m,\infty}$, $m \in \mathcal{N}$; this measurable space $(E,\mathcal{L})$ is
visually represented by the examples from Sec. 4.4. For the f.a. measure
$\eta \in (\mathrm{add})_+ [\mathcal{L}]$, we set $\forall L \in \mathcal{L}$

$$\eta(L) \triangleq \sum_{k=1}^{\infty} \frac{1}{2^k} \delta_k(L); \tag{7.2.1}$$

in other words, η is the sum of the series composed of Dirac measures $(\delta_k|\mathcal{L})$
with appropriate coefficients that provide convergence. Taking account of
(7.2.1), it is easy to check that (as in Example 4.4.1) $\mathfrak{N} = \{\varnothing\}$, $\eta(L) > 0$ for
$L \in \mathcal{L}$, $L \neq \varnothing$. Let us consider the following two (convergent) sequences,

$$\pi \triangleq \left(-\frac{1}{k}\right)_{k \in \mathcal{N}} \in B(E,\mathcal{L}); \qquad s \triangleq \left(\frac{1}{k^2}\right)_{k \in \mathcal{N}} \in B(E,\mathcal{L}),$$

which we shall use for deriving a criterion (objective functional) and for
formulating constraints, respectively. Let the function $g^0 : \mathbf{R} \to \mathbf{R}$ be, by
definition, such that $\forall \xi \in \mathbf{R}$

$$g^0(\xi) \triangleq \sup(\{\xi; -1\}).$$

Let us consider the problem

$$g^0\left(\int_E \pi f\, d\eta\right) \to \inf,$$

$$f \in B_0^+(E, \mathcal{L}), \qquad \int_E s f\, d\eta \leqslant 0, \tag{7.2.2}$$

and also its relaxations that differ from (7.2.2) by a perturbation of the right-hand side of the inequality by ε, $\varepsilon > 0$. Any relaxed problem of this kind will be called an ε-problem. Finally, let us consider the generalized problem

$$g^0\left(\int_E \pi\, d\mu\right) \to \inf,$$

$$\mu \in (\mathrm{add})^+[\mathcal{L}, \eta], \qquad \int_E s\, d\mu \leqslant 0. \tag{7.2.3}$$

Note that the objective functionals of problems (7.2.2) and (7.2.3) are connected by the following obvious relation: if $\mu = f * \eta$, where $f \in B_0^+(E, \mathcal{L})$, then

$$g^0\left(\int_E \pi f\, d\eta\right) = g^0\left(\int_E \pi\, d\mu\right). \tag{7.2.4}$$

Relation (7.2.4) allows us to interpret the criterion of the conventional problem (7.2.2) as a trace of the objective functional of the generalized problem; in the subsequent extension constructions this property will play an important role. Since s is a strictly positive sequence, it is easy to check that the only admissible element for problem (7.2.2) is the function f, which is identically zero: $f(k) \equiv 0$. Problem (7.2.2) is consistent and its value is 0.

From (7.2.3) we immediately obtain that the zero measure $\mathbf{O}$ is an admissible element of the given generalized problem. Let now $\mu_\partial \in (\mathrm{add})^+[\mathcal{L}, \eta]$ be an arbitrary admissible element for problem (7.2.3). Then, again taking into account the strict positiveness property of s, we get $\mu_\partial(\{k\}) = 0$ for $k \in \mathcal{N}$. This means that $\forall\, k \in \mathcal{N}$, $l \in \mathcal{N}$: $\mu_\partial(\overline{k, l}) = 0$. But in this case for this generalized element μ_∂ we necessarily have the equality

$$\int_E \pi\, d\mu_\partial = 0. \tag{7.2.5}$$

Indeed, if $\alpha > 0$, then, by virtue of the convergence of the sequence π to 0, we can find a number $m \in \mathcal{N}$ such that $\forall\, k \in \overrightarrow{m, \infty}$

$$|\pi(k)| \leqslant \frac{\alpha}{2(\mu_\partial(E) + 1)}.$$

Then we necessarily have

$$\int\limits_{E} \pi \, d\mu_\partial = \int\limits_{\overline{1,m}} \pi \, d\mu_\partial + \int\limits_{\overrightarrow{m+1,\infty}} \pi \, d\mu_\partial = \int\limits_{\overrightarrow{m+1,\infty}} \pi \, d\mu_\partial. \qquad (7.2.6)$$

But the last integral admits the estimate

$$\left| \int\limits_{\overrightarrow{m+1,\infty}} \pi \, d\mu_\partial \right| \leqslant \frac{\alpha}{2} \frac{\mu_\partial(E)}{1 + \mu_\partial(E)} < \alpha. \qquad (7.2.7)$$

Since $\alpha > 0$ was arbitrary, (7.2.6) and (7.2.7) yield equality (7.2.5). However, in this case the value of problem (7.2.3) necessarily vanishes because μ_∂ was also an arbitrary element from the admissible set of the generalized problem. Thus, problems (7.2.2) and (7.2.3) are equivalent with respect to the result and have zero value. Let now $\forall\, r \in \mathcal{N}$

$$f_r \triangleq (r2^r)\chi_{\{r\}}$$

(f_r is a sequence in $\mathbf{R}$ for which $f_r(i) = 0$ when $i \in \mathcal{N} \setminus \{r\}$, $f_r(r) = r2^r$). Then the sequence

$$k \mapsto f_k \colon \mathcal{N} \to B_0^+(E, \mathcal{L}) \qquad (7.2.8)$$

has the following easily verifiable properties. Namely, $\forall\, r \in \mathcal{N}$ we have

$$\left(\int\limits_{E} s f_r \, d\eta = \frac{1}{r} \right) \& \left(\int\limits_{E} \pi f_r \, d\eta = -1 \right). \qquad (7.2.9)$$

This property follows immediately from the definition of the integral considered in Chap. 3 with due account of (7.2.1) and the definitions of the sequences π and s. Taking (7.2.9) into account, we find that for almost all $k \in \mathcal{N}$ the control is f_k-admissible for $\varepsilon > 0$ in the ε-problem differing from (7.2.2) only by the ε-perturbation of the right-hand side of the last inequality. Therefore, the value of every ε-problem of this kind, $\varepsilon > 0$, is equal to -1. For the same reason, -1, as the limit of the values of the ε-problems, $\varepsilon > 0$, can be called an asymptotic value, which, consequently, is strictly smaller than both the conventional value and the generalized value. Sequence (7.2.8) is an approximate sequential solution, which has, however, the meaning of a latent "chaos"; as one can easily check, this sequence does not contain $*$-weakly convergent subsequences but can observe, with a growing accuracy, the constraints of problem (7.2.2). We note that the given unbounded problem (7.2.2) provides a typical example of a convex programming problem with a duality discontinuity (for details, see [7, 10, 12, 33]).

7.3.   Generalized Extremal Problem and Asymptotic Values

We shall now specify the assertions of Sec. 2.3 using the results of Chap. 5. Throughout this section, we assume Condition 5.4.1 to be fulfilled (see Sec. 5.4). Next, we shall suppose that the elements K_0, ε_0, c_0 (5.4.5) are selected with allowance for condition (5.4.6) and fixed. Finally, in this section, we shall use families (5.4.10) and (5.4.11) corresponding to the 3-tuple $(K_0, \varepsilon_0, c_0)$ defined by relations (5.4.5) and (5.4.6). However, these families of sets play in the sequel an auxiliary role; the assertions obtained with their aid actually characterize asymptotics (5.2.9) and (5.2.11). Thus, we adhere to the agreement on the nonnegativeness of the conventional controls f and the generalized elements μ. The alternating case will be considered in the next section. Throughout this section, we shall assume that the natural condition is satisfied.

CONDITION 7.3.1 (of generalized consistency). $\widetilde{\Omega}_+ \neq \varnothing$.

We have now from (5.2.15) that $\mathcal{T}_+$ is a filter-base for $B_0^+(E, \mathcal{L})$; moreover, $\forall\, \Gamma_* \in (\mathrm{Step})\,[\Gamma]$ the family $\mathcal{T}_0^+[\Gamma_*]$ is a filter-base for $B_0^+(E, \mathcal{L})$. Theorem 5.4.3 now implies that

$$(\varnothing \notin \widetilde{\mathcal{T}}_+)\,\&\,(\forall\, \Gamma^* \in (\mathrm{Step})\,[\Gamma]:\ \varnothing \notin \mathcal{T}_0^+[\Gamma^*]). \qquad (7.3.1)$$

Taking into account (7.3.1), it is now easy to check that $\widetilde{\mathcal{T}}_+$ is a filter-base for $M_{c_0}^+$, and exactly likewise $\forall\, \Gamma_* \in (\mathrm{Step})\,[\Gamma]$ the family $\widetilde{\mathcal{T}}_0^+[\Gamma_*]$ is a filter-base for $M_{c_0}^+$. Here, of course, we take into account that according to (5.2.4) and (5.4.10), the family $\widetilde{\mathcal{T}}_+$ has the semimultiplicativity property discussed in Sec. 2.2 (see (2.2.7)). The same is true for the families $\widetilde{\mathcal{T}}_0^+[\Gamma_*]$, $\Gamma_* \in (\mathrm{Step})\,[\Gamma]$, introduced in Chap. 5. Thereby, for these families we also find from (7.3.1) their useful characteristic as bases of the respective filters. Recall that it was this case of constraints of an asymptotic character that was the main one in Sec. 2.3. Now, making use of this circumstance, we shall specify the assertions of Sec. 2.3. We shall need the notion of the H-compactificator in the case where

$$H \triangleq \{(g(f * \eta))_{f \in M_{c_0}^+}:\ g \in \mathbf{C}((\mathrm{add})^+[\mathcal{L}, \eta], \tau_\eta^*(\mathcal{L}))\}; \qquad (7.3.2)$$

throughout this section, H is understood in the sense of (7.3.2); the elements of H will be regarded as restrictions to $M_{c_0}^+$ of functionals continuous on the topological space (6.2.3). We should observe, however, that (7.3.2) plays an auxiliary role; as objective functionals of the problem under consideration we shall use the elements of the set

$$G \triangleq \{(g(f * \eta))_{f \in B_0^+(E,\mathcal{L})} \colon g \in \mathbf{C}((\mathrm{add})^+[\mathcal{L}, \eta], \tau_\eta^*(\mathcal{L}))\} \tag{7.3.3}$$

(the symbol G will be used only in the sense of (7.3.3)). However, the relationship between (7.3.2) and (7.3.3) is quite obvious. Namely,

$$H = \{(s|M_{c_0}^+) \colon s \in G\},$$

so that in the sequel we may not distinguish between G and H from the standpoint of asymptotic optimization. Note that the choice of the elements $s \in G$ for objective functionals is not exotic. On the contrary, it is typical of many problems of practice, and, in particular, of many optimal control problems (see examples in Chap. 1). We shall now point out only one typical special case of s: $\forall f \in B_0^+(E, \mathcal{L})$

$$s(f) \triangleq S\left(\int_E \pi_1 f \, d\eta, \ldots, \int_E \pi_k f \, d\eta\right), \tag{7.3.4}$$

where $k \in \mathcal{N}$, S is a continuous real-valued function of k variables, $\pi_1 \in B(E, \mathcal{L}), \ldots, \pi_k \in B(E, \mathcal{L})$. Indeed, to represent s (7.3.4) in the form of a restriction to $B_0^+(E, \mathcal{L})$ of a functional g continuous on space (6.2.3), it is sufficient to define it by the condition

$$g(\mu) \triangleq S\left(\int_E \pi_1 \, d\mu, \ldots, \int_E \pi_k \, d\mu\right), \tag{7.3.5}$$

$\mu \in (\mathrm{add})^+[\mathcal{L}, \eta]$ and then use relation (3.4.11); as a result, we find for the functional s (7.3.4) that $s \in G$. Note that a number of settings of linear problems of controlling system (1.3.1) with the so-called terminal criterion (see Sec. 1.3) reduce to the form (7.3.4), (7.3.5); in this case, we should, of course, properly specify space (4.2.1). We shall not consider this issue in detail now and shall refer the interested reader to Secs. 1.3 and 6.5, where one can find relevant substantive arguments. Observe the following very important property related to (4.2.23): $\forall g \in \mathbf{C}((\mathrm{add})^+[\mathcal{L}, \eta], \tau_\eta^*(\mathcal{L}))$

$$(g|\Xi_+[c_0]) = (g(\mu))_{\mu \in \Xi_+[c_0]} \in \mathbf{C}(\Xi_+[c_0], \tilde{\tau}_{c_0}^*(\mathcal{L})). \tag{7.3.6}$$

Functionals (7.3.6) can, of course, be used for forming elements of H by natural restriction from $\Xi_+[c_0]$ to $M_{c_0}^+$. Taking into account the compactness of the topological spaces (4.2.25), we obviously have that every functional (7.3.6) is bounded, and this, in turn, implies that every functional from H is bounded. At the same time, it should be pointed out that $\mathbf{C}((\mathrm{add})^+[\mathcal{L}, \eta], \tau_\eta^*(\mathcal{L}))$ is a linear subspace of the space of all functionals on set (4.2.2). Therefore, H (7.3.2) is, obviously, a linear subspace of $\mathbf{B}(M_{c_0}^+)$, so that, in complete accordance with Sec. 2.3, we can introduce the notion of the H-compactificator.

In this connection, let us consider the integration operator restricted to $M_{c_0}^+$, i.e., the operator

$$f \mapsto f * \eta: \ M_{c_0}^+ \to \Xi_+[c_0]. \tag{7.3.7}$$

Lemma 4.3.2 implies, in particular, that $\widetilde{M_{c_0}^+}$ is dense in $\Xi_+[c_0]$ with respect to topological space (4.2.6), but from (7.3.3), (7.3.6), and the definition of H, it follows immediately that $\forall h \in H \ \exists g \in \mathbf{C}(\Xi_+[c_0], \tilde{\tau}_{c_0}^*(\mathcal{L})) \ \forall f \in M_{c_0}^+$: $h(f) = g(f * \eta)$; here, of course, we have to make allowance for Lemma 4.3.2. The latter makes it possible to treat the first space in (6.3.2) and operator (7.3.7) as an H-compactificator. Its further application corresponds to Sec. 2.3. In particular, in using set (2.3.4), we have to take into account Theorem 5.2.1 and (5.4.12). The specification of the functionals $R_H^0[Y, \tau, m](h)$ (see Sec. 2.3) is also rather simple in this case. Indeed, if $h \in H$ and the functional g continuous on the topological space (6.2.3) generates h in the sense of the representation $h(f) = g(f * \eta)$, $f \in M_{c_0}^+$, then functional (7.3.6) provides the required representation of h in the form of a superposition with operator (7.3.7), so that it is (7.3.6) that coincides with $R_H^0[Y, \tau, m](h)$ in our case. Thus, problem (2.3.5) can be easily specified, and we shall not consider its realization now in detail. We have thus transferred, as a matter of fact, the techniques of studying the general problems of asymptotic optimization of Sec. 2.3 to a concrete version of such a setting. We shall use this specification to extend the problem of asymptotic optimization of an arbitrary functional from (7.3.3).

Let us take an element $W \in G$, assuming in the sequel that this functional W is an objective one; with due account of (7.3.3), we choose a functional

$$w \in \mathbf{C}((\mathrm{add})^+[\mathcal{L}, \eta], \tau_\eta^*(\mathcal{L})), \tag{7.3.8}$$

which has the property that

$$W = (w(f * \eta))_{f \in B_0^+(E, \mathcal{L})}. \tag{7.3.9}$$

By means of (7.3.8) and (7.3.9), we actually specify the basic extremal problem in the form

$$W(f) \to \inf, \qquad f \in B_0^+(E,\mathcal{L}),$$

$$\left(\int_E S_\gamma f \, d\eta\right)_{\gamma \in \Gamma} \in \mathbf{Y}. \tag{7.3.10}$$

It stands to reason that the main object of our consideration is not problem (7.3.10) itself in its rigid setting, but its relaxations, whose versions were already discussed in Chap. 5. Note that (by analogy with (7.3.6)) it follows from (7.3.8), by virtue of (4.2.23), that $\forall \, b \in [0,\infty)$

$$(w|\Xi_+[b]) \in \mathbf{C}(\Xi_+[b], \tilde{\tau}_b^*(\mathcal{L})); \tag{7.3.11}$$

every functional (7.3.11) is necessarily bounded. We thus have the boundedness property for the functionals $(W|M_b^+)$, $b \geqslant 0$. In particular, this property holds if $b = c_0$. As a consequence, the functional

$$T \mapsto \inf_{f \in T} W(f): \ \widetilde{\mathcal{T}}_+ \to \mathbf{R}$$

is bounded. Similarly, $\forall \, \Gamma_* \in (\text{Step})\,[\Gamma]$ the functional

$$T \mapsto \inf_{f \in T} W(f): \ \widetilde{\mathcal{T}}_0^+[\Gamma_*] \to \mathbf{R}$$

is bounded. But then the quantities

$$\left(V^+ \triangleq \sup_{T \in \widetilde{\mathcal{T}}_+} \inf_{f \in T} W(f) \in \mathbf{R}\right)$$

$$\&\left(\forall \, \Gamma_* \in (\text{Step})\,[\Gamma]: \ V_0^+[\Gamma_*] \triangleq \sup_{T \in \widetilde{\mathcal{T}}_0^+[\Gamma_*]} \inf_{f \in T} W(f) \in \mathbf{R}\right) \tag{7.3.12}$$

are, obviously, asymptotic values of the form (2.3.2). In this case, the main parameters of (2.3.2) can be specified as follows. In the capacity of the set X of Sec. 2.3, we can take $M_{c_0}^+$, in the capacity of the filter-base $\mathcal{X}$, the family $\widetilde{\mathcal{T}}_+$ or the family $\widetilde{\mathcal{T}}_0^+[\Gamma_*]$ for $\Gamma_* \in (\text{Step})\,[\Gamma]$, and in the capacity of the (bounded) functional s used in (2.3.2), the functional $(W|M_{c_0}^+)$. Now the theorem on the coincidence of all quantities (7.3.12) is practically clear. To formulate it, we have only to notice that (in this case) $\widetilde{\Omega}_+$ is a nonempty closed subset of space (4.2.6), which, by virtue of Theorem 5.4.2, is a subset of the compact space $\Xi_+(c_0)$ (see (4.2.25)). Therefore, $\widetilde{\Omega}_+$ is a closed, in the sense of $\tilde{\tau}_{c_0}^*(\mathcal{L})$, subset of $\Xi_+[c_0]$, which follows from (4.2.23). Consequently, this set is nonempty and compact in the sense of $\tilde{\tau}_{c_0}^*(\mathcal{L})$. Therefore, the functional $(w|\Xi_+[c_0])$ attains, by (7.3.11), its minimum on $\tilde{\Omega}_+$; the same is clearly true for w (7.3.8).

THEOREM   7.3.1. *The generalized problem*

$$w(\mu) \to \min, \qquad \mu \in \tilde{\Omega}_+, \qquad (7.3.13)$$

is equivalent, with respect to the result, to the asymptotic versions of problem (7.3.10) under conditions where the families of admissible sets of relaxed problems are defined as in Theorem 5.2.1; namely, the quantity

$$\tilde{V}_+ \triangleq \min_{\mu \in \tilde{\Omega}_+} w(\mu) \qquad (7.3.14)$$

characterizes the asymptotic values

$$(\tilde{V}_+ = V^+)\&(\forall\, \Gamma_* \in (\text{Step})\,[\Gamma]: \tilde{V}_+ = V_0^+[\Gamma_*]).$$

The proof reduces to using the constructions of Sec. 2.3 under the conditions of specifying an H-compactificator. We omit the elementary arguments and turn to a discussion of one possible application of Theorem 7.3.1. We shall speak about a certain regularization of the value function of an extremal problem similar to (7.3.10) under the perturbation of a part of its constraints. For example, we may consider the problem

$$W(f) \to \inf, \quad f \in B_0^+(E, \mathcal{L}), \quad \left\{ y \in \mathbf{Y} \,\middle|\, \left(\forall\, \gamma \in \Gamma_0: \left| \int_E S_\gamma f \, d\eta - y(\gamma) \right| \leqslant \delta \right) \right.$$

$$\left. \&\left(\forall\, \gamma \in \Gamma \setminus \Gamma_0: \int_E S_\gamma f \, d\eta = y(\gamma) \right) \right\} \neq \varnothing, \qquad (7.3.15)$$

where $\delta \geqslant 0$ and $\Gamma_0 \in \mathcal{P}(\Gamma)$. In (7.3.15), the number δ is the accuracy parameter. In a number of cases, it is of interest to consider the dependence of the value of (7.3.15) on δ, which, as we know from Chap. 1, can be rather irregular. Now, on the basis of the generalized construction (7.3.13), (7.3.14), we shall try to modify the notion of the value of problem (7.3.15) itself in order to eliminate the pathologies similar to those mentioned in Chap. 1. In doing so, we shall proceed from the hypothesis that the true optimum depends on the parameter δ, $\delta \geqslant 0$, continuously. We shall discuss one (regularizing) procedure of this kind, which uses the idea of some additional perturbations having the meaning of weakening the constraints (for a substantial discussion, see [40, Chap. 4]). We first note that for $\delta = 0$ problem (7.3.15) reduces to (7.3.10). We shall now replace, for every δ, $\delta \geqslant 0$, problem (7.3.15) by an infinite system of relaxed problems of the form

$$W(f) \to \inf, \quad f \in B_0^+(E, \mathcal{L}), \quad \left\{ y \in \mathbf{Y} \;\middle|\; \left(\forall \gamma \in K \cap \Gamma_0: \right.\right.$$

$$\left| \int_E S_\gamma f \, d\eta - y(\gamma) \right| \leqslant \delta \Big)$$

$$\& \left(\forall \gamma \in K \setminus \Gamma_0: \left| \int_E S_\gamma f \, d\eta - y(\gamma) \right| \leqslant \varepsilon \right) \right\} \neq \varnothing, \tag{7.3.16}$$

$K \in \operatorname{Fin}(\Gamma)$, $\varepsilon > 0$. Of course, nothing will change from the standpoint of describing the corresponding asymptotics, provided that in (7.3.16) we confine ourselves to the natural case, where $K_0 \subset K$ and $\varepsilon \leqslant \varepsilon_0$. In the same way, we can assume the parameter δ to be sufficiently small since we are interested in the regularization of the value function in a neighborhood of the point $\delta = 0$; taking this into account, we shall confine ourselves in (7.3.15) and in problems of the type (7.3.16) to the case $\delta \leqslant \varepsilon_0$. It is crucial, however, to establish which sets Γ_0 are used in (7.3.15) and (7.3.16). Here, we again confine ourselves to family (5.2.13), this condition being essential (it can be easily verified by the use of examples from Chap. 1). We suppose that $\forall \Gamma_* \in \mathcal{P}(\Gamma)$, $K \in \operatorname{Fin}(\Gamma)$, $\varepsilon \in [0, \infty)$, $\delta \in [0, \infty)$

$$\Omega_+^0[\Gamma_*, K, \varepsilon, \delta] \triangleq \left\{ f \in B_0^+(E, \mathcal{L}) \;\middle|\; \exists y \in \mathbf{Y}: \right.$$

$$\left(\forall \gamma \in K \cap \Gamma_*: \left| \int_E S_\gamma f \, d\eta - y(\gamma) \right| \leqslant \delta \right)$$

$$\& \left(\forall \gamma \in K \setminus \Gamma_*: \left| \int_E S_\gamma f \, d\eta - y(\gamma) \right| \leqslant \varepsilon \right) \right\}, \tag{7.3.17}$$

obtaining each time the admissible set of a problem similar to (7.3.16). Sets (7.3.17) are, in a way, similar to (5.2.6); the latter can be used for estimating (7.3.17). Namely, under the conditions defining (7.3.17), we have

$$\Omega_0^+(\Gamma_*, K, \varepsilon) \subset \Omega_+^0[\Gamma_*, K, \varepsilon, \delta] \subset \Omega^+(K, \sup(\{\varepsilon, \delta\})). \tag{7.3.18}$$

It is now clear that, in principle, in order to obtain the required assertions, it suffices to combine (7.3.18) and Theorem 5.2.1 and use a "limiting process" for (K, ε). For more exact assertions, we introduce $\forall \Gamma_* \in$ (Step) $[\Gamma]$, $\delta \in [0, \varepsilon_0]$ the nonempty family

$$\tilde{\mathcal{T}}_+^0[\Gamma_*, \delta] \triangleq \{\Omega_+^0[\Gamma_*, K, \varepsilon, \delta]: (K, \varepsilon) \in \mathrm{Fin}\,(\Gamma) \times (0, \varepsilon_0], \; K_0 \subset K\}; \quad (7.3.19)$$

the elements of (7.3.19) are nonempty subsets of $M_{c_0}^+$, which can be easily checked with due allowance for (7.3.18). However, $(W|M_{c_0}^+) \in \mathbf{B}(M_{c_0}^+)$; consequently, we obtain, in particular, $\forall \Gamma_* \in$ (Step) $[\Gamma]$, $\delta \in [0, \varepsilon_0]$ the upper boundedness property for the set

$$\{\inf_{f \in T} W(f): T \in \tilde{\mathcal{T}}_+^0[\Gamma_*, \delta]\} \in 2^{\mathbf{R}}.$$

Taking account of this property, we have $\forall \Gamma_* \in$ (Step) $[\Gamma]$, $\delta \in [0, \varepsilon_0]$

$$V_+^0[\Gamma_*, \delta] \triangleq \sup_{T \in \tilde{\mathcal{T}}_+^0[\Gamma_*, \delta]} \inf_{f \in T} W(f) \in \mathbf{R}; \quad (7.3.20)$$

we are going to associate quantity (7.3.20), as the value, with the respective problem (7.3.15). With respect to the case $\delta = 0$, we can make at once the following important stipulation: $\forall \Gamma_* \in$ (Step) $[\Gamma]$

$$\tilde{\mathcal{T}}_+^0[\Gamma_*, 0] = \tilde{\mathcal{T}}_0^+[\Gamma_*]. \quad (7.3.21)$$

Combining (7.3.20) and (7.3.21), we can obviously interpret the quantity $V_+^0[\Gamma_*, 0]$ as $V_0^+[\Gamma_*]$, for which purpose, it suffices to make allowance for (7.3.12). In other words, according to (7.3.14) and Theorem 7.3.1, we have in (7.3.20), for $\delta = 0$, the value of the generalized problem (7.3.13). For the entire collection of quantities (7.3.20) we have obtained a dependence defined on the interval $[0, \varepsilon_0]$. It is easy to see that it is monotone nonincreasing, i.e., $\forall \Gamma_* \in$ (Step) $[\Gamma]$, $\delta_1 \in [0, \varepsilon_0]$, $\delta_2 \in [\delta_1, \varepsilon_0]$

$$V_+^0[\Gamma_*, \delta_2] \leqslant V_+^0[\Gamma_*, \delta_1]; \quad (7.3.22)$$

inequality (7.3.22) easily follows from (7.3.17) and (7.3.19).

THEOREM 7.3.2. *Let* $\Gamma_* \in$ (Step) $[\Gamma]$. *Then the function* $V_+^0[\Gamma_*, \cdot]$ *is continuous at the point* $\delta = 0$: $\forall \varkappa \in (0, \infty) \; \exists \delta \in (0, \infty] \; \forall \theta \in [0, \delta) \cap [0, \varepsilon_0]$

$$V_+^0[\Gamma_*, 0] - \varkappa < V_+^0[\Gamma_*, \theta] \leqslant V_+^0[\Gamma_*, 0].$$

The proof reduces to using Theorem 7.3.1 and relation (7.3.18); analogous arguments can be found in [40]. In this presentation, we omit this rather obvious proof.

To conclude this section, we want to point out that the condition $\Gamma_* \in$ (Step) $[\Gamma]$ is essential for the validity of Theorem 7.3.2. The fact is that the construction on the basis of (7.3.19) and (7.3.20) could also be realized in a more general case $\Gamma_* \in \mathcal{P}(\Gamma)$. However, the property of continuity of the dependence, similar to the $V_+[\Gamma_*, \cdot]$, may not hold in this case. To verify this, it suffices to turn to the examples of Chap. 1, and we shall do this at an informal level. Thus, for the minimization problem considered in Sec. 1.2 with constraints on the choice of $f \in F$ (the notation is that of Sec. 1.2) defined by (1.2.8), we could identify Γ with the two-element set $\{1,2\} = \overline{1,2}$ and specify the functions S_1 and S_2 as an identity function and a linear one ($S_2(t) = t$ for $t \in [0,1)$), respectively. Let us try to use as Γ_* the one-element set $\{2\}$, i.e., $\Gamma_* = \{2\}$, and then make use of a construction similar to (7.3.19), (7.3.20). However, since Γ is finite, the construction becomes more elementary, and, instead of (7.3.17), it suffices to consider the constraints

$$
f \in F, \quad \int_0^1 f(t)\, dt \leqslant 1 + \varepsilon, \quad \int_0^1 t f(t)\, dt \leqslant \delta, \qquad (7.3.23)
$$

where $\varepsilon > 0$, $\delta \geqslant 0$. The quality functional is, as in Sec. 1.2, the Riemannian integral of f taken with the opposite sign. Then, for $\delta = 0$, we have, as an admissible element, a unique element, namely, the identically vanishing function. The value of the minimization problem under constraints (7.3.23) coincides, in the case $\delta = 0$, with 0, for any value of the parameter $\varepsilon > 0$. If, however, $\delta > 0$, then the value of this problem equals $-(1 + \varepsilon)$, where $\varepsilon > 0$. As a result, if $\delta > 0$, then, in the limit, as $\varepsilon \downarrow 0$, we get the convergence of the values of constrained problems (7.3.23) to -1. Consequently, for a dependence similar to (7.3.20) we have a discontinuity at the point $\delta = 0$; the corresponding jump is equal to 1. Thus, we did not succeed here in regularizing the value function by performing an additional limiting process for ε, $\varepsilon > 0$. This is due to the fact that for $\gamma \in \Gamma_*$ the function S_γ is not, in this case, a step function in the natural sense. We shall not give now exact formulations with a specification of space (4.2.1) for our example and a description of admissible sets (7.3.17); the reader can easily do this himself by using the examples of Chap. 1 and the definitions from Chap. 4. However, some specifications concerning the case of a finite set Γ will be considered in Chap. 9.

7.4. Value Asymptotics for Strongly Bounded Problems

We now consider the asymptotic optimization problems whose constraints are connected with a perturbation of condition (5.5.1). Conceptually, the material of this section is in large measure similar to the respective constructions of Sec. 7.3, and therefore we shall not give detailed proofs in the cases where they can be easily executed by analogy with the arguments of the preceding section or by analogy with general statements of Sec. 2.3. At the same time, we shall give a general scheme of the investigation, admitting, perhaps, some repetitions. The corresponding constructions connected with a regularization of the value function of an extremal problem and admitting a natural analogy with Theorem 7.3.2 will be considered in Sec. 7.5, so that the question will be of asymptotic values and their generalized representation. Recall that constraints similar to (5.5.1) frequently arise in optimal control problems. Therefore, it should be pointed out that the version of linear control problem considered in [30] and concerning, in particular, the problem of minimization of the terminal criterion considered in Sec. 1.3 on the paths of system (1.3.1) that are realized under the action of alternating control programs with bounded total impulse is a special case of the assertions presented in what follows. Furthermore, a problem may have constraints on the vector integrand of the control, similar to (1.3.3). Many issues of this kind were already considered in Sec. 6.5, and we shall not consider them any more. In this section, we shall use, without additional explanations, the constructions and assertions of Sec. 2.3, since we speak about one of the specifications of the compactificator from Sec. 2.3. Throughout this section, we fix the (energy) parameter $c \in [0, \infty]$ from Sec. 5.5. Moreover, we use notation from Sec. 5.5 such as (5.5.4), (5.5.5), (5.5.8), (5.5.9), (5.5.11), (5.5.12). Finally, throughout this section we shall assume the following condition.

CONDITION 7.4.1 (of generalized consistency). $\tilde{\Omega}_*^{(c)} \neq \varnothing$.

Theorem 5.5.1 and relation (5.2.1) imply that $\tilde{\Omega}_*^{(c)} \in \mathcal{F}_{\tau_*(\mathcal{L})}$. But from (5.5.11) obviously follows the property of strong boundedness of the set $\tilde{\Omega}_*^{(c)}$; here, of course, we have to take into account (4.2.22). Thus, according to (3.4.14), the set $\tilde{\Omega}_*^{(c)}$ is compact in space (4.2.6) and nonempty by virtue of the above condition; hence (see (2.2.3)),

$$\tau_*(\mathcal{L})|_{\tilde{\Omega}_*^{(c)}} = \{\tilde{\Omega}_*^{(c)} \cap G \colon G \in \tau_*(\mathcal{L})\} \tag{7.4.1}$$

is a topology turning (5.5.11) into a nonempty compact space. However,

(5.5.11) may also be regarded as a subspace of the topological space

$$(\mathbf{A}_\eta[\mathcal{L}], \tau_*^{(\eta)}[\mathcal{L}]), \tag{7.4.2}$$

and, by virtue of (4.2.16) and the transitivity of the operation of passing to a subspace, we obviously get the coincidence of the compact topology (7.4.1) and the topology induced in (5.5.11) from (7.4.2). But then $\forall\, g \in \mathbf{C}(\mathbf{A}_\eta[\mathcal{L}], \tau_*^{(\eta)})[\mathcal{L}]$

$$(g|\tilde{\Omega}_*^{(c)}) = (g(\mu))_{\mu\in\tilde{\Omega}_*^{(c)}} \tag{7.4.3}$$

is a functional on (5.5.11), continuous in the sense of topology (7.4.1). Every functional of this kind necessarily attains its minimum. This means that every functional from $\tilde{\Omega}_*^{(c)}$ attains its minimum on $\mathbf{C}(\mathbf{A}_\eta[\mathcal{L}], \tau_*^{(\eta)}[\mathcal{L}])$ of (5.5.11). In other words, $\forall\, g \in \mathbf{C}(\mathbf{A}_\eta[\mathcal{L}], \tau_*^{(\eta)}[\mathcal{L}])$ the problem

$$g(\mu) \to \min, \qquad \mu \in \tilde{\Omega}_*^{(c)}, \tag{7.4.4}$$

is solvable in the traditional sense. Problems of the type (7.4.4) will be called generalized. We notice here the obvious fact that according to (4.2.16), (4.2.22), and (4.2.24), $\forall\, g \in \mathbf{C}(\mathbf{A}_\eta[\mathcal{L}], \tau_*^{(\eta)}[\mathcal{L}])$, $b \in [0, \infty)$

$$(g|\Xi[b]) = (g(\mu))_{\mu\in\Xi[b]} \in \mathbf{C}(\Xi[b], \tilde{\tau}_*^{(b)}[\mathcal{L}]); \tag{7.4.5}$$

functionals of the type (7.4.5) will be needed for specifying the compactificator of Sec. 2.3 in the case $b \geqslant c$. Every functional of this kind is necessarily bounded by virtue of the compactness of the second topological space used in (4.2.25). In turn, we obviously get from (4.2.20) and Lemma 4.3.3 the boundedness property for every functional

$$f \mapsto g(f * \eta)\colon M_b \to \mathbf{R}, \tag{7.4.6}$$

where g is a continuous functional on (7.4.2), $b \geqslant 0$. This actually allows us to use the constructions from Sec. 2.3 concerning the problems of asymptotic optimization of bounded functionals. Here we shall have to introduce an analog of (7.3.2) for the alternating case considered here; for this, we shall use functionals (7.4.6). Taking this into account, we introduce the indicated analog (analogs, to be more precise) in the following form: throughout this section we shall assume that

$$\begin{aligned}
H_1 &\triangleq \{(g(f * \eta))_{f\in M_c}\colon g \in \mathbf{C}(\mathbf{A}_\eta[\mathcal{L}], \tau_*^{(\eta)}[\mathcal{L}])\}, \\
H_2 &\triangleq \{(g(f * \eta))_{f\in M_{c+1}}\colon g \in \mathbf{C}(\mathbf{A}_\eta[\mathcal{L}], \tau_*^{(\eta)}[\mathcal{L}])\};
\end{aligned} \tag{7.4.7}$$

every set (7.4.7) has as its elements only bounded functionals:

$$H_1 \subset \mathbf{B}(M_c), \qquad H_2 \subset \mathbf{B}(M_{c+1}).$$

Here we should observe that H_1 and H_2 are linear spaces, and therefore we can use them as H from Sec. 2.3 by setting $(H = H_1) \vee (H = H_2)$ depending on the type of asymptotics of the perturbed constraints. The case $H = H_1$ corresponds to asymptotics (5.5.9); $H = H_2$ will be used for studying asymptotics (5.5.12), and, in the final analysis, asymptotics (5.5.8). Thus, we consider below two versions of compactifications from Sec. 2.3; such a dichotomy of the generalized construction is, of course, not necessary, but it is convenient, and we shall use it. Consider two operators:

$$f \mapsto f * \eta \colon M_c \to \Xi[c], \tag{7.4.8}$$

$$f \mapsto f * \eta \colon M_{c+1} \to \Xi[c+1]. \tag{7.4.9}$$

Operators (7.4.8) and (7.4.9) realize the embeddings of M_c and M_{c+1}, respectively, into the space of f.a. measures of bounded variation. The images of M_c and M_{c+1} are (see Theorem 4.3.3) everywhere dense sets in $\Xi[c]$ and $\Xi[c+1]$, respectively, relative to the topological space (4.2.6). Making use of (4.2.24), it is easy to check that (for $b \in [0, \infty)$)

$$\mathrm{cl}\,(\widetilde{M_b}, \tilde{\tau}_*^{(b)}[\mathcal{L}]) = \mathrm{cl}\,(\widetilde{M_b}, \tau_*(\mathcal{L})) \cap \Xi[b] = \Xi[b], \tag{7.4.10}$$

so that in (7.4.8) and (7.4.9) we actually get operators of everywhere dense embedding into compact spaces of the type (4.2.25). Taking into account (7.4.5), we also have $\forall\, h \in H_1$

$$\{g \in \mathbf{C}(\Xi[c], \tilde{\tau}_*^{(c)}[\mathcal{L}]) \mid h = (g(f * \eta))_{f \in M_c}\} \neq \varnothing.$$

By virtue of (7.4.10), this means that the triple

$$(\Xi[c], \tilde{\tau}_*^{(c)}[\mathcal{L}], m_1),$$

where m_1 is defined by (7.4.8), is an H-compactificator from Sec. 2.3 for $H = H_1$. In the same way, by complementing the compact space (5.5.18) by the operator m_2 (7.4.9), we have, as the resulting tuple, the H-compactificator from Sec. 2.3 for $H = H_2$. Now it is not difficult to use Theorem 2.3.1, where we should only take into account the limit representations of admissible sets obtained in Sec. 5.5 (we mean Theorem 5.5.8 and relations (5.5.17)), which makes it possible to treat the admissible set of the version of problem

(2.3.5) used here as $\tilde{\Omega}_*^{(c)}$ from Sec. 5.5. For definiteness, we choose and fix, throughout the remainder of this section, the functional

$$w \in \mathbf{C}(\mathbf{A}_\eta[\mathcal{L}], \tau_*^{(\eta)}[\mathcal{L}]); \tag{7.4.11}$$

the notation w is understood here in the sense of (7.4.11). Then, as one can see from (7.4.7),

$$W \triangleq (w(f * \eta))_{f \in B_0(E,\mathcal{L})} \in \mathbf{R}^{B_0(E,\mathcal{L})},$$

$$W_1 \triangleq (W|M_c) \in H_1, \qquad W_2 \triangleq (W|M_{c+1}) \in H_2,$$

$$w_1 \triangleq (w|\Xi[c]) \in \mathbf{C}(\Xi[c], \tilde{\tau}_*^{(c)}[\mathcal{L}]),$$

$$w_2 \triangleq (w|\Xi[c+1]) \in \mathbf{C}(\Xi[c+1], \tilde{\tau}_*^{(c+1)}[\mathcal{L}]),$$

$$R_{H_1}^0[\Xi[c], \tilde{\tau}_*^{(c)}[\mathcal{L}], m_1](W_1) = w_1,$$

$$R_{H_2}^0[\Xi[c+1], \tilde{\tau}_*^{(c+1)}[\mathcal{L}], m_2](W_2) = w_2.$$

Thus, we have sufficiently specified the generalized problem (2.3.5). Note that when studying asymptotics (5.5.8), we shall use the "whole" functional W. It should be noticed that (see (7.4.5)) $\forall b \in [0, \infty)$

$$(W|M_b) \in \mathbf{B}(M_b). \tag{7.4.12}$$

Therefore, as one can see from (5.5.4) and (5.5.8), any set

$$\{W(f): f \in T\}, \qquad T \in \widehat{\mathcal{T}}_c, \tag{7.4.13}$$

is bounded. Now we have to take into account Theorem 5.5.2 and Condition 7.4.1, so that in the case under consideration (7.4.13) are nonempty bounded subsets of $\mathbf{R}$, as also are the similar sets of values of W on the elements of families (5.5.9). Taking account of these circumstances, we introduce the number sets

$$\{\inf_{f \in T} W(f): T \in \widehat{\mathcal{T}}_c\}, \qquad \{\inf_{f \in T} W(f): T \in \widehat{\mathcal{T}}_c^{(1)}\}. \tag{7.4.14}$$

The second set in (7.4.14) is necessarily bounded from above because every set from $\widehat{\mathcal{T}}_c^{(1)}$ is contained in M_{c+1} and we can use property (7.4.12) for $b = c + 1$. Therefore, the second set in (7.4.14) has a finite upper bound. But then, by virtue of (5.5.14), the first set is also bounded from above, and

$$\widehat{V}_c \triangleq \sup_{T \in \widehat{\mathcal{T}}_c} \inf_{f \in T} W(f) = \sup_{T \in \widehat{\mathcal{T}}_c^{(1)}} \inf_{f \in T} W(f) \in \mathbf{R}. \tag{7.4.15}$$

Using the same argument and applying Theorem 5.5.2 and relation (7.4.12), we come to the conclusion that $\forall \Gamma_* \in (\text{Step})\,[\Gamma]$, $T \in \widehat{T}_c^0[\Gamma_*]$ the set $\{W(f): f \in T\}$ is nonempty and bounded in $(\mathbf{R}, |\cdot|)$; this time we use (7.4.12) for $b = c$. Moreover, it easily follows from (7.4.12) that $\forall \Gamma_* \in (\text{Step})\,[\Gamma]$

$$\{\inf_{f \in T} W(f): T \in \widehat{T}_c^0[\Gamma_*]\}$$

is a bounded subset of $\mathbf{R}$, so that its least upper bound is well defined:

$$\widehat{V}_c^0[\Gamma_*] \triangleq \sup_{T \in \widehat{T}_c^0} \inf_{f \in T} W(f) \in \mathbf{R}. \tag{7.4.16}$$

Relations (7.4.15) and (7.4.16) actually provide us with some versions of asymptotic values from Sec. 2.3. To verify this fact, we observe that, in accordance with Theorem 5.5.2, in our case $\widehat{T}_c^{(1)}$ is a filter-base for M_{c+1}; of course, here we have to use the semimultiplicativity properties from Sec. 2.2 which hold for the families $\widehat{T}_c$ and $\widehat{T}_c^{(1)}$ (see Sec. 5.5). We can also observe that $\widehat{T}_c^0[\Gamma_*]$ is a filter-base for M_c provided that $\Gamma_* \in (\text{Step})\,[\Gamma]$. These two circumstances, together with (7.4.12), allow us to determine, following the prescriptions of Sec. 2.3, the values

$$((\widehat{T}_c^{(1)} - \min)[(W|M_{c+1})] \in \mathbf{R})$$

$$\&(\forall \Gamma_* \subset (\text{Step})\,[\Gamma]: (\widehat{T}_c^0[\Gamma_*] - \min)[(W|M_c)] \in \mathbf{R}). \tag{7.4.17}$$

In turn, (7.4.15) and (7.4.16) are connected with quantities (7.4.17) by obvious relations that follow immediately from the general definitions of Sec. 2.3. Namely, the first of the quantities in (7.4.17) coincides with $\widehat{V}_c$, and the second one coincides with $\widehat{V}_c^0[\Gamma_*]$, where $\Gamma_* \in (\text{Step})\,[\Gamma]$. Now, to determine (7.4.15) and (7.4.16), we can directly use Theorem 2.3.1. In addition, here we have to take into account Theorem 5.5.1 and the remarks made at the beginning of the section concerning the attainability of minimum on the values of functionals of the type (7.4.3); in this case, we must apply these remarks to functional (7.4.11).

THEOREM 7.4.1. *The asymptotic values (7.4.15) and (7.4.16) and the value of the generalized problem*

$$w(\mu) \to \min, \qquad \mu \in \widetilde{\Omega}_*^{(c)}, \tag{7.4.18}$$

coincide, so that

$$\left(\widehat{V}_c = \min_{\mu \in \widetilde{\Omega}_*^{(c)}} w(\mu)\right) \& \left(\forall \Gamma_* \in (\text{Step})\,[\Gamma]: \widehat{V}_c^0[\Gamma_*] = \min_{\mu \in \widetilde{\Omega}_*^{(c)}} w(\mu)\right).$$

The proof reduces to the determination of (7.4.17) on the basis of the prescriptions of Theorem 2.3.1; here problem (2.3.5) is transformed to the form (7.4.18) in both cases of determining values (7.4.17). We must bear in mind that, strictly speaking, the objective functional of problem (2.3.5) is transformed either into the functional w_1 or into the functional w_2, as was pointed out above, but, in view of (5.5.11), the two generalized problems are equivalent and correspond to the minimum problem for the values of w, i.e., to problem (7.4.18).

Theorem 7.4.1 provides an important property of computational equivalence for asymptotics (5.5.8) and (5.5.9). Now we shall consider (following conceptually the approach put forward in the preceding section) the issue of a peculiar regularization of the value function. The case in point is the minimization problem

$$W(f) \to \inf, \qquad f \in M_{c+\delta},$$

$$\left\{ y \in \mathbf{Y} \,\middle|\, \left(\forall \gamma \in \Gamma_0 \colon \left| \int_E S_\gamma f \, d\eta - y(\gamma) \right| \leqslant \delta \right) \right.$$

$$\left. \& \left(\forall \gamma \in \Gamma \setminus \Gamma_0 \colon \int_E S_\gamma f \, d\eta = y(\gamma) \right) \right\} \neq \varnothing, \qquad \delta \geqslant 0. \tag{7.4.19}$$

Here we assume that $\Gamma_0 \in \mathcal{P}(\Gamma)$. The values of problem (7.4.19) can be regarded as a function of the parameter δ, $\delta \geqslant 0$, which, however, can be quite irregular (in the sense of its limit properties) at the point $\delta = 0$. Note that for $\delta = 0$ the admissible set of problem (7.4.19) is defined by condition (5.5.1); it is weakened when $\delta > 0$. As in the preceding section, we shall try to replace, for every $\delta \geqslant 0$, the real value of problem (7.4.19) by some other quantity, which, nevertheless, can also, for some reason, be interpreted as a value taking into account various minor errors in realizing the constraints. For brevity, we shall introduce admissible sets for the relaxed problems; the essential meaning can be easily restored by analogy with the arguments used in the preceding section. Thus, $\forall \Gamma_* \in \mathcal{P}(\Gamma)$, $K \in \mathrm{Fin}\,(\Gamma)$, $\varepsilon \in [0, \infty)$, $\delta \in [0, \infty)$

$$\widehat{\Omega}_c^0[\Gamma_*, K, \varepsilon, \delta] \triangleq \left\{ f \in M_{c+\delta} \,\middle|\, \exists y \in \mathbf{Y} \colon \right.$$

$$\left(\forall \gamma \in K \cap \Gamma_* \colon \left| \int_E S_\gamma f \, d\eta - y(\gamma) \right| \leqslant \delta \right)$$

$$\left. \& \left(\forall \gamma \in K \setminus \Gamma_* \colon \left| \int_E S_\gamma f \, d\eta - y(\gamma) \right| \leqslant \varepsilon \right) \right\} \tag{7.4.20}$$

is the new admissible set corresponding to the relaxation of condition (7.4.19) toward their weakening, provided that $\Gamma_* = \Gamma_0$. Let us observe, as a special case, that $\forall K \in \mathrm{Fin}\,(\Gamma)$, $\varepsilon \in [0,\infty)$, $\delta \in [0,\infty)$

$$\widehat{\Omega}^0_c[\varnothing, K, \varepsilon, \delta] = \left\{ f \in M_{c+\delta} \;\middle|\; \exists\, y \in \mathbf{Y}:\; \sup_{\gamma \in K} \left| \int_E S_\gamma f \, d\eta - y(\gamma) \right| \leqslant \varepsilon \right\}. \qquad (7.4.21)$$

This case deserves a separate discussion in connection with problem (7.4.19), too. Indeed, if we set $\Gamma_0 = \varnothing$ in (7.4.19), then (in this problem) there arises the question of dependence of the value of the extremal problem with constraints (5.5.1) under a perturbation of the resource parameter c toward increase, i.e., with the replacement of c by $c + \delta$, $\delta > 0$. The examples from Sec. 1.2 (after their elementary modification) show that even such a dependence of the value function may have a discontinuity at the point $\delta = 0$; moreover, the extremal problem (7.4.19) in this simplest case $\Gamma_0 = \varnothing$ may be transformed from a consistent to an inconsistent one when we pass from the case $\delta > 0$ to the case $\delta = 0$. The construction based on (7.4.21), where $\varepsilon > 0$, eliminates this nuisance. Let us return to the general case. As one can easily verify, $\forall \Gamma_* \in \mathcal{P}(\Gamma)$, $K \in \mathrm{Fin}\,(\Gamma)$, $\varepsilon \in [0,\infty)$, $\delta \in [0,\infty)$,

$$\widehat{\Omega}_0[\Gamma_*, K, \varepsilon | c] \subset \widehat{\Omega}^0_c[\Gamma_*, K, \varepsilon, \delta] \subset \widehat{\Omega}[K; \sup(\{\varepsilon, \delta\}) | c]. \qquad (7.4.22)$$

When establishing (7.4.22), one ought to take into account (5.2.5), (5.2.6), (5.5.4), and (5.5.5); (7.4.22) defines a peculiar "fork" for admissible sets (7.4.20). Indeed, the sets bounding (7.4.20) from above and from below are, for $\varepsilon > 0$, elements of families (5.5.8) and (5.5.9) respectively, and we have already established for $\Gamma_* \in (\mathrm{Step})\,[\Gamma]$ asymptotic properties common for these two families (see Theorems 5.5.1 and 7.4.1). Here we also observe that $\forall \Gamma_* \in \mathcal{P}(\Gamma)$, $K \in \mathrm{Fin}\,(\Gamma)$, $\varepsilon \in [0,\infty)$: $\widehat{\Omega}^0_c[\Gamma_*, K, \varepsilon, 0] = \widehat{\Omega}_0[\Gamma_*, K, \varepsilon | c]$; this equality follows from a straightforward juxtaposition of (5.2.5), (5.2.6), (5.5.5), and (7.4.20). The properties connected with the monotonic dependence of sets (7.4.20) on (K, ε) and δ are practically clear, and we therefore omit their discussion. Let us now introduce $\forall \Gamma_* \in \mathcal{P}(\Gamma)$, $\delta \in [0,\infty)$:

$$\widehat{\mathcal{T}}^{(c)}_0[\Gamma_*, \delta] \triangleq \{ \widehat{\Omega}^0_c[\Gamma_*, K, \varepsilon, \delta]\colon (K, \varepsilon) \in \mathrm{Fin}\,(\Gamma) \times (0, \infty) \}, \qquad (7.4.23)$$

getting each time a nonempty set, possessing, as one can easily check, the semimultiplicativity property from Sec. 2.2. Of special interest for us is the case where in (7.4.20) and (7.4.23) the set Γ_* is an element of family (5.2.13). In this case, as one can see from (7.4.22), family (7.4.23) is a filter-base of $M_{c+\delta}$ if only $\Gamma_* \in (\mathrm{Step})\,[\Gamma]$ and $\delta \in [0,\infty)$. Thus, $\forall \Gamma_* \in (\mathrm{Step})\,[\Gamma]$,

$\delta \in [0, \infty)$,

$$\widehat{T}_0^{(c)}[\Gamma_*, \delta] \subset 2^{M_{c+\delta}}. \tag{7.4.24}$$

Hence, we have in (7.4.23) nonempty families composed of nonempty subsets of $M_{c+\delta}$, if only Γ_* satisfies the condition of step-valuedness of operator (5.1.1) on this set. In this case, as one can see from (7.4.12), $(W|M_{c+\delta}) \in \mathbf{B}(M_{c+\delta})$ for $\delta \geqslant 0$. Taking account of this circumstance, we obviously have $\forall \, \Gamma_* \in (\text{Step})\,[\Gamma]$, $\delta \in [0, \infty)$ a bounded nonempty set

$$\{ \inf_{f \in T} W(f) \colon T \in \widehat{T}_0^{(c)}[\Gamma_*, \delta]\},$$

for which the finite least upper bound is well defined. Let $\forall \, \Gamma_* \in (\text{Step})\,[\Gamma]$, $\delta \in [0, \infty)$

$$\widehat{V}_0^{(c)}[\Gamma_*, \delta] \triangleq \sup_{T \in \widehat{T}_0^{(c)}[\Gamma_*, \delta]} \inf_{f \in T} W(f). \tag{7.4.25}$$

For a fixed $\Gamma_* \in (\text{Step})\,[\Gamma]$, we obtain from (7.4.25) a real-valued function

$$\widehat{V}_0^{(c)}[\Gamma_*, \cdot\,] \triangleq (\widehat{V}_0^{(c)}[\Gamma_*, \delta])_{\delta \in [0, \infty)} \tag{7.4.26}$$

on the semiaxis $[0, \infty)$, which we shall regard as the new value function. Namely, for any fixed δ, $\delta \geqslant 0$, number (7.4.25) is assigned to problem (7.4.19) as its value. The main advantage of function (7.4.26) is the continuity property at the point $\delta = 0$. To establish this fact, we must first of all take into account (7.4.22); this relation implies, in view of (5.5.9) and (7.4.23), that $\forall \, \Gamma_* \in (\text{Step})\,[\Gamma]$

$$\widehat{T}_0^{(c)}[\Gamma_*, 0] = \widehat{T}_c^0[\Gamma_*]. \tag{7.4.27}$$

Relation (7.4.27) allows us to invoke Theorem 7.4.1, so that $\forall \, \Gamma_* \in (\text{Step})\,[\Gamma]$

$$\widehat{V}_0^{(c)}[\Gamma_*, 0] = \widehat{V}_c^0[\Gamma_*] = \min_{\mu \in \widetilde{\Omega}_*^{(c)}} w(\mu). \tag{7.4.28}$$

Let us observe here the practically obvious property of monotonicity of function (7.4.26): $\forall \, \Gamma_* \in (\text{Step})\,[\Gamma]$, $\delta_1 \in [0, \infty)$, $\delta_2 \in [\delta_1, \infty)$,

$$\widehat{V}_0^{(c)}[\Gamma_*, \delta_2] \leqslant \widehat{V}_0^{(c)}[\Gamma_*, \delta_1].$$

Taking this property into account, one can easily establish the validity of the following general assertion on regularization of a value function.

THEOREM 7.4.2. *Let* $\Gamma_* \in$ (Step) $[\Gamma]$. *Function* (7.4.26) *is continuous at the point* $\delta = 0$, *so that* $\forall \varepsilon \in (0, \infty) \; \exists \tilde{\delta} \in (0, \infty) \; \forall \delta \in [0, \tilde{\delta}]$

$$\widehat{V}_0^{(c)}[\Gamma_*, 0] - \varepsilon \leqslant \widehat{V}_0^{(c)}[\Gamma_*; \delta] \leqslant \widehat{V}_0^{(c)}[\Gamma_*, 0].$$

This assertion implies, in particular, an important practical assertion on regularization of the dependence on the resource parameter. To obtain it, we have to set $\Gamma_* = \varnothing$ in Theorem 7.4.2. However, it makes sense to take into account (7.4.21) and to reformulate this special case in a somewhat different way, using families of the form (5.5.9). Here we have to take into account that, according to (7.4.21), we get, in the form $\widehat{\Omega}_c^0[\varnothing, K, \varepsilon, \delta]$, a set differing from (5.5.4) only by the replacement of the number c by $c + \delta$. If we turn to (7.4.19), then we have now, as a version of this setting, the family of problems

$$W(f) \to \inf, \qquad f \in M_{c+\delta}, \qquad \left(\int_E S_\gamma f \, d\eta \right)_{\gamma \in \Gamma} \in \mathbf{Y}, \qquad (7.4.29)$$

$\delta \geqslant 0$. It is for the purpose of regularizing the value function of problems (7.4.29) as δ varies in a neighborhood of the point $\delta = 0$ that the introduction of relaxations (7.4.21) is subject to and, in the final analysis, becomes the dependence $\widehat{V}_0^{(c)}[\varnothing, \cdot]$. Thus, to every problem (7.4.29), i.e., to every number $\delta \in [0, \infty)$ playing the role of perturbation of the resource parameter, we assign, as the value of the δ-problem, the number $\widehat{V}_0^{(c)}[\varnothing, \delta]$; it seems worthwhile to note that the dependence $\widehat{V}_0^{(c)}[\varnothing, \cdot]$ possesses some properties which are natural for the value function of the family of problems (7.4.29). Theorem 7.4.2 provides, in turn, the required continuity property of this dependence at the point $\delta = 0$, so that we always have some structural stability of the problem with respect to the resource parameter.

7.5. On an Extremal Problem with Inequality-Type Constraints

Here we consider a special case of setting used in Sec. 7.3, namely, an infinite-dimensional problem of linear programming [7, 10, 12], which, for reasons of a methodological character, should be represented, in our mind, in the following, somewhat unusual form:

$$\int_E \pi f \, d\eta \to \inf, \quad f \in M_c^+, \quad \int_E S_\gamma f \, d\eta \leqslant r(\gamma) \quad (\gamma \in \Gamma); \qquad (7.5.1)$$

here, as in the preceding section, we fix $c \in [0, \infty)$ as the resource parameter of the problem, $r \in \mathbf{R}^\Gamma$, $\pi \in B(E, \mathcal{L})$. Taking account of the definitions from Sec. 4.2, it is easy to check that all constraints in (7.5.1) have the inequality character, so that we deal with a typical linear programming problem [24]. At the same time, we shall consider this setting only as an example of the problem considered in Secs. 7.2 and 7.3. Therefore, we shall adhere to a certain notation scheme coordinated with Secs. 5.2 and 7.3. We shall omit some obvious details concerning, for example, some modification of the set Γ and the operator of Sec. 5.1 in order to make allowance for the resource constraint in (7.5.1) (we mean the reduction of the system of constraints (7.5.1) to the form (5.1.4) with the addition of the nonnegativeness condition). In the general part, we omit justifications since they can be easily derived from the assertions of Sec. 7.3; at the same time, we shall pay attention to a construction logically connected with the generalized duality theorem for problems of convex programming [7, 10] and presented in a form, admitting, in principle, a transition in (7.5.1) to the use of the convex (and not necessarily linear) criterion. As for the set Γ and operator (5.5.1) involved in (7.5.1), we do not make any special assumptions (for a more extensive exposition, see [51]). Before passing to the general case, we shall formulate one almost obvious result concerning the problems of "unconditional" optimization. We shall suppose in this section the f.a. measure η to be nonzero, $\eta \neq \mathbf{O}$. Consider the following two problems:

$$\int_E sf\, d\eta \to \inf, \qquad f \in M_c^+, \qquad (7.5.2)$$

$$\int_E s\, d\mu \to \min, \qquad \mu \in \Xi_+[c], \qquad (7.5.3)$$

where $s \in B(E, \mathcal{L})$. Let $\mathcal{L}_0 \triangleq \{L \in \mathcal{L} \mid \eta(L) \neq 0\} = \mathcal{L} \setminus \mathfrak{N}$. Then, as one can easily check,

$$b_0[s] \triangleq \inf \left(\left\{ \frac{(s * \eta)(L)}{\eta(L)} : L \in \mathcal{L}_0 \right\} \right) \in \mathbf{R},$$

$$\beta_0[s] \triangleq \inf(\{0, b_0[s]\}) \in (-\infty, 0]. \qquad (7.5.4)$$

Quantity (7.5.4) plays an important role since, as one can easily check [51, pp. 36–39], it coincides, up to a factor, with the value of each one of the problems (7.5.2) and (7.5.3), and in the latter case it is attained on a generalized element:

$$\inf_{f \in M_c^+} \int_E sf \, d\eta = \min_{\mu \in \Xi_+[c]} \int_E s \, d\mu = c\beta_0[s]. \tag{7.5.5}$$

Let us observe here that [51, pp. 39, 40]

$$b_0[s] = \inf_{x \in E} s(x) \tag{7.5.6}$$

if only $\forall L \in \mathcal{L}$ the conditions $\eta(L) = 0$ and $L = \varnothing$ are equivalent (here the functional s may be any element of $B(E, \mathcal{L})$); in Chap. 4 and Sec. 7.2, we gave an example in which this condition is satisfied. In the subsequent constructions, the functional s can assume various values determined by the requirements of problem (7.5.1), which we begin to consider now. Here we introduce our notation for relaxations, although we could, with some transformations, use the definitions of Chap. 5. We shall confine ourselves to the case where

$$\{S_\gamma \colon \gamma \in \Gamma\} \subset B_0(E, \mathcal{L});$$

this condition will be assumed throughout the remainder of this section, although a number of the main propositions do not require it. Here we shall confine ourselves to studying only one asymptotics of relaxed constraints, namely, the asymptotics of finite subsystems; for other settings, see [51]. We set $\forall m \in \mathcal{N}$, $(\gamma_i)_{i \in \overline{1,m}} \in \Gamma^m$

$$\Lambda_m[(\gamma_i)_{i \in \overline{1,m}}] \triangleq \left\{ f \in M_c^+ \,\middle|\, \forall j \in \overline{1, m} \colon \int_E S_{\gamma_j} f \, d\eta \leqslant r(\gamma_j) \right\} \tag{7.5.7}$$

(for some reasons, it is more convenient here to use ordered systems $(\gamma_1, \ldots, \gamma_m)$ but not the elements of Fin (Γ) as before). It is easy to see that (7.5.7) is an admissible set for the problem

$$\int_E \pi f \, d\eta \to \inf, \qquad f \in M_c^+,$$

$$\int_E S_{\gamma_i} f \, d\eta \leqslant r(\gamma_i) \qquad (i \in \overline{1, m}).$$

The indicated reduction of problem (7.5.1) defines the main type of relaxations [51] (some differences in notation compared to [51] are inessential). We denote the family of all sets (7.5.7) by Λ, so that

$$\Lambda \triangleq \{ T \in \mathcal{P}(M_c^+) \,|\, \exists\, m \in \mathcal{N}, \, (\gamma_i)_{i \in \overline{1,m}} \in \Gamma^m \colon T = \Lambda_m[(\gamma_i)_{i \in \overline{1,m}}] \} \tag{7.5.8}$$

is a nonempty family of sets possessing the property of semimultiplicativity discussed in Sec. 2.2 (see (2.2.7)): together with every pair of its sets family (7.5.8) contains a subset of their intersection. If (7.5.8) consists only of nonempty sets, then we obviously have a filter-base of M_c^+. Furthermore, let us introduce a natural version of the generalized problem of linear programming

$$\int_E \pi \, d\mu \to \inf, \qquad \mu \in \Xi_+[c],$$

$$\int_E S_\gamma \, d\mu \leqslant r(\gamma) \qquad (\gamma \in \Gamma). \tag{7.5.9}$$

It stands to reason that problem (7.5.9) can actually be extracted from construction (7.3.13) with a respective change of the notation of Sec. 7.3. We denote the admissible set of problem (7.5.9) by $\widehat{\Xi}_+[c]$, so that

$$\widehat{\Xi}_+[c] \triangleq \left\{ \mu \in \Xi_+[c] \,\middle|\, \forall \gamma \in \Gamma \colon \int_E S_\gamma \, d\mu \leqslant r(\gamma) \right\}. \tag{7.5.10}$$

Set (7.5.10) is closed in (4.2.6) but may turn out to be empty. An elementary verification, with due allowance for Lemma 4.3.2, shows that

$$(\widehat{\Xi}_+[c] \neq \varnothing) \Leftrightarrow (\varnothing \notin \Lambda); \tag{7.5.11}$$

this simple statement is a special case of the respective assertions of Chap. 5 on the conditions of consistency. Here, Lemma 4.3.2 is used for approximating in the sense of the topology $\tau_0(\mathcal{L}) \in \mathfrak{M}$. We omit now the proof of (7.5.11) since the indicated arguments can easily be realized by the interested reader himself.

Throughout the remainder of this section, we assume that the following condition holds.

CONDITION 7.5.1. The constraints of problem (7.5.9) are consistent: $\widehat{\Xi}_+[c] \neq \varnothing$.

With due account of (7.5.11), we now have, in the form of Λ, a filter-base of M_c^+. Moreover,

$$\Pi \triangleq \left(\int_E \pi f \, d\eta \right)_{f \in M_c^+} \in \mathbf{B}(M_c^+).$$

This follows from the obvious fact that

$$\mu \mapsto \int_E \pi \, d\mu \colon \Xi_+[c] \to \mathbf{R} \tag{7.5.12}$$

is a functional continuous on the compact topological space

$$(\Xi_+[c], \tilde{\tau}_c^*(\mathcal{L})) \tag{7.5.13}$$

(see (4.2.25)). This assertion on the continuity of (7.5.12) as a functional on the compact space (7.5.13) can be easily checked by analogy with (7.3.6): in the form (7.5.12) we have the trace of a functional continuous on space (4.2.6). But then, in accordance with the prescriptions given in Chap. 2, we can introduce the asymptotic value

$$(\Lambda - \min)[\Pi] = \sup_{\Lambda \in \boldsymbol{\Lambda}} \inf_{f \in \Lambda} \Pi(f) \in \mathbf{R}, \tag{7.5.14}$$

which is of fundamental interest to us. The fact is that (7.5.14) defines (see Sec. 2.3) the true optimum for which the general representation in terms of problem (7.5.9) holds true. Of course, here we have to make allowance for assertions similar to Theorem 5.2.1 in the part of it that is connected with the use of $\Gamma_* \in$ (Step) $[\Gamma]$ (of course, we ought to modify the notation in order to interpret the constraints of problem (7.5.1) in the form (5.1.4)). As a result of these rather elementary constructions, we get the equality

$$\bigcap_{\Lambda \in \boldsymbol{\Lambda}} \mathrm{cl}\,(\{f * \eta \colon f \in \Lambda\}, \tilde{\tau}_c^*(\mathcal{L})) = \hat{\Xi}_+[c].$$

Now, a very simple specification of the propositions from Sec. 2.3 (see Theorem 2.3.1) yields the equality

$$(\Lambda - \min)[\Pi] = \min_{\mu \in \hat{\Xi}_+[c]} \int_E \pi \, d\mu, \tag{7.5.15}$$

characterizing the asymptotic value (7.5.14) in terms of the generalized problem (7.5.9), which can also be considered as a conditional extremum problem. Now, however, we can further specify the asymptotic value (7.5.14), without confining ourselves to equality (7.5.15), and making use of the familiar techniques of convex programming (we shall confine ourselves to the most obvious construction, making use of the Kuhn–Tucker theorem; see [12, 17] and others). For this, we shall first of all consider, along with (7.5.9), generalized problems with relaxation of constraints, setting $\forall K \in \mathrm{Fin}\,(\Gamma)$, $\varepsilon \in (0, \infty)$,

$$\Lambda^*[K, \varepsilon] \triangleq \left\{ \mu \in \Xi_+[c] \,\middle|\, \forall \gamma \in K \colon \int_E S_\gamma \, d\mu \leqslant r(\gamma) + \varepsilon \right\}. \tag{7.5.16}$$

We denote the family of all sets (7.5.16) by Λ^*, so that

$$\Lambda^* \triangleq \{T \in \mathcal{P}(\Xi_+[c])|\ \exists\, K \in \text{Fin}\,(\Gamma),\ \varepsilon \in (0,\infty)\colon T = \Lambda^*[K,\varepsilon]\}. \quad (7.5.17)$$

Family (7.5.17) is a filter-base of $\Xi_+[c]$, which easily follows from the assumption that Condition 7.5.1 is fulfilled. Making use of the continuity of functional (7.5.12), which implies its boundedness, and of the easily verifiable equality

$$\widehat{\Xi}_+[c] = \bigcap_{T \in \Lambda^*} T$$

(and the fact that in (7.5.13) every set (7.5.16) is closed) we can show that

$$(\Lambda^* - \min)\left[\left(\int_E \pi\, d\mu\right)_{\mu \in \Xi_+[c]}\right] = \min_{\mu \in \widehat{\Xi}_+[c]} \int_E \pi\, d\mu; \quad (7.5.18)$$

in turn, we can use (7.5.18) in (7.5.15) for determining the asymptotic value:

$$(\Lambda - \min)[\Pi] = \sup_{T \in \Lambda^*} \min_{\mu \in T} \int_E \pi\, d\mu. \quad (7.5.19)$$

The fact is that every problem

$$\int_E \pi\, d\mu \to \min, \qquad \mu \in \Lambda^*[K,\varepsilon], \quad (7.5.20)$$

where $K \in \text{Fin}\,(\Gamma)$ and $\varepsilon \in (0,\infty)$, is a standard problem of convex programming satisfying the Slater condition (actually, (7.5.20) is even a problem of linear programming) and having a finite number of inequality-type constraints; problem (7.5.20) always has an optimal solution. In the sequel, we suppose that $\forall\, m \in \mathcal{N}$

$$\mathbf{R}_+^m \triangleq \{(x_i)_{i\in\overline{1,m}} \in \mathbf{R}^m|\ \forall\, j \in \overline{1,m}\colon 0 \leqslant x_j\}$$

(the positive cone in $\mathbf{R}^m$). Furthermore, we assume that $\forall\, m \in \mathcal{N}$

$$(\lambda_i)_{i\in\overline{1,m}} \in \Gamma^m, \qquad (l_i)_{i\in\overline{1,m}} \in \mathbf{R}_+^m, \qquad \varepsilon \in [0,\infty),$$

$$\mathbf{L}_m[(\gamma_i)_{i\in\overline{1,m}}, (l_i)_{i\in\overline{1,m}}, \varepsilon] \triangleq \left(\int_E \left(\pi + \sum_{i=1}^{m} l_i S_{\gamma_i}\right) d\mu\right.$$

$$\left. - \sum_{i=1}^{m} l_i\big(r(\gamma_i) + \varepsilon\big)\right)_{\mu \in \Xi_+[c]} \quad (7.5.21)$$

Relation (7.5.21) defines for $\varepsilon > 0$ the Lagrange function [7, 10] for problem (7.5.20) under the conditions where $K = \{\gamma_i\colon i \in \overline{1,m}\}$. Function (7.5.21) also plays an important role when $\varepsilon = 0$ in connection with the use of dual representations allowing one to perform the limiting process in ε, $\varepsilon > 0$. The role of function (7.5.21) for problem (7.5.20) is defined by the obvious (by virtue of the Kuhn–Tucker theorem) representation that for $m \in \mathcal{N}$, $(\gamma_i)_{i \in \overline{1,m}} \in \Gamma^m$, $\varepsilon \in (0, \infty)$, $K = \{\gamma_i\colon i \in \overline{1,m}\}$,

$$\min_{\mu \in \Lambda^*[K,\varepsilon]} \int_E \pi\, d\mu = \max_{(l_i)_{i \in \overline{1,m}} \in \mathbf{R}_+^m} \min_{\mu \in \Xi_+[c]} \mathbf{L}_m[(\gamma_i)_{i \in \overline{1,m}}, (l_i)_{i \in \overline{1,m}}, \varepsilon](\mu) \quad (7.5.22)$$

(all maxima and minima in (7.5.22) are attained); for the proof of relations similar to (7.5.22) see, e.g., [4, Chap. 3; 17, Chap. 1]. It is connected with the familiar rule of Lagrange multipliers; we shall omit these arguments in the present brief exposition. Now we can directly use (7.5.19) for finding the asymptotic value in which we are interested, because, in view of (7.5.22),

$$(\Lambda - \min)[\Pi]$$

$$= \sup_{\varepsilon \in (0,\infty)} \sup_{m \in \mathcal{N}} \sup_{(\gamma_i)_{i \in \overline{1,m}} \in \Gamma^m} \max_{(l_i)_{i \in \overline{1,m}} \in \mathbf{R}_+^m} \min_{\mu \in \Xi_+[c]} \mathbf{L}_m[(\gamma_i)_{i \in \overline{1,m}}, (l_i)_{i \in \overline{1,m}}, \varepsilon](\mu)$$

$$= \sup_{m \in \mathcal{N}} \sup_{(\gamma_i)_{i \in \overline{1,m}} \in \Gamma^m} \sup_{(l_i)_{i \in \overline{1,m}} \in \mathbf{R}_+^m} \sup_{\varepsilon \in (0,\infty)} \min_{\mu \in \Xi_+[c]} \mathbf{L}_m[(\gamma_i)_{i \in \overline{1,m}}, (l_i)_{i \in \overline{1,m}}, \varepsilon](\mu)$$

$$= \sup_{m \in \mathcal{N}} \sup_{(\gamma_i)_{i \in \overline{1,m}} \in \Gamma^m} \sup_{(l_i)_{i \in \overline{1,m}} \in \mathbf{R}_+^m} \min_{\mu \in \Xi_+[c]} \mathbf{L}_m[(\gamma_i)_{i \in \overline{1,m}}, (l_i)_{i \in \overline{1,m}}, 0](\mu). \quad (7.5.23)$$

In (7.5.23), we used the following fact, which follows from (7.5.21): $\forall\, m \in \mathcal{N}$, $(\gamma_i)_{i \in \overline{1,m}} \in \Gamma^m$,

$$(l_i)_{i \in \overline{1,m}} \in \mathbf{R}_+^m, \qquad \varepsilon \in (0, \infty), \qquad \mu \in \Xi_+[c]:$$

$$\mathbf{L}_m[(\gamma_i)_{i \in \overline{1,m}}, (l_i)_{i \in \overline{1,m}}, \varepsilon](\mu)$$

$$= \mathbf{L}_m[(\gamma_i)_{i \in \overline{1,m}}, (l_i)_{i \in \overline{1,m}}, 0](\mu) - \varepsilon \sum_{i=1}^m l_i. \quad (7.5.24)$$

Representation (7.5.24) allows us to determine in (7.5.23) the inner supremum with respect to ε, $\varepsilon > 0$; the other arguments used in (7.5.23) are trivial. Now, let us observe that $\forall\, m \in \mathcal{N}$

$$(\gamma_i)_{i \in \overline{1,m}} \in \Gamma^m, \qquad (l_i)_{i \in \overline{1,m}} \in \mathbf{R}_+^m:$$

$$\pi + \sum_{i=1}^{m} l_i S_{\gamma_i} \in B(E, \mathcal{L}),$$

so that, in accordance with (7.5.4), the quantity

$$\beta_0\left[\pi + \sum_{i=1}^{m} l_i S_{\gamma_i}\right] \in (-\infty, 0]$$

is well defined. Combining (7.5.5), (7.5.21), and (7.5.23), we obtain the required dual representation of the asymptotic value:

$$(\Lambda - \min)[\Pi]$$

$$= \sup_{m \in \mathcal{N}} \sup_{(\gamma_i)_{i \in \overline{1,m}} \in \Gamma^m} \sup_{(l_i)_{i \in \overline{1,m}} \in \mathbf{R}_+^m} \left(c\beta_0\left[\pi + \sum_{i=1}^{m} l_i S_{\gamma_i}\right] - \sum_{i=1}^{m} l_i r(\gamma_i) \right). \quad (7.5.25)$$

Here we make use of the auxiliary problems (7.5.2) and (7.5.3) with an appropriate specification of their objective functionals; we have to bear in mind the possibility defined by (7.5.6) under the imposition of a relevant condition on the measure η. We restricted ourselves to constructing one specific duality with the aim of representing the asymptotic value (7.5.14). We could consider other versions of introducing perturbations, making use of the prescriptions from Chap. 5, but we shall not do so now because the main object of study in this monograph is nonconvex extremal problems: we keep in mind that the set $\mathbf{Y}$ of Chap. 5 was not assumed to be convex until this section. However, we ought to take into account that adding some other convexity conditions provides new possibilities for the dual representation of asymptotic values as such; this was noticed earlier in various settings of problems of mathematical programming (see, e.g., [7, 10]).

For problems similar to (7.5.1) in the asymptotic setting (7.5.14), we point out [47], and also [14–16], where the problem of consecutive control with constraints on the travel points was studied.

Chapter 8

Asymptotic Efficiency

8.1. Introduction

This chapter is a natural continuation of the preceding one, but it is devoted to the study of a more complicated problem of optimization with respect to the cone corresponding to the pointwise ordering of the space of estimates. We discussed general issues concerning the extension of these problems, including statements with constraints of an asymptotic character, in Chap. 2. However, starting with Chap. 5, we have been studying problems in which we used only one concrete version of compactification of the space of conventional solutions by means of their integration with respect to a specified f.a. measure η. Let us consider a utilization of this compactificator for studying more complicated problems of optimization with respect to the cone corresponding to the pointwise ordering of the space of estimates. An example of such a problem was given in Chap. 1. We shall now present general constructions of this kind, confining ourselves to the case where the conventional solutions are identified with functionals on the set E. A relevant generalization embracing the example mentioned in Chap. 1 will be considered later. Our purpose in this chapter is to investigate the conditions of asymptotic efficiency both when studying the values and when constructing approximate solutions. A major role is played here by generalized problems, to which diverse asymptotic settings lead.

8.2. Generalized Problem of Cone Optimization. I

We shall now discuss one specification of the construction considered in Sec. 2.6, connected with the relaxations of constraints defined in Chap. 5. We shall deal with perturbations of condition (5.1.4) and their influence on the result, which is defined, as in Sec. 2.6, as the set of minimal elements.

Consider operator (5.1.1) and set (5.1.3); these objects define a system of constraints (recall that space (4.2.1) is fixed throughout the book). In this section, we discuss the case where the system of constraints contains a condition of nonnegativity similar to that used in Sec. 5.2, and it is not perturbed. We shall use the notation and conventions of Sec. 5.2 without additional explanation; we shall need, in particular, definitions (5.2.3)–(5.2.6), (5.2.9), (5.2.11), (5.2.13), and (5.2.14). Finally, throughout this section we assume that Condition 5.4.1 is fulfilled. In this connection, we shall use, without additional explanation, the 3-tuple $(K_0, \varepsilon_0, c_0)$ (5.4.5) satisfying condition (5.4.6), and, moreover, families (5.4.10) and (5.4.11) corresponding to the choice of (5.4.5) and (5.4.6) and playing the role of auxiliary truncated asymptotics of perturbed constraints. It should be emphasized, however, that the resultant assertions of this section relate to "asymptotics" (5.2.9), (5.2.11) and do not depend on the subjective choice of (5.4.5) and (5.4.6), so that families (5.4.10) and (5.4.11) are used only at the intermediate stages of reasoning where a need arises of applying the procedures of Sec. 2.6. Throughout this section, we fix the nonempty set Q (see Sec. 2.6) and the operator

$$w\colon Q \to \mathbf{C}((\mathrm{add})^+[\mathcal{L}, \eta], \tau_\eta^*(\mathcal{L})); \tag{8.2.1}$$

also, throughout this section the symbol w is understood only in the sense of (8.2.1). In the sequel, we shall use, without additional explanation, space (2.6.1) and the relevant notation of Sec. 2.6; in particular, we shall need a set of the type (2.6.2) and (2.6.6). By means of (8.2.1) we introduce the objective operator

$$W\colon Q \to \mathbf{R}^{B_0^+(E,\mathcal{L})}, \tag{8.2.2}$$

by setting $\forall\, q \in Q$

$$W(q) \triangleq (w(q)(f * \eta))_{f \in B_0^+(E,\mathcal{L})}. \tag{8.2.3}$$

Throughout this section, the symbol W is understood only in the sense of (8.2.2) and (8.2.3); here, of course, $\forall\, q \in Q,\ f \in B_0^+(E, \mathcal{L})$:

$$W(q)(f) = w(q)(f * \eta). \tag{8.2.4}$$

Note that we already frequently used the definitions of criterion in terms of the elements of the function space $\mathbf{C}((\text{add})^+[\mathcal{L}, \eta], \tau_\eta^*(\mathcal{L}))$ in Chap. 7. Therefore, we shall not discuss in detail various concrete settings of extremal problems for which such a way of defining the objective operator (criterion) is acceptable. We only observe that a number of control problems for system (1.3.1) of Chap. 1 can be reduced to this form. At the same time, strictly speaking, the problem considered in Sec. 1.4 does not conform to (8.2.2)– (8.2.4) for the only reason that the case discussed in Sec. 1.4 uses a "vector" control f, whereas in (8.2.2)–(8.2.4) we used functionals as f. However, conceptually, the scheme of investigation of the example from Sec. 1.4 is in full agreement with the one considered here (later, in a separate section we shall give one rather general setting, embracing, in particular, the example from Sec. 1.4). For brevity, we introduce, following Sec. 2.6, the operators

$$\mathcal{W} \triangleq (\text{term})[w], \qquad \mathbf{W} \triangleq (\text{term})[W], \tag{8.2.5}$$

thus obtaining maps mapping $(\text{add})_+[\mathcal{L}, \eta]$ and $B_0^+(E, \mathcal{L})$, respectively, into $\mathbf{R}^Q$. The first operator in (8.2.5) will be used in the generalized problem as an objective one. Recall (see 2.6) that, in accordance with (8.2.5), $\forall \mu \in (\text{add})^+[\mathcal{L}, \eta]$, $q \in Q$,

$$\mathcal{W}(\mu)(q) = w(q)(\mu).$$

In just the same way, $\forall f \in B_0^+(E, \mathcal{L})$, $q \in Q$,

$$\mathbf{W}(f)(q) = W(q)(f) = \mathcal{W}(f * \eta)(q).$$

The last equality means that $\forall f \in B_0^+(E, \mathcal{L})$

$$\mathbf{W}(f) = \mathcal{W}(f * \eta) \in \mathbf{R}^Q. \tag{8.2.6}$$

Here we have a natural analogy with (7.3.9). However, in this case the quality criterion is more complicated. Note that the difference between (w, W) and $(\mathcal{W}, \mathbf{W})$ is inessential; it is merely a difference in the form of notation and nothing else. Here (see 2.6),

$$W \in \mathcal{Q}_{B_0^+(E, \mathcal{L})}, \qquad w \in \mathcal{Q}^0_{(\text{add})^+[\mathcal{L}, \eta]}(\tau_\eta^*(\mathcal{L})) \tag{8.2.7}$$

is a concordant pair of operators, just like $(\mathbf{W}, \mathcal{W})$ in (8.2.6). For the purposes of compactification of an essential part of the solution space by the prescriptions given in Sec. 2.6, we introduce a number of auxiliary operators with due allowance for (7.3.6). As the functional g in (7.3.6), we shall use

$w(q)$, $q \in Q$. All these auxiliary definitions are connected with the requirements of compactification and are not used in the final assertions. Thus, with due allowance for (7.3.6), we introduce the operator

$$w_0 \colon Q \to \mathbf{C}(\Xi_+[c_0], \tilde{\tau}^*_{c_0}(\mathcal{L})), \tag{8.2.8}$$

setting $\forall q \in Q$

$$w_0(q) \triangleq (w(q)|\Xi_+[c_0]). \tag{8.2.9}$$

It is easy to see that operator (8.2.8), (8.2.9) is in full agreement with the conditions from Sec. 2.6 (see (2.6.6)), so that

$$w_0 \in \mathcal{Q}^0_{\Xi_+[c_0]}(\tilde{\tau}^*_{c_0}(\mathcal{L})). \tag{8.2.10}$$

However, the most important fact connected with operator (8.2.10) is the fact that its images are continuous functionals on a compact space (see (4.2.25)). Of course, under our conditions, the transition from w to w_0 can be regarded as an inessential operation caused by the requirement of formalization by the prescriptions from Sec. 2.6. By analogy with this replacement (of w by w_0), we also introduce a respective shortened operator for handling the conventional solutions. Namely, let, by definition,

$$W_0 \colon Q \to \mathbf{R}^{M^+_{c_0}}$$

be an operator such that $\forall q \in Q$

$$W_0(q) \triangleq (W(q)|M^+_{c_0}). \tag{8.2.11}$$

Here again we use a restriction of the values of (8.2.2); then,

$$W_0 \in \mathcal{Q}_X|_{X=M^+_{c_0}}. \tag{8.2.12}$$

In the sequel, operator (8.2.11), (8.2.12) plays the role of the objective operator s from Sec. 2.6. Here, $W_0(q)(f) = W(q)(f)$ for $q \in Q$, $f \in M^+_{c_0}$. We also have

$$\forall q \in Q, \quad \mu \in \Xi_+[c_0]\colon \quad w(q)(\mu) = w_0(q)(\mu).$$

Taking account of (8.2.4), we get a natural relation of truncated operators (8.2.10) and (8.2.12), namely, $\forall q \in Q$, $f \in M^+_{c_0}$,

$$W_0(q)(f) = w_0(q)(f * \eta); \tag{8.2.13}$$

in the sequel, we shall need (8.2.13) for checking the conditions of pointwise consistency (see Sec. 2.6). Finally, with due allowance for (8.2.10) and

(8.2.12), we naturally introduce the transformations of the indicated operators of the type (2.6.8):

$$\mathcal{W}_0 \triangleq (\text{term})\,[w_0], \qquad \mathbf{W}_0 \triangleq (\text{term})\,[W_0],$$

mapping the sets $\Xi_+[c_0]$ and $M_{c_0}^+$, respectively, into $\mathbf{R}^Q$. Here, according to (2.6.4), $\forall\,\mu \in \Xi_+[c_0]$, $q \in Q$,

$$\mathcal{W}_0(\mu)(q) = w_0(q)(\mu) = w(q)(\mu) = \mathcal{W}(\mu)(q).$$

This means that $\mathcal{W}_0(\mu) = \mathcal{W}(\mu)$ for $\mu \in \Xi_+[c_0]$. In other words,

$$\mathcal{W}_0 = (\mathcal{W}|\Xi_+[c_0]). \tag{8.2.14}$$

Next, we have (see Sec. 2.6) $\forall\,f \in M_{c_0}^+$, $q \in Q$,

$$\mathbf{W}_0(f)(q) = W_0(q)(f) = W(q)(f) = \mathbf{W}(f)(q).$$

Thus, $\mathbf{W}_0(f) = \mathbf{W}(f)\ \forall\,f \in M_{c_0}^+$. But this means that

$$\mathbf{W}_0 = (\mathbf{W}|M_{c_0}^+). \tag{8.2.15}$$

Relations (8.2.14) and (8.2.15) give a visual characterization of auxiliary construction (8.2.8), (8.2.10), and (8.2.12) as a restriction of the corresponding problem, and, obviously, (8.2.14) corresponds to a certain compactification of the space of conventional solutions in the problem with criterion (8.2.15); see the assertions of Sec. 4.3. Let us now introduce the truncated integration operator

$$\mathcal{J} \triangleq (f * \eta)_{f \in M_{c_0}^+}, \tag{8.2.16}$$

acting, obviously, from $M_{c_0}^+$ into $\Xi_+[c_0]$. It is easy to see that $\forall\,q \in Q$

$$W_0(q) = w_0(q) \circ \mathcal{J} \tag{8.2.17}$$

(see (8.2.13)). Relation (8.2.17) defines in exact form the conditions of pointwise consistency from Sec. 2.6 (see, for example, the conditions of Theorem 2.6.1). We now observe that each of the families $\widetilde{\mathcal{T}}_+$, $\widetilde{\mathcal{T}}_0^+[\Gamma_*]$ ($\Gamma_* \in (\text{Step})\,[\Gamma]$) has the semimultiplicativity property from Sec. 2.2; this fact was already mentioned in Sec. 7.3. Now we are going to make use of a direct specification of Theorem 2.6.1 under the conditions where (in this theorem) $X = M_{c_0}^+$, $\mathcal{X}$ being defined as $\widetilde{\mathcal{T}}_+$ (5.4.10) or $\widetilde{\mathcal{T}}_0^+[\Gamma_*]$ (5.4.11) for some $\Gamma_* \in (\text{Step})\,[\Gamma]$, $Y = \Xi_+[c_0]$, $\tau = \tilde{\tau}_{c_0}^*(\mathcal{L})$, $s = W_0$, $m = \mathcal{J}$, $g = w_0$. Here, operators (2.6.8) are specified, as one can easily see, as follows: $S = \mathbf{W}_0$, $G = \mathcal{W}_0$. From (5.4.12) we find, in terms of operator (8.2.16), that

$$\bigcap_{U \in \tilde{\mathcal{T}}_+} \mathrm{cl}\,(\mathcal{J}^1(U), \tilde{\tau}^*_{c_0}(\mathcal{L})) = \tilde{\Omega}_+;$$

moreover, in view of (5.4.13), we get $\forall\, \Gamma_* \in$ (Step) $[\Gamma]$

$$\bigcap_{U \in \tilde{\mathcal{T}}_0^+[\Gamma_*]} \mathrm{cl}\,(\mathcal{J}^1(U), \tilde{\tau}^*_{c_0}(\mathcal{L})) = \tilde{\Omega}_+.$$

Now, Theorem 2.6.1 implies that

$$\left(\bigcap_{T \in \tilde{\mathcal{T}}_+} \mathrm{cl}\,(\mathbf{W}_0^1(T), \otimes^Q(\tau_{\mathbf{R}})) = \mathcal{W}_0^1(\tilde{\Omega}_+) \right)$$

$$\&\left(\forall\, \Gamma_* \in \text{(Step)}\,[\Gamma]: \bigcap_{T \in \tilde{\mathcal{T}}_0^+[\Gamma_*]} \mathrm{cl}\,(\mathbf{W}_0^1(T), \otimes^Q(\tau_{\mathbf{R}})) = \mathcal{W}_0^1(\tilde{\Omega}_+) \right). \quad (8.2.18)$$

Next, taking account of (8.2.14) and Theorem 5.4.2, we obtain the relation

$$\mathcal{W}_0^1(\tilde{\Omega}_+) = \{\mathcal{W}_0(\mu)\colon \mu \in \tilde{\Omega}_+\} = \{\mathcal{W}(\mu)\colon \mu \in \tilde{\Omega}_+\} = \mathcal{W}^1(\tilde{\Omega}_+). \quad (8.2.19)$$

However, by virtue of (5.4.10) and (5.4.11), we have, by definition,

$$(\forall\, T \in \tilde{\mathcal{T}}_+\colon \mathbf{W}^1(T) = \mathbf{W}_0^1(T))$$

$$\&(\forall\, \Gamma_* \in \text{(Step)}\,[\Gamma], \tilde{T} \in \tilde{\mathcal{T}}_0^+[\Gamma_*]\colon \mathbf{W}^1(\tilde{T}) = \mathbf{W}_0^1(\tilde{T})). \quad (8.2.20)$$

From (8.2.18)–(8.2.20), we immediately obtain (see (5.2.9), (5.2.12), (5.4.10), (5.4.11)) the following theorem.

THEOREM 8.2.1. *The sets of asymptotically attainable estimates admit the following generalized representation:*

$$\left(\bigcap_{T \in \mathcal{T}_+} \mathrm{cl}\,(\mathbf{W}^1(T), \otimes^Q(\tau_{\mathbf{R}})) = \mathcal{W}^1(\tilde{\Omega}_+) \right)$$

$$\&\left(\forall\, \Gamma_* \in \text{(Step)}\,[\Gamma]: \bigcap_{T \in \mathcal{T}_0^+[\Gamma_*]} \mathrm{cl}\,(\mathbf{W}^1(T), \otimes^Q(\tau_{\mathbf{R}})) = \mathcal{W}^1(\tilde{\Omega}_+) \right).$$

COROLLARY. *Asymptotic values have a generalized representation in terms of the $\mathcal{W}$-image of set (5.2.14):*

$$\left((\leqq - \mathrm{MIN}\,)\left[\bigcap_{T \in \mathcal{T}_+} \mathrm{cl}\,(\mathbf{W}^1(T), \otimes^Q(\tau_{\mathbf{R}})) \right] = (\leqq - \mathrm{MIN}\,)[\mathcal{W}^1(\tilde{\Omega}_+)] \right)$$

$$\& \left(\forall \, \Gamma_* \in (\text{Step})\,[\Gamma] : (\underset{=}{\leq} - \text{MIN}\,) \left[\bigcap_{T \in \mathcal{T}_0^+[\Gamma_*]} \text{cl}\,(\mathbf{W}^1(T), \otimes^Q(\tau_{\mathbf{R}})) \right] \right.$$

$$\left. = (\underset{=}{\leq} - \text{MIN}\,)[\mathcal{W}^1(\tilde{\Omega}_+)] \right).$$

Among other things, let us observe that the corollary to Theorem 8.2.1 implies, in particular, that $\forall \, \Gamma_* \in (\text{Step})\,[\Gamma]$,

$$(\underset{=}{\leq} - \text{MIN}\,) \left[\bigcap_{T \in \mathcal{T}_+} \text{cl}\,(\mathbf{W}^1(T), \otimes^Q(\tau_{\mathbf{R}})) \right]$$

$$= (\underset{=}{\leq} - \text{MIN}\,) \left[\bigcap_{T \in \mathcal{T}_0^+[\Gamma_*]} \text{cl}\,(\mathbf{W}^1(T), \otimes^Q(\tau_{\mathbf{R}})) \right]; \qquad (8.2.20)'$$

relation $(8.2.20)'$ can be interpreted as asymptotic equivalence with respect to the result for families (5.2.9) and (5.2.11). In the next chapter, we shall study some special cases, in which the inference mentioned above acquires the meaning of a distinctive computational stability. We shall not discuss these cases in detail for the situation of the finite set Γ.

We shall consider, in the general case of the system of constraints of Secs. 5.1 and 5.2, the issue of seeking asymptotically efficient approximate solutions. Here, we could, in principle, use the construction of Theorem 2.6.3. However, the approximate solutions obtained by its prescriptions have a very complicated structure and can hardly be used for practical purposes. Let us consider a somewhat different structure, being guided by assertions about the approximation of f.a. measures by indefinite integrals, stated in Sec. 4.3. In particular, we shall use Lemma 4.3.1, but we shall first specify some propositions from Sec. 2.6 concerning generalized problems. Recall that (see (2.6.11), (5.2.14), and (8.2.5))

$$(\mathcal{W} - \text{min})[\tilde{\Omega}_+] = \{ \mu \in \tilde{\Omega}_+ \, | \, \mathcal{W}(\mu) \in (\underset{=}{\leq} - \text{MIN}\,)[\mathcal{W}^1(\tilde{\Omega}_+)] \} \qquad (8.2.21)$$

is the set of all efficient, in the traditional sense, solutions of the following generalized minimization problem:

$$\mathcal{W}(\mu) \to [\underset{=}{\leq} - \text{MIN}\,], \qquad \mu \in \tilde{\Omega}_+; \qquad (8.2.22)$$

it is easy to check that (8.2.22) is a specification of problem (2.6.9). Taking account of (2.6.12) and (8.2.21), we obviously have the equality

$$\mathcal{W}^1((\mathcal{W} - \text{min})[\tilde{\Omega}_+]) = \{ \mathcal{W}(\mu) : \mu \in (\mathcal{W} - \text{min})[\tilde{\Omega}_+] \} = (\underset{=}{\leq} - \text{MIN}\,)[\mathcal{W}^1(\tilde{\Omega}_+)].$$

With due account of (8.2.1), (8.2.5), and the definitions from Sec. 2.4, we get

$$(w - \text{Min})[\tilde{\Omega}_+] = (\mathcal{W} - \min)[\tilde{\Omega}_+]; \tag{8.2.23}$$

$$\mathcal{W}^1((w - \text{Min})[\tilde{\Omega}_+]) = (\leqq - \text{MIN})[\mathcal{W}^1(\tilde{\Omega}_+)]. \tag{8.2.24}$$

Relation (8.2.24), together with the corollary to Theorem 8.2.1, provides the important property of exhaustibility of asymptotic values by the functionals $\mathcal{W}(\mu)$, where μ runs over the set (8.2.23) of all efficient solutions of problem (8.2.22), namely,

$$\left((\leqq - \text{MIN})\left[\bigcap_{T \in \mathcal{T}_+} \text{cl}\,(\mathbf{W}^1(T), \otimes^Q(\tau_{\mathbf{R}}))\right] = \{\mathcal{W}(\mu)\colon \mu \in (w - \text{Min})[\tilde{\Omega}_+]\}\right)$$

$$\&\left(\forall\, \Gamma_* \in (\text{Step})[\Gamma]\colon (\leqq - \text{MIN})\left[\bigcap_{T \in \mathcal{T}_0^+[\Gamma_*]} \text{cl}\,(\mathbf{W}^1(T), \otimes^Q(\tau_{\mathbf{R}}))\right]\right.$$

$$= \{\mathcal{W}(\mu)\colon \mu \in (w - \text{Min})[\tilde{\Omega}_+]\}\Big). \tag{8.2.25}$$

In turn, (8.2.25) will provide, in the sequel, the property of sufficiency of the approximating construction on the basis of Lemma 4.3.1. By analogy with (8.2.21) and (8.2.23), we introduce

$$(\mathcal{W}_0 - \min)[\tilde{\Omega}_+] \in \mathcal{P}(\tilde{\Omega}_+),$$

$$(w_0 - \text{Min})[\tilde{\Omega}_+] = (\mathcal{W}_0 - \min)[\tilde{\Omega}_+],$$

where we take into account the natural relationship between w_0 and $\mathcal{W}_0$. However, Theorem 5.4.2 and relations (8.2.19) and (8.2.21) yield

$$(w_0 - \text{Min})[\tilde{\Omega}_+] = \{\mu \in \tilde{\Omega}_+ \mid \mathcal{W}_0(\mu) \in (\leqq - \text{MIN})[\mathcal{W}^1(\tilde{\Omega}_+)]\}$$

$$= (\mathcal{W} - \min)[\tilde{\Omega}_+] = (w - \text{Min})[\tilde{\Omega}_+]. \tag{8.2.26}$$

THEOREM 8.2.2. *Let* $\tilde{\Omega}_+ \neq \varnothing$. *Then*

$$(w - \text{Min})[\tilde{\Omega}_+] \neq \varnothing.$$

The proof is obvious, but nevertheless let us consider a brief scheme of it, turning to operator (8.2.8) and to the operator $\mathcal{W}_0$ obtained with its aid. We shall use the specification of property (2.6.13); for this purpose, we ought to take into account (8.2.10) and the fact that under the conditions of the theorem $\tilde{\Omega}_+$ is a nonempty subset of $\Xi_+[c_0]$ closed in the sense of $\tilde{\tau}_{c_0}^*(\mathcal{L})$

$$\tilde{\Omega}_+ \in \mathcal{F}_\tau|_{\tau = \tilde{\tau}_{c_0}^*(\mathcal{L})}, \qquad \tilde{\Omega}_+ \neq \varnothing.$$

Now, relations (2.6.13) and (8.2.10) imply that

$$(w_0 - \text{Min})[\tilde{\Omega}_+] \neq \varnothing;$$

taking into account (8.2.26), we get the assertion of the theorem.

From Theorem 8.2.2 and relations (8.2.21), (8.2.23), it follows that

$$(\tilde{\Omega}_+ \neq \varnothing) \Rightarrow ((w - \text{Min})[\tilde{\Omega}_+] \in 2^{\tilde{\Omega}_+}).$$

This relation is quite natural from the standpoint of the general assertions from Sec. 2.6. We now observe that, in view of (5.2.14), set (8.2.23) is a subset of cone (4.2.2). This fact allows us to associate every element (8.2.23) with a net in $B_0^+(E, \mathcal{L})$ using the prescriptions from Sec. 4.3; here $\mu \in (\text{DIR})[\mathcal{D}]$ (the notation is that of Secs. 2.4 and 4.3). Then $\forall \mu \in (w - \text{Min})[\tilde{\Omega}_+]$

$$(\mathcal{D}, \prec, \Theta_\mu^+[\cdot]) \tag{8.2.27}$$

is a net in $B_0^+(E, \mathcal{L})$. The last inference can be made somewhat more precise by using Theorem 5.4.2. Namely, taking account of the indicated theorem and of relations (8.2.21) and (8.2.23), we have

$$(w - \text{Min})[\tilde{\Omega}_+] \in \mathcal{P}(\Xi_+[c_0]).$$

Now, from (4.3.8), we get the following important inference concerning nets (8.2.27). Namely, if $\mu \in (w - \text{Min})[\tilde{\Omega}_+]$, then (8.2.27) is a net in $M_{c_0}^+$ for which (see Lemma 4.3.1) net (4.3.11) converges to μ in the sense of any topology $\tau \in \mathfrak{M}$. Note that for any choice of μ from set (8.2.23), the directed set in (8.2.27) is fixed. In view of this circumstance, it is expedient to consider the set

$$\{\Theta_\mu^+[\cdot] : \mu \in (w - \text{Min})[\tilde{\Omega}_+]\} \tag{8.2.28}$$

(which is a subset of the set of all operators acting from $\mathcal{D}$ into $B_0^+(E, \mathcal{L})$) as a description of the whole totality of approximate net-solutions related by the prescriptions from Sec. 4.3 with the generalized extremal set (8.2.23).

The following theorem holds true.

THEOREM 8.2.3. *Let* $\mu \in (w - \text{Min})[\tilde{\Omega}_+]$. *Then*

$$\mathcal{W}(\mu) \in (\leqq - \text{MIN})\left[\bigcap_{T \in \mathcal{T}_+} \text{cl}(\mathbf{W}^1(T), \otimes^Q(\tau_{\mathbf{R}}))\right].$$

Net (8.2.27) is an asymptotically efficient approximate solution under the conditions of relaxation of the constraints corresponding to the family

$\mathcal{T}_+$ which realizes in the limit $\mathcal{W}(\mu)$ the values of the objective operator $\mathbf{W}$. Namely, we have: (1) the convergence of the net

$$(\mathcal{D}, \prec, \mathbf{W} \circ \Theta_\mu^+[\cdot]) = (\mathcal{D}, \prec, (\mathbf{W}(\Theta_\mu^+[\mathcal{K}]))_{\mathcal{K} \in \mathcal{D}}) \tag{8.2.29}$$

to $\mathcal{W}(\mu)$ in the sense of the topology $\otimes^Q(\tau_{\mathbf{R}})$, (2) $\forall T \in \mathcal{T}_+ \; \exists \mathcal{K}_0 \in \mathcal{D} \; \forall \widetilde{\mathcal{K}} \in \mathcal{D}$,

$$(\mathcal{K}_0 \prec \widetilde{\mathcal{K}}) \Rightarrow (\Theta_\mu^+[\widetilde{\mathcal{K}}] \in T).$$

THEOREM 8.2.4. *Let* $\mu \in (w - \text{Min})[\Omega_+]$. *Then* $\forall \Gamma_* \in (\text{Step})[\Gamma]$

(1) $\mathcal{W}(\mu) \in (\underline{\leqq} - \text{MIN})\left[\bigcap_{T \in \mathcal{T}_0^+[\Gamma_*]} \text{cl}\,(\mathbf{W}^1(T), \otimes^Q(\tau_{\mathbf{R}})) \right],$

(2) *net* (8.2.29) *converges to* $\mathcal{W}(\mu)$ *in the sense of* $\otimes^Q(\tau_{\mathbf{R}})$,

(3) $\forall T \in \mathcal{T}_0^+[\Gamma_*] \; \exists \mathcal{K}_0 \in \mathcal{D} \; \forall \widetilde{\mathcal{K}} \in \mathcal{D}$,

$$(\mathcal{K}_0 \prec \widetilde{\mathcal{K}}) \Rightarrow (\Theta_\mu^+[\widetilde{\mathcal{K}}] \in T).$$

The proofs of the two assertions are similar, and therefore it is sufficient to consider only the scheme of the proof of Theorem 8.2.4. The simplest is the proof of assertion (2) of the last theorem because Lemma 4.3.1 implies, in particular, the convergence of net (4.3.11) to μ in the topological space (4.2.6). From (8.2.1) and (8.2.5) it follows, however, that (see (2.6.7))

$$\mathcal{W} \in C((\text{add})^+[\mathcal{L}, \eta], \tau_\eta^*(\mathcal{L}), \mathbf{R}^Q, \otimes^Q(\tau_{\mathbf{R}})). \tag{8.2.30}$$

Since, in particular, $\mu \in (\text{add})^+[\mathcal{L}, \eta]$ and (4.3.11) is a net in $(\text{add})^+[\mathcal{L}, \eta]$, it follows, according to (4.2.15) that we have a convergence of (4.3.11) to μ in the sense of $\tau_\eta^*(\mathcal{L})$. From (8.2.30) we now have the convergence in the sense of $\otimes^Q(\tau_{\mathbf{R}})$ of the net

$$(\mathcal{D}, \prec, (\mathbf{W}(\Theta_\mu^+[\mathcal{K}] * \eta))_{\mathcal{K} \in \mathcal{D}}) \tag{8.2.31}$$

to $\mathcal{W}(\mu)$. However, it follows from (8.2.6) and (8.2.31) that

$$\mathbf{W} \circ \Theta_\mu^+[\cdot] = (\mathbf{W}(\Theta_\mu^+[\mathcal{K}]))_{\mathcal{K} \in \mathcal{D}} = (\mathcal{W}(\Theta_\mu^+[\mathcal{K}] * \eta))_{\mathcal{K} \in \mathcal{D}}$$

and, as a consequence, we have property (2), i.e., the convergence of (8.2.29). Assertion (1) follows from Theorem 8.2.1 and relation (8.2.24). The proof of (3) uses the convergence of net (4.3.11) to μ in the sense of the topology $\tau_0(\mathcal{L})$. Here we make essential use of the fact that $S_\gamma \in B_0(E, \mathcal{L})$ for $\gamma \in \Gamma_*$, as one can see from (5.2.13). In addition, we ought to bear in mind (5.2.5) and (5.2.6), and the representation of topology (4.2.14), mentioned in Sec. 4.2, inherited by the elementary integrals from Chap. 3. We shall not dwell on the indicated proofs, which are elementary in principle but rather unwieldy; the interested reader can carry out the proof himself, using the assertions of Chap. 4. We note one simple consequence of Theorems 8.2.3 and 8.2.4, which was discussed earlier. The fact is that by virtue of (8.2.25) and Theorems 8.2.3 and 8.2.4, set (8.2.28) can be regarded as a collection of approximate solutions sufficient for realizing the entire asymptotic value. Indeed, the generalized limits of values along the operators from (8.2.28) exhaust, with μ running over the set (8.2.23), each of the asymptotic values. Here, the directed set $(\mathcal{D}, \prec)$ for the approximate solutions obtained is fixed, and therefore it just becomes an element of the construction of the indicated solutions. This remark actually makes it possible to interpret set (8.2.28) as exhaustive in the sense of the possibilities of asymptotic optimization; the operators themselves, being elements of (8.2.28), can now be identified with approximate solutions. At the same time, this may allow us to infer that the set (8.2.23) of all generalized efficient solutions is exhaustive from the standpoint of asymptotic optimization.

8.3.   Generalized Problem of Cone Optimization. II

Here we study a problem which is conceptually similar to that considered in Sec. 8.2 but which has a different system of constraints. We shall deal with an analog of the extremal problem from Sec. 7.4, whereas Sec. 8.2 was devoted to studying a setting generalizing the problem with the system of constraints (5.1.4) complemented with the additional condition of nonnegativity considered in Sec. 7.3. Thus, in the sequel, we shall again fix $c \in [0, \infty)$ and consider condition (5.5.1) and its relaxations. In this connection, we observe that the notation of Sec. 5.5 will be used in the sequel without additional explanations (see, in particular, (5.5.4), (5.5.5), (5.5.8), (5.5.9), (5.5.11), (5.5.12)). At the same time, in contrast to Sec. 7.4, we shall speak about an optimization problem with respect to the cone corresponding to the pointwise ordering of the estimate space. In this connection, we shall need the notation and conventions of Sec. 2.6. In particular, as in Sec. 8.2, we fix a nonempty set Q throughout this chapter; thus, in the sequel, we shall use the conforming space (2.6.1). We shall also use the notation (2.6.2), (2.6.3), (2.6.6), and others. We shall refrain from a more detailed discussion of relations with Sec. 2.6, referring the reader to Sec. 8.2, where we gave similar arguments, though they concerned a somewhat different case.

Throughout this section, we fix an operator

$$w\colon Q \to \mathbf{C}(\mathbf{A}_\eta[\mathcal{L}], \tau_*^{(\eta)}[\mathcal{L}]); \qquad (8.3.1)$$

the symbol w is used throughout the remainder of this chapter only in the sense of (8.3.1). Note that (2.6.6) and (8.3.1), in particular, imply

$$w \in \mathcal{Q}^0_{\mathbf{A}_\eta[\mathcal{L}]}(\tau_*^{(\eta)}[\mathcal{L}]), \qquad (8.3.2)$$

and from (8.3.2) we find, in turn, that (8.3.1) is an element of (2.6.2) for $M = \mathbf{A}_\eta[\mathcal{L}]$. Therefore, according to (2.6.3) and (2.6.4), we introduce the operator

$$\mathcal{W} \triangleq (\text{term})\,[w] \qquad (8.3.3)$$

that maps $\mathbf{A}_\eta[\mathcal{L}]$ into $\mathbf{R}^Q$; the symbol $\mathcal{W}$ in this section is used only in the sense of (8.3.3). On the basis of the (generalized) operator (8.3.1), (8.3.2), we define the objective operator

$$W\colon Q \to \mathbf{R}^{B_0(E,\mathcal{L})} \qquad (8.3.4)$$

by a relation similar to (8.2.3): $\forall\, q \in Q$, $f \in B_0(E, \mathcal{L})$,

$$W(q)(f) = w(q)(f * \eta). \qquad (8.3.5)$$

Here, of course, by virtue of (2.6.2) and (8.3.4), we have

$$W \in \mathcal{Q}_{B_0(E,\mathcal{L})}, \qquad (8.3.6)$$

so that, according to (2.6.3), (2.6.4), (8.3.4)–(8.3.6), we can define a natural (from the standpoint of Sec. 2.6) transformation of the indicated operator:

$$\mathbf{W} \triangleq (\text{term})\,[W] \in (\mathbf{R}^Q)^{B_0(E,\mathcal{L})}; \qquad (8.3.7)$$

the symbols W and $\mathbf{W}$ are used throughout the remainder of this section only in the sense of (8.3.4)–(8.3.6) and (8.3.7) respectively. As for operators (8.3.3) and (8.3.7), we recall their explicit expressions using (2.6.3) and (2.6.4). Namely, $\forall\, \mu \in \mathbf{A}_\eta[\mathcal{L}]$, $q \in Q$: $\mathcal{W}(\mu)(q) = w(q)(\mu)$. In addition, $\forall\, f \in B_0(E, \mathcal{L})$, $q \in Q$,

$$\mathbf{W}(f)(q) = W(q)(f) = w(q)(f * \eta) = \mathcal{W}(f * \eta)(q).$$

This clearly means that $\forall\, f \in B_0(E, \mathcal{L})$

$$\mathbf{W}(f) = \mathcal{W}(f * \eta). \qquad (8.3.8)$$

Relation (8.3.8) is similar to the representation of the criterion in Sec. 7.4 but concerns a more general situation. In the form of $(\mathbf{W}, \mathcal{W})$ we again have a consistent pair of operators, each of which is defined on an unbounded set. The last circumstance hinders their study by means of the extension constructions from Sec. 2.6. To improve the situation, we again introduce truncated operators, acting by analogy with Sec. 7.4. Consider, though it is somewhat redundant, two types of such operators, one of which will be oriented to the work with asymptotics (5.5.9) and the other of which will correspond, in some natural sense, to asymptotics (5.5.12), and, hence, to asymptotics (5.5.8). In principle, we could use a unique compactificator, but, for methodological reasons, we shall use the scheme of arguments from Sec. 7.4. For this purpose, we shall observe, in view of (7.4.5) and (8.3.1), that $\forall\, q \in Q$

$$((w(q)|\Xi[c]) \in \mathbf{C}(\Xi[c], \tilde{\tau}_*^{(c)}[\mathcal{L}]))$$
$$\&((w(q)|\Xi[c+1]) \in \mathbf{C}(\Xi[c+1], \tilde{\tau}_*^{(c+1)}[\mathcal{L}])). \tag{8.3.9}$$

Making use of (8.3.9), we introduce two truncated generalized operators. Namely, let, by definition,

$$w_1\colon Q \to \mathbf{C}(\Xi[c], \tilde{\tau}_*^{(c)}[\mathcal{L}]) \tag{8.3.10}$$

be an operator such that $\forall\, q \in Q$

$$w_1(q) \triangleq (w(q)|\Xi[c]) = (w(q)(\mu))_{\mu \in \Xi[c]}. \tag{8.3.11}$$

Similarly, we assume, by definition, that

$$w_2\colon Q \to \mathbf{C}(\Xi[c+1], \tilde{\tau}_*^{(c+1)}[\mathcal{L}]) \tag{8.3.12}$$

is an operator such that $\forall\, q \in Q$

$$w_2(q) \triangleq (w(q)|\Xi[c+1]) = (w(q)(\mu))_{\mu \in \Xi[c+1]}. \tag{8.3.13}$$

The symbols w_1 and w_2 are understood in this section only in the sense of (8.3.10), (8.3.11) and (8.3.12), (8.3.13) respectively. Operators (8.3.10)–(8.3.13) are auxiliary; they actually play the same role as operator (8.2.8)–(8.2.10). The concrete choice of compact sets to which we restrict the functionals $w(q)$, may, of course, be different; this operation is inessential and is given in the interest of reduction to the procedures of Sec. 2.6. By analogy with (8.3.10)–(8.3.13), we introduce truncated operators on the spaces of conventional solutions. Namely, we assume, by definition, the operator

$$W_1\colon Q \to \mathbf{R}^{M_c} \tag{8.3.14}$$

to be such that $\forall q \in Q$: $W_1(q) \triangleq (W(q)|M_c)$. Moreover, we assume, by definition, that

$$W_2: Q \to \mathbf{R}^{M_{c+1}} \tag{8.3.15}$$

is an operator for which $\forall q \in Q$: $W_2(q) \triangleq (W(q)|M_{c+1})$. Throughout the remainder of this section, the symbols W_1 and W_2 have the meaning of (8.3.14) and (8.3.15) respectively. Let us consider the relationship between operators (8.3.10)–(8.3.15). For this purpose, it is convenient to introduce operators similar to (8.3.3) and (8.3.7). Let us first note that

$$w_1 \in \mathcal{Q}^0_{\Xi[c]}(\tilde{\tau}^{(c)}_*[\mathcal{L}]), \tag{8.3.16}$$

$$w_2 \in \mathcal{Q}^0_{\Xi[c+1]}(\tilde{\tau}^{(c+1)}_*[\mathcal{L}]); \tag{8.3.17}$$

(8.3.16) and (8.3.17) are in good agreement with Sec. 2.6 in part of the compactification constructions. Making use of (2.6.3), (2.6.4), (2.6.6), (8.3.16), and (8.3.17), we introduce the operators

$$\mathcal{W}_1 \triangleq (\text{term})\,[w_1], \qquad \mathcal{W}_2 \triangleq (\text{term})\,[w_2], \tag{8.3.18}$$

acting, respectively, from $\Xi[c]$ and $\Xi[c+1]$ into $\mathbf{R}^Q$. Here, of course, $\forall \mu \in \Xi[c]$, $q \in Q$,

$$\mathcal{W}_1(\mu)(q) = w_1(q)(\mu) = w(q)(\mu) = \mathcal{W}(\mu)(q).$$

The last assertion means that $\mathcal{W}_1(\mu) = \mathcal{W}(\mu)$ for $\mu \in \Xi[c]$. But in this case

$$\mathcal{W}_1 = (\mathcal{W}|\Xi[c]) = (\mathcal{W}(\mu))_{\mu \in \Xi[c]}. \tag{8.3.19}$$

Similarly, we have $\forall \mu \in \Xi[c+1]$, $q \in Q$,

$$\mathcal{W}_2(\mu)(q) = w_2(q)(\mu) = w(q)(\mu) = \mathcal{W}(\mu)(q).$$

This means that

$$\mathcal{W}_2 = (\mathcal{W}|\Xi[c+1]) = (\mathcal{W}(\mu))_{\mu \in \Xi[c+1]}; \tag{8.3.20}$$

relations (8.3.19) and (8.3.20) characterize the truncated generalized operators in terms of the restriction understood in the traditional sense. Taking into account that (see (8.3.14), (8.3.15))

$$W_1 \in Q_{M_c}, \qquad W_2 \in Q_{M_{c+1}},$$

we introduce, in accordance with Sec. 2.6, the operators

$$\mathbf{W}_1 \triangleq (\text{term})\,[W_1], \qquad \mathbf{W}_2 \triangleq (\text{term})\,[W_2], \tag{8.3.21}$$

operating here with relations (2.6.3) and (2.6.4). Then operators (8.3.21) are defined by the following obvious relations. Namely, $\mathbf{W}_1$ acts from M_c into $\mathbf{R}^Q$ and is defined by the condition $\mathbf{W}_1(f)(q) = \mathbf{W}(f)(q)$, $f \in M_c$, $q \in Q$. But then, obviously, $\mathbf{W}_1 = (\mathbf{W}|M_c)$. Likewise, we find, for $\mathbf{W}_2$, that this operator acting from M_{c+1} into $\mathbf{R}^Q$ satisfies the condition $\mathbf{W}_2(f)(q) = W_2(q)(f) = W(q)(f) = \mathbf{W}(f)(q)$. This means that $\mathbf{W}_2 = (\mathbf{W}|M_{c+1})$. Thus,

$$(\mathbf{W}_1 = (\mathbf{W}|M_c)) \& (\mathbf{W}_2 = (\mathbf{W}|M_{c+1})). \tag{8.3.22}$$

It follows from (8.3.8), (8.3.19), (8.3.20), and (8.3.22) that

$$(\forall f \in M_c: \mathbf{W}_1(f) = \mathcal{W}_1(f * \eta))$$

$$\& (\forall f \in M_{c+1}: \mathbf{W}_2(f) = \mathcal{W}_2(f * \eta)). \tag{8.3.23}$$

Here we have also taken into account (4.2.20) and Theorem 4.3.3. Relations (8.3.23), which are obvious, though, by virtue of (8.3.8), play an essential role because they are directly connected with the compactification procedures from Sec. 2.6. To be more precise, these relations characterize the conditions of pointwise consistency. In order to formulate the corresponding statement, we introduce the operators

$$\mathcal{J}_1 \triangleq (f * \eta)_{f \in M_c}, \qquad \mathcal{J}_2 \triangleq (f * \eta)_{f \in M_{c+1}}. \tag{8.3.24}$$

Then $\forall q \in Q$, $f \in M_c$,

$$W_1(q)(f) = w_1(q)(f * \eta) = (w_1(q) \circ \mathcal{J}_1)(f).$$

This means that $\forall q \in Q: W_1(q) = w_1(q) \circ \mathcal{J}_1$. Likewise, we have $W_2(q)(f) = w_2(q)(f * \eta) = (w_2(q) \circ \mathcal{J}_2)(f)$ for $q \in Q$, $f \in M_{c+1}$. This yields $\forall q \in Q: W_2(q) = w_2(q) \circ \mathcal{J}_2$. We have established the validity of the pointwise consistency, which is, however, also expressed by an obvious consequence of relations (8.3.23) and (8.3.24):

$$\mathbf{W}_1 = \mathcal{W}_1 \circ \mathcal{J}_1, \qquad \mathbf{W}_2 = \mathcal{W}_2 \circ \mathcal{J}_2.$$

At the same time, let us now note, as a consequence of (2.6.7), that

$$\mathcal{W} \in C(\mathbf{A}_\eta[\mathcal{L}], \tau_*^{(\eta)}[\mathcal{L}], \mathbf{R}^Q, \otimes^Q(\tau_\mathbf{R})),$$

$$\mathcal{W}_1 \in C(\Xi[c], \tilde{\tau}_*^{(c)}[\mathcal{L}], \mathbf{R}^Q, \otimes^Q(\tau_\mathbf{R})),$$

$$\mathcal{W}_2 \in C(\Xi[c+1], \tilde{\tau}_*^{(c+1)}[\mathcal{L}], \mathbf{R}^Q, \otimes^Q(\tau_\mathbf{R})).$$

Now we can directly use Theorem 2.6.1, which should, however, be considered in two versions, each one corresponding to a selected concrete definition of the compactificator. The set X in Sec. 2.6 should here be identified either with M_c or with M_{c+1}, the operator s either with W_1 or with W_2, the operator g either with w_1 or with w_2, (Y, τ) is identified with a respective copy of (4.2.25) for $b = c$ and $b = c + 1$, m is replaced by one of the operators (8.3.24); here (in Theorem 2.6.1) S is identified with $\mathbf{W}_1$ or $\mathbf{W}_2$, G with $\mathcal{W}_1$ or $\mathcal{W}_2$ respectively. In addition, we have to take into account Theorem 5.5.1, relation (5.5.17), and a similar representation for asymptotics (5.5.9). We shall start with the latter representations in terms of (8.3.24), so that

$$\left(\forall\, \Gamma_* \in (\text{Step})\,[\Gamma]\colon \ \tilde{\Omega}_*^{(c)} = \bigcap_{T \in \widehat{\mathcal{T}}_c^0[\Gamma_*]} \mathrm{cl}\,(\mathcal{J}_1^1(T), \tilde{\tau}_*^{(c)}[\mathcal{L}]) \right)$$

$$\&\left(\tilde{\Omega}_*^{(c)} = \bigcap_{T \in \widehat{\mathcal{T}}_c^{(1)}} \mathrm{cl}\,(\mathcal{J}_2^1(T), \tilde{\tau}_*^{(c+1)}[\mathcal{L}]) \right). \tag{8.3.25}$$

Now, Theorem 2.6.1 and relation (8.3.25) imply that

$$\left(\forall\, \Gamma_* \in (\text{Step})\,[\Gamma]\colon \ \bigcap_{T \in \widehat{\mathcal{T}}_c^0[\Gamma_*]} \mathrm{cl}\,(\mathbf{W}_1^1(T), \otimes^Q(\tau_\mathbf{R})) = \mathcal{W}_1^1(\tilde{\Omega}_*^{(c)}) \right)$$

$$\&\left(\bigcap_{T \in \widehat{\mathcal{T}}_c^{(1)}} \mathrm{cl}\,(\mathbf{W}_2^1(T), \otimes^Q(\tau_\mathbf{R})) = \mathcal{W}_2^1(\tilde{\Omega}_*^{(c)}) \right). \tag{8.3.26}$$

This immediate consequence of the assertions from Sec. 2.6 admits elementary transformations to a form that is characterized only in terms of the pair $(\mathbf{W}, \mathcal{W})$ and families (5.5.8) and (5.5.9). Similar transformations were also used in the preceding section but without additional explanation. Now we shall briefly turn our attention to this question by examining the indicated chain of arguments. First of all, we observe that, by virtue of (5.5.11), (8.3.19), and (8.3.20), the following equalities hold true:

$$\mathcal{W}_1^1(\tilde{\Omega}_*^{(c)}) = \mathcal{W}_2^1(\tilde{\Omega}_*^{(c)}) = \mathcal{W}^1(\tilde{\Omega}_*^{(c)}). \tag{8.3.27}$$

Furthermore, from the constructions of Sec. 5.5, it easily follows, in view of (8.3.22), that $\forall\, \Gamma_* \in (\text{Step})\,[\Gamma],\ T \in \widehat{\mathcal{T}}_c^0[\Gamma_*]$

$$\mathbf{W}_1^1(T) = \mathbf{W}^1(T). \tag{8.3.28}$$

Similarly, we have $\forall\, T \in \mathcal{T}_c^{(1)}$,

$$\mathbf{W}_2^1(T) = \mathbf{W}^1(T). \tag{8.3.29}$$

Now it remains to note that (see (5.5.13), (5.5.14))

$$\bigcap_{T\in\widehat{\mathcal{T}}_c^{(1)}} \mathrm{cl}\,(\mathbf{W}^1(T),\otimes^Q(\tau_{\mathbf{R}})) = \bigcap_{T\in\widehat{\mathcal{T}}_c} \mathrm{cl}\,(\mathbf{W}^1(T),\otimes^Q(\tau_{\mathbf{R}})). \tag{8.3.30}$$

From (8.3.26)–(8.3.30) easily follows

THEOREM 8.3.1. *The following general limit representation for the sets of all asymptotically attainable estimates is valid:*

$$\left(\forall\,\Gamma_* \in (\mathrm{Step})\,[\Gamma]\colon\ \mathcal{W}^1(\widetilde{\Omega}_*^{(c)}) = \bigcap_{T\in\widehat{\mathcal{T}}_c^0[\Gamma_*]} \mathrm{cl}\,(\mathbf{W}^1(T),\otimes^Q(\tau_{\mathbf{R}}))\right)$$

$$\&\left(\mathcal{W}^1(\widetilde{\Omega}_*^{(c)}) = \bigcap_{T\in\widehat{\mathcal{T}}_c} \mathrm{cl}\,(\mathbf{W}^1(T),\otimes^Q(\tau_{\mathbf{R}}))\right).$$

Note that this theorem, just as Theorem 8.2.1, admits natural analogs with the propositions from Chap. 6 and, in principle, can be deduced from the latter directly. Theorem 8.3.1 implies that

$$\left(\forall\,\Gamma_* \in (\mathrm{Step})\,[\Gamma]\colon\ (\leqq - \mathrm{MIN}\,)[\mathcal{W}^1(\widetilde{\Omega}_*^{(c)})]\right.$$

$$= (\leqq - \mathrm{MIN}\,)\left[\bigcap_{T\in\widehat{\mathcal{T}}_c^0[\Gamma_*]} \mathrm{cl}\,(\mathbf{W}^1(T),\otimes^Q(\tau_{\mathbf{R}}))\right]\right)$$

$$\&\left((\leqq - \mathrm{MIN}\,)[\mathcal{W}^1(\widetilde{\Omega}_*^{(c)})] = (\leqq - \mathrm{MIN}\,)\left[\bigcap_{T\in\widehat{\mathcal{T}}_c} \mathrm{cl}\,(\mathbf{W}^1(T),\otimes^Q(\tau_{\mathbf{R}}))\right]\right);$$

$$\tag{8.3.31}$$

relation (8.3.31) expresses, in particular, the asymptotic equivalence with respect to the result in the problem of $\mathbf{W}$-minimization of families (5.5.8) and (5.5.9): $\forall\,\Gamma^* \in (\mathrm{Step})\,[\Gamma]$

$$(\leqq - \mathrm{MIN}\,)\left[\bigcap_{T\in\widehat{\mathcal{T}}_c^0[\Gamma_*]} \mathrm{cl}\,(\mathbf{W}^1(T),\otimes^Q(\tau_{\mathbf{R}}))\right]$$

$$= (\leqq - \mathrm{MIN}\,)\left[\bigcap_{T\in\widehat{\mathcal{T}}_c} \mathrm{cl}\,(\mathbf{W}^1(T),\otimes^Q(\tau_{\mathbf{R}}))\right]. \tag{8.3.32}$$

In turn, (8.3.32) can be interpreted, with due allowance for the definitions of Sec. 5.5, as asymptotic nonsensitivity to perturbation of a part of the constraints, for which purpose it suffices to compare (5.5.4) and (5.5.5). Indeed, the relaxed sets forming families (5.5.8) and (5.5.9) differ by the requirements of approximate and exact matching, respectively, of the f-integrand with an element of $\mathbf{Y}$ on finite subsets of Γ^*. The constraints defined in (5.5.5) are much stronger. To this we should also add the fact that in (5.5.4) the constraint of the "energy" character is weakened. Relation (8.3.32) under the condition of finiteness of Γ, provides, under some natural conditions, properties of computational stability; these issues will be considered in the next chapter. But in the general case considered here, (8.3.32) testifies to a certain structural stability of the problem to some types of perturbation. We mean the requirement to matching the f-integrand with an element of $\mathbf{Y}$ at the points of $\gamma \in \Gamma^*$, $\Gamma^* \in (\text{Step})\,[\Gamma]$ (the last condition is essential).

Chapter 9

Issues of Computational Stability

9.1. Introduction

In this chapter, we consider a very important special case of the problems investigated in the preceding two chapters. We mean extremal problems with constraints in the form of the inclusion

$$\left(\int\limits_{E} s_1 f \, d\eta, \ldots, \int\limits_{E} s_n f \, d\eta \right) \in Y, \qquad (9.1.1)$$

where $s_1, \ldots, s_n$ are stratum functionals and Y is a closed subset in an n-dimensional space. These problems were already illustrated by examples in Chap. 1 connected with settings typical of the optimal control theory. Condition (9.1.1) can, in an obvious way, be reduced to form (5.1.4) if we use the finite set Γ in the last relation. It is this case that will be the subject of our subsequent consideration. The main facts can be easily derived from the corresponding theorems of Chap. 8, so that sometimes we shall omit detailed proofs, emphasizing the main concepts. One of them is the property of computational stability that occurs for problems with constraints (9.1.1) under the assumption of step-valuedness of $s_1, \ldots, s_n$. We should, however, bear in mind that this concept is nothing but a consequence of assertions similar to Theorem 8.2.1. As a matter of fact, "asymptotics" of the form (5.2.11) were introduced in a form that makes it possible, under complete step-valuedness of the operator (5.1.1), to include easily an unperturbed problem into a family of its respective relaxations. Moreover, in this case a family of the type (5.2.11) actually reduces to an unperturbed problem. Other relaxations become inessential because their system of constraints is weakened compared with (9.1.1). In this connection, we shall concentrate on the problem of comparing the asymptotic and the conventional value of the problem. Moreover, at the end of the chapter we shall consider an extremal problem with constraints that are relevant, by their meaning, to (9.1.1) but embrace an important (in particular, for control problems) case, where f is a vector-function; this issue is not considered in full generality. The example from Sec. 1.4 gives an idea of the possible applications of this abstract problem.

9.2.   Constraints on the Vector Integrand

Throughout this and the subsequent two sections, we shall suppose the integer $n \in \mathcal{N}$ to be given and postulate, if the contrary is not stipulated, the equality

$$\Gamma = \overline{1, n}. \tag{9.2.1}$$

Relation (9.2.1) will be the main simplifying point in the settings considered in the sequel. The fact is that the set $\mathbf{Y}$ (5.1.3) is, in this case, a closed subset of $\mathbf{R}^n$. For definiteness, we shall provide the n-dimensional arithmetic space $\mathbf{R}^n$ with the sup-norm $\| \cdot \|$, assigning to any vector $y \triangleq (y_i)_{i \in \overline{1,n}} \in \mathbf{R}^n$ the number

$$\|y\|_n \triangleq \sup(\{|y_i|: \ i \in \overline{1, n}\}) \in [0, \infty).$$

Thus, in the case of (9.2.1), $\mathbf{Y}$ is, obviously, a subset of $\mathbf{R}^n$, closed in the sense of $\| \cdot \|_n$. We shall assume throughout this chapter that $\mathbf{Y} \neq \varnothing$. Taking account of the well-known property of bounded compactness of $(\mathbf{R}^n, \| \cdot \|_n)$, we then have $\forall y \in \mathbf{R}^n \ \exists \tilde{y} \in \mathbf{Y}$:

$$\|y - \tilde{y}\|_n = \inf(\{\|y - z\|_n: z \in \mathbf{Y}\}).$$

As a consequence, we get $\forall \varepsilon \in [0, \infty)$

$$\mathbf{Y}_\varepsilon \triangleq \{z \in \mathbf{R}^n | \ \inf(\{\|y - z\|_n: y \in \mathbf{Y}\}) \leqslant \varepsilon\}$$

$$= \{z \in \mathbf{R}^n | \ \exists y \in \mathbf{Y}: \ \|y - z\|_n \leqslant \varepsilon\}. \tag{9.2.2}$$

In view of (9.2.2), condition (5.2.2) reduces to the form

$$f \in B_0(E, \mathcal{L}), \qquad \left(\int_E S_i f \, d\eta \right)_{i \in \overline{1,n}} \in \mathbf{Y}_\varepsilon$$

only if $K = \Gamma = \overline{1,n}$, $\varepsilon > 0$. Transformations of this kind can also be easily realized for other conditions of Chap. 5, and the assumption $K = \overline{1,n}$ is clearly not restrictive. It follows from (9.2.2) that

$$\mathbf{Y}_0 = \mathbf{Y}_\varepsilon|_{\varepsilon=0} = \mathbf{Y}. \tag{9.2.3}$$

REMARK. In particular, taking into account (9.2.3), we shall consider the issue concerning the conditions of computational stability as $\varepsilon \downarrow 0$. The fact is that if $S_\gamma \in B_0(E, \mathcal{L})$, $\gamma \in \Gamma$, we can identify Γ_* with Γ in (5.2.3) and (5.2.6). In this case, family (5.2.11) will obviously include the admissible set corresponding to the unperturbed condition

$$f \in B_0^+(E, \mathcal{L}), \qquad \left(\int_E S_i f \, d\eta\right)_{i\in\overline{1,n}} \in \mathbf{Y}.$$

Arguments of this sort can also be expressed with respect to other conditions on the choice of f in the specification of (9.2.1). If the condition of full step-valuedness of the operator (5.1.1) is not fulfilled, then we cannot speak about identification with the unperturbed problem of the respective asymptotics ((5.2.11), (5.2.12) or any other, including the dependence on $\Gamma_* \in (\mathrm{Step})\,[\Gamma]$), which can be seen from the simplest examples of Chap. 1.

Relation (9.2.1) leads, in a natural way, to the necessity of considering the one-parameter family of problems

$$\mathbf{W}(f) \to [\leqq - \mathrm{MIN}], \qquad f \in B_0^+(E, \mathcal{L}),$$

$$\left(\int_E S_i f \, d\eta\right)_{i\in\overline{1,n}} \in \mathbf{Y}_\varepsilon, \tag{9.2.4}$$

$\varepsilon \geqslant 0$, in the specification of the settings from Sec. 8.2 (throughout this section the symbols $w, \mathcal{W}, W, \mathbf{W}$ have the sense of Sec. 8.2). Actually, in these problems the asymptotics corresponding to the family $\mathcal{T}_+$ of (9.2.2) is realized if we consider the case $\varepsilon > 0$ in (9.2.4). Throughout this section we assume, in this connection, $\forall \varepsilon \in [0, \infty)$

$$F_\varepsilon \triangleq \left\{ f \in B_0^+(E, \mathcal{L}) \Big| \left(\int_E S_i f \, d\eta\right)_{i\in\overline{1,n}} \in \mathbf{Y}_\varepsilon \right\}. \tag{9.2.5}$$

In (9.2.5), we introduced the admissible sets of problems (9.2.4). Here, $F_0 = F_\varepsilon|_{\varepsilon=0}$ is, obviously, the admissible set of the unperturbed problem (see condition (5.1.4) in the specification of (9.2.1)). However, $\forall \varepsilon \in (0, \infty)$

$$F_\varepsilon = \Omega^+(\overline{1,n}, \varepsilon) \in \mathcal{T}_+. \tag{9.2.6}$$

This equality can be easily checked with due regard for (5.2.4), (9.2.1), and (9.2.2). Moreover, sets (9.2.6) are sufficient, in our case, for characterizing the whole family (5.2.9) since

$$\forall K \in \mathrm{Fin}\,(\Gamma), \quad \varepsilon \in (0,\infty): \quad F_\varepsilon \subset \Omega^+(K,\varepsilon). \tag{9.2.7}$$

Indeed, in all asymptotic settings the family $\{F_\varepsilon: \varepsilon \in (0,\infty)\}$ is, by virtue of (9.2.6) and (9.2.7), adequate to $\mathcal{T}_+$, so that we shall use the corresponding substitution without additional explanation (we mean the consideration, where it is convenient, of the family of all sets F_ε, $\varepsilon > 0$, instead of $\mathcal{T}_+$). By analogy with (9.2.4), we introduce the one-parameter family of generalized problems

$$\mathcal{W}(\mu) \to [\underset{=}{\leq} - \mathrm{MIN}\,], \qquad \mu \in (\mathrm{add})^+\,[\mathcal{L}, \eta], \qquad \left(\int_E S_i\, d\mu\right)_{i \in \overline{1,n}} \in \mathbf{Y}_\varepsilon,$$

$\varepsilon \geqslant 0$. For $\varepsilon = 0$, this problem reduces to the generalized extremal problem from Sec. 8.2 (see (8.2.22)), since its admissible set degenerates in this case into (5.2.14). Above we confined ourselves to considering this special case. However, in the problem under consideration it makes sense to discuss this issue somewhat more extensively. Let $\forall \varepsilon \in [0,\infty)$

$$\tilde{\Omega}_\varepsilon^+ \triangleq \left\{\mu \in (\mathrm{add})^+\,[\mathcal{L}, \eta] \,\middle|\, \left(\int_E S_i\, d\mu\right)_{i \in \overline{1,n}} \in \mathbf{Y}_\varepsilon\right\}.$$

Here, as one can easily see,

$$\tilde{\Omega}_0^+ = \tilde{\Omega}_\varepsilon^+|_{\varepsilon=0} = \tilde{\Omega}_+;$$

this case will be of interest to us when we study the family of generalized problems.

Throughout the remainder of this section, we shall assume Condition 5.4.1 to be fulfilled. Then (see (9.2.7), as one can easily see,

$$\exists \varepsilon \in (0,\infty), \quad c \in [0,\infty): \quad \left\{\int_E f(x)\, \eta(dx): f \in F_\varepsilon\right\} \subset [0,c]. \tag{9.2.8}$$

Taking into account (9.2.8), we choose and fix numbers

$$\varepsilon^* \in (0,\infty), \qquad a^* \in [0,\infty), \tag{9.2.9}$$

for which we have $\forall f \in F_{\varepsilon^*}$

$$\int_E f\, d\eta \leqslant a^* \tag{9.2.10}$$

(of course, the choice of (9.2.9) with the fulfillment of condition (9.2.10) can be realized in several ways; we fix an arbitrary pair of this kind, proceeding from condition (9.2.8), which is, by the way, equivalent, in the case of (9.2.1), to Condition 5.4.1, which follows from (9.2.6) and (9.2.7)).

LEMMA 9.2.1. *Let $\varepsilon \in [0, \varepsilon^*)$. Then $\tilde{\Omega}_\varepsilon^+ \subset \Xi_+[a^*]$.*

The proof is practically an obvious consequence of Theorem 4.3.1 and Definition (9.2.5); it also uses (9.2.9) and (9.2.10). Let us note here the property of monotone dependence of the set (9.2.7) on the parameter ε, $\varepsilon \geqslant 0$:

$$(0 \leqslant \varepsilon_1 \leqslant \varepsilon_2) \Rightarrow (\tilde{\Omega}_{\varepsilon_1}^+ \subset \tilde{\Omega}_{\varepsilon_2}^+). \tag{9.2.11}$$

Let us also note that the intersection of all sets $\tilde{\Omega}_\varepsilon^+$, $\varepsilon \in (0, \infty)$, coincides with $\tilde{\Omega}_+$. This equality easily follows from (9.2.2).

THEOREM 9.2.1. *Let H_* be a neighborhood of the set $\tilde{\Omega}_+$ in $(\mathbf{A}(\mathcal{L})$, $\tau_*(\mathcal{L}))$. Then $\exists \varepsilon \in (0, \infty) \; \forall \delta \in [0, \varepsilon]: \tilde{\Omega}_\delta^+ \subset H_*$.*

SCHEME OF THE PROOF. Let $G_* \in \tau_*(\mathcal{L})$: $\tilde{\Omega}_+ \subset G_* \subset H_*$. Suppose that $\forall \varepsilon \in (0, \infty)$

$$\left(\bigcup_{\delta \in [0,\varepsilon]} \tilde{\Omega}_\delta^+ \right) \setminus G_* \neq \varnothing. \tag{9.2.12}$$

Taking account of (9.2.11) and (9.2.12), we then obtain $\forall \varepsilon \in (0, \infty)$

$$\tilde{\Omega}_\varepsilon^+ \setminus G_* \neq \varnothing.$$

Every set (9.2.2) is closed in $(\mathbf{R}^n, \|\cdot\|_n)$, so that every generalized admissible set is also closed in the sense of $\tau_*(\mathcal{L})$. In this case, $\mathbf{A}(\mathcal{L}) \setminus G_*$ is also closed in the sense of $\tau_*(\mathcal{L})$. But this means that $\forall \varepsilon \in (0, \infty)$

$$\tilde{\Omega}_\varepsilon^+ \setminus G_* = \tilde{\Omega}_\varepsilon^+ \cap (\mathbf{A}(\mathcal{L}) \setminus G_*) \tag{9.2.13}$$

is a nonempty $*$-weakly closed set. In particular, this property holds for $\varepsilon \in (0, \varepsilon^*)$. Under this condition, however, (9.2.13) is a subset of $\Xi_+[a^*]$. As a result we find that $\forall \varepsilon \in (0, \varepsilon^*)$

$$\tilde{\Omega}_\varepsilon^+ \setminus G_* \in \mathfrak{F}_{\tilde{\tau}_b^*(\mathcal{L})}|_{b=a^*}. \tag{9.2.14}$$

Taking account of (9.2.11) and (9.2.14), we infer that $\{\tilde{\Omega}_\varepsilon^+ \setminus G_*: \varepsilon \in (0, \varepsilon^*)\}$ is a family of closed (in the sense of (9.2.14)) sets with a finite intersection property in a compact space. This implies (see Secs. 2.2 and 4.2) the nonemptiness property for the intersection of all sets of the indicated family. Let μ^* be an element of this intersection, i.e., $\mu^* \in \tilde{\Omega}_\varepsilon^+ \setminus G_*$ for $0 < \varepsilon < \varepsilon^*$.

Then we have the following two assertions: (1) on one hand, $\mu^* \in \tilde{\Omega}_\varepsilon^+$ for $\varepsilon > 0$ (see (9.2.11)); (2) $\mu^* \notin G_*$. But from (1) it follows that $\mu^* \in \tilde{\Omega}_+$, and this contradicts (2); this contradiction shows that our assumption was erroneous. Thus, $\exists \varepsilon \in (0, \infty] \ \forall \delta \in [0, \varepsilon]: \tilde{\Omega}_\delta^+ \subset G_*$. The further reasoning is obvious.

Theorem 9.2.1 actually establishes the property of monotonic convergence of the admissible sets of the family of relaxed generalized problems under the condition $\varepsilon \downarrow 0$ to the similar set corresponding to the limit problem. If we now introduce $\forall \varepsilon \in [0, \infty)$

$$\tilde{F}_\varepsilon \triangleq \{f * \eta: \ f \in F_\varepsilon\} \in \mathcal{P}(\tilde{\Omega}_\varepsilon^+), \tag{9.2.15}$$

then, generally speaking, the dependence $\varepsilon \mapsto \tilde{F}_\varepsilon: [0, \infty) \to \mathcal{P}((\text{add})^+ [\mathcal{L}, \eta])$ does not possess a similar property; this was illustrated well enough by the examples of Sec. 1.2. At the same time, it makes sense to observe the following obvious consequence of Theorem 5.2.1.

THEOREM   9.2.2. *Let* $\tau \in \mathfrak{M}$. *Then*

$$\tilde{\Omega}_+ = \bigcap_{\varepsilon > 0} \text{cl} \, (\tilde{F}_\varepsilon, \tau).$$

The proof easily follows from Theorem 5.2.1 with due account of (9.2.6) and (9.2.7). We see that the set $\tilde{\Omega}_+$ is limiting with respect not only to generalized relaxed admissible sets but also to the corresponding relaxations of the admissible set in the class of conventional solutions, so that $\tilde{\Omega}_+$ provides a sort of regularization of the admissible set corresponding to condition (9.2.4). We have not considered the optimization problem but confined ourselves to examining asymptotically admissible sets. In the next section, we shall consider in detail the issue of conditions of computational stability. Now we shall briefly consider the specification of the general case of the problem from Sec. 8.2 corresponding to condition (9.2.1). From the standpoint of the constructions of Chaps. 5–8, this case is rather special (recall that we also assume (9.2.8), and, moreover, adhere to conventions (9.2.9) and (9.2.10)) but is inherent in a large class of problems of applied character. The substantive examples of Chap. 1 give an idea of the problems of this kind. Throughout the remainder of this section, we assume that

$$\Gamma_0 \triangleq \{i \in \overline{1, n} | \ S_i \in B_0(E, \mathcal{L})\}, \tag{9.2.16}$$

obtaining the largest element of (Step) $[\Gamma]$. We use (9.2.16) as the parameter Γ_* from Sec. 5.2; here we do not exclude the case where (9.2.16) is an empty set. In many cases, however, $\Gamma_0 \neq \varnothing$. We assume that

$$\mathbf{Y}_\varepsilon^0 \triangleq \{(z_i)_{i \in \overline{1,n}} \in \mathbf{R}^n \mid \exists (y_i)_{i \in \overline{1,n}} \in \mathbf{Y} \colon (\forall k \in \Gamma_0 \colon y_k = z_k)$$

$$\&(\forall r \in \overline{1,n} \setminus \Gamma_0 \colon |y_r - z_r| \leqslant \varepsilon)\}. \tag{9.2.17}$$

Actually, we realize in (9.2.17) (under the conditions of (9.2.1)) the definitions of Sec. 5.2. Furthermore, we set, $\forall \varepsilon \in [0, \infty)$

$$F_\varepsilon^0 \triangleq \left\{ f \in B_0^+(E, \mathcal{L}) \,\middle|\, \left(\int_E S_i f \, d\eta \right)_{i \in \overline{1,n}} \in \mathbf{Y}_\varepsilon^0 \right\}. \tag{9.2.18}$$

For $\varepsilon = 0$, we again have in (9.2.8) the admissible set of the unperturbed problem. However, we shall now give much attention to considering sets (9.2.18) under the condition $\varepsilon > 0$. Of course, here it easily follows from (9.2.2) and (9.2.17) that $\mathbf{Y}_\varepsilon^0 \subset \mathbf{Y}_\varepsilon$ for $\varepsilon \geqslant 0$. In turn, from (9.2.5) and (9.2.18), it follows that $\forall \varepsilon \in [0, \infty)$

$$F_\varepsilon^0 \subset F_\varepsilon. \tag{9.2.19}$$

Thus, we have obtained a family $\{F_\varepsilon^0 \colon \varepsilon \in (0, \infty)\}$ of relaxations of the admissible set that are, generally speaking, more severe than (9.2.5); every one of these sets (9.2.18) corresponds to the case of realization of the f-integrand within the "asymmetric" neighborhood (9.2.17) of the set $\mathbf{Y}$. In connection with (5.2.6), (9.2.1), and (9.2.16), we observe that in our case $\forall \varepsilon \in [0, \infty)$

$$\Omega_0^+(\Gamma_0, \overline{1,n}, \varepsilon) = \left\{ f \in B_0^+(E, \mathcal{L}) \,\middle|\, \exists (y_i)_{i \in \overline{1,n}} \in \mathbf{Y} \colon \right.$$

$$\left(\forall k \in \Gamma_0 \colon \int_E S_{\gamma_k} f \, d\eta = y_k \right)$$

$$\left. \&\left(\forall r \in \overline{1,n} \setminus \Gamma_0 \colon \left| \int_E S_r f \, d\eta - y_r \right| \leqslant \varepsilon \right) \right\} = F_\varepsilon^0. \tag{9.2.20}$$

Thus, an essential part of asymptotics (5.2.11) is realized (see (9.2.20)) by means of (9.2.18) under the conditions where $\Gamma_* = \Gamma_0$. Indeed, by virtue of the monotonicity of (5.2.6) with respect to K, $K \in \mathrm{Fin}\,(\Gamma)$, we obviously have that $\forall K \in \mathrm{Fin}\,(\Gamma)$, $\tilde\varepsilon \in [0, \infty)$,

$$F_{\tilde\varepsilon}^0 \subset \Omega_0^+(\Gamma_0, \widetilde{K}, \tilde\varepsilon). \tag{9.2.21}$$

Combining (5.2.11), (9.2.20), and (9.2.21), we find that in the case under consideration

$$(\forall \varepsilon \in (0,\infty): F_\varepsilon^0 \in \mathcal{T}_0^+[\Gamma_0])$$

$$\&(\forall T \in \mathcal{T}_0^+[\Gamma_0] \,\exists\, \delta \in (0,\infty): F_\delta^0 \subset T). \tag{9.2.22}$$

In turn, by monotonicity of the closure and image operators, we get from (9.2.22) the property of asymptotic equivalence of the families $\mathcal{T}_0^+[\Gamma_0]$ and $\{F_\varepsilon^0 : \varepsilon \in (0,\infty)\}$ with respect to the result:

$$\bigcap_{\varepsilon \in (0,\infty)} \mathrm{cl}\,(\mathbf{W}^1(F_\varepsilon^0), \otimes^Q(\tau_\mathbf{R})) = \bigcap_{T \in \mathcal{T}_0^+[\Gamma_0]} \mathrm{cl}\,(\mathbf{W}^1(T), \otimes^Q(\tau_\mathbf{R})). \tag{9.2.23}$$

Taking account of (9.2.5)–(9.2.7), we also have a similar representation for the asymptotics $\mathcal{T}_+$ (5.2.9) since

$$(\forall \varepsilon \in (0,\infty): F_\varepsilon \in \mathcal{T}_+)\&(\forall T \in \mathcal{T}_+ \,\exists\, \delta \in (0,\infty): F_\delta \subset T).$$

To be more precise, the respective set of all asymptotically attainable estimates has the form

$$\bigcap_{\delta \in (0,\infty)} \mathrm{cl}\,(\mathbf{W}^1(F_\varepsilon), \otimes^Q(\tau_\mathbf{R})) = \bigcap_{T \in \mathcal{T}_+} \mathrm{cl}\,(\mathbf{W}^1(T), \otimes^Q(\tau_\mathbf{R})). \tag{9.2.24}$$

Combining (9.2.23), (9.2.24), and Theorem 8.2.1, we get the equality

$$\bigcap_{\varepsilon \in (0,\infty)} \mathrm{cl}\,(\mathbf{W}^1(F_\varepsilon), \otimes^Q(\tau_\mathbf{R})) = \bigcap_{\varepsilon \in (0,\infty)} \mathrm{cl}\,(\mathbf{W}^1(F_\varepsilon^0), \otimes^Q(\tau_\mathbf{R})) = \mathcal{W}^1(\tilde{\Omega}_+).$$

$$\tag{9.2.25}$$

Here, the set $\tilde{\Omega}_+$ is determined by means of $\tilde{\Omega}_\varepsilon^+$ for $\varepsilon = 0$, in other words, by a relation (see (9.2.3)) differing from $\tilde{\Omega}_\varepsilon^+$ only by the replacement of $\mathbf{Y}_\varepsilon$ by $\mathbf{Y}$. From (9.2.25), we get the following representation of asymptotic values (a specific consequence of Theorem 8.2.1):

$$(\leqq - \mathrm{MIN}\,)\left[\bigcap_{\varepsilon \in (0,\infty)} \mathrm{cl}\,(\mathbf{W}^1(F_\varepsilon), \otimes^Q(\tau_\mathbf{R})) \right]$$

$$= (\leqq - \mathrm{MIN}\,)\left[\bigcap_{\varepsilon \in (0,\infty)} \mathrm{cl}\,(\mathbf{W}^1(F_\varepsilon^0), \otimes^Q(\tau_\mathbf{R})) \right]$$

$$= (\leqq - \mathrm{MIN}\,)[\mathcal{W}^1(\tilde{\Omega}_+)]. \tag{9.2.26}$$

We now omit a similar specification of the construction of asymptotically efficient approximate solutions from Sec. 8.2. Let us only observe that under conditions where the set Q is finite, the class of sequential approximate solutions, i.e., sequence-solutions $(f_1, f_2, \ldots)$, is sufficient from the conceptual point of view; see [2, 49]. However, here it is also advisable to use the construction (8.2.27) of an approximate net-solution at the intermediate stages, thinning it out afterwards to a sequence, following the scheme of [49], without loss of the asymptotic efficiency property. Substantial progress in studying properties of the stability type for the family of extremal problems (9.2.4) can be achieved for the case of scalar optimization, which, in turn, is treated as a special case in [50]. We shall not dwell on this in detail, but shall observe that for a family of relaxed generalized problems stability was established in [50] and weakened (or more precise, regularized) stability was established under some conditions for problems of the type (9.2.4). We shall touch upon issues of this kind at the end of this chapter, where we shall consider a problem generalizing the setting of Sec. 1.4.

9.3. Computational Stability

Here we consider an important special case of the problem from Sec. 8.2 for which Theorem 8.2.1 and its corollary reduce to asserting computational stability. In this section we postulate (9.2.1) and (9.2.8), and, moreover, assume the following condition.

CONDITION 9.3.1. $\forall\, i \in \overline{1, n}$: $S_i \in B_0(E, \mathcal{L})$.

Without additional explanations, we use the numbers ε^*, a^* defined in (9.2.9) and satisfying (9.2.10). Thus, a new circumstance, compared to that in Sec. 9.2, is the assumption that $S_1, \ldots, S_n$ are $\mathcal{L}$-step-valued maps on E, so that

$$\Gamma = \overline{1, n} \in (\text{Step})\,[\Gamma]. \tag{9.3.1}$$

Taking account of (9.3.1), we can specify the choice of Γ_* of Sec. 5.2 in the form $\Gamma_* = \Gamma$, and this will be done in the sequel. The fact is that for this specification the corresponding family of relaxed constraints reduces to the constraints of the unperturbed problem. We shall now explain in what sense we understand this reduction procedure. The case in point is that by virtue of (9.1.1) and (9.3.1), we can speak of the sets $\Omega_0^+(\overline{1, n}, \overline{1, n}, \varepsilon) = \Omega_0^+(\Gamma, \overline{1, n}, \varepsilon)$, where $\varepsilon > 0$. In view of (5.2.11), any such set of this kind is an element of the family $\mathcal{T}_0^+[\Gamma] = \mathcal{T}_0^+[\Gamma_*]$. However, it follows from (5.2.5), (5.2.6), and (9.2.1) that $\forall \varepsilon \in (0, \infty)$

$$\Omega_0^+(\overline{1, n}, \overline{1, n}, \varepsilon) = \left\{ f \in B_0^+(E, \mathcal{L}) \Big| \left(\int_E S_i f\, d\eta \right)_{i \in \overline{1, n}} \in \mathbf{Y} \right\}. \tag{9.3.2}$$

In other words, every set (9.3.2) coincides with $F_0 = F_\varepsilon|_{\varepsilon=0}$ (9.2.5), as one can see from (9.2.3). Thus

$$F_0 \in \mathcal{T}_0^+[\Gamma]. \tag{9.3.3}$$

Moreover, F_0 is, as one can easily see, the least-by-inclusion element of $\mathcal{T}_0^+[\Gamma]$, which follows directly from (5.2.5), (5.2.6), and (5.2.11). Taking into account the monotonicity of the closure and image operators, we obtain, by virtue of Theorem 8.2.1 and relation (9.3.3), the relation

$$\mathcal{W}^1(\tilde{\Omega}_+) = \mathrm{cl}\,(\mathbf{W}^1(F_0), \otimes^Q(\tau_{\mathbf{R}})).$$

For the same reason, in view of (9.2.26), we obviously have the relation

$$\bigcap_{\varepsilon\in(0,\infty)} \mathrm{cl}\,(\mathbf{W}^1(F_\varepsilon), \otimes^Q(\tau_{\mathbf{R}})) = \mathrm{cl}\,(\mathbf{W}^1(F_0), \otimes^Q(\tau_{\mathbf{R}})). \tag{9.3.4}$$

Actually, (9.3.4) is a special case of (9.2.25) because Condition 9.3.1 (assumed to be fulfilled) provides the coincidence of sets (9.2.1) and (9.2.16). From (9.3.4) we obviously get the relation

$$(\leqq - \mathrm{MIN})\left[\bigcap_{\varepsilon\in(0,\infty)} \mathrm{cl}\,(\mathbf{W}^1(F_\varepsilon), \otimes^Q(\tau_{\mathbf{R}}))\right]$$

$$= (\leqq - \mathrm{MIN})[\mathrm{cl}\,(\mathbf{W}^1(F_0), \otimes^Q(\tau_{\mathbf{R}}))]. \tag{9.3.5}$$

Relation (9.3.5) shows that the conventional and asymptotic values coincide. In other words, (9.3.5) expresses the property of computational stability under the perturbations

$$\mathbf{Y} \to \mathbf{Y}_\varepsilon, \qquad \varepsilon > 0.$$

9.4.   A Problem of the Extremal Choice of Probability Density

Let us consider the setting of a problem actually corresponding to Sec. 9.2 but concerning a somewhat specific situation. The question is the choice of the probability density with the aim of optimizing the system of expectations for a given family of random variables under the constraints on a finite (for simplicity) system of expectations of some other random variables. In the interests of greater consistency with traditional settings of probabilistic problems, we shall assume throughout this section that space (4.2.1) is standard, namely (although it is not essential from the conceptual point of view), that $\mathcal{L}$ is a σ-algebra of subsets of E and η is a f.a. nonnegative measure on $\mathcal{L}$. As is well known, in this case $B(E, \mathcal{L})$ is the set of all $\mathcal{L}$-measurable functions from $\mathbf{B}(E)$; see Sec. 3.4. Throughout this section, the elements of $B(E, \mathcal{L})$ will be called random variables (strictly speaking, bounded random variables; here we presume, of course, that a probability is given on $\mathcal{L}$, i.e., a nonnegative normalized f.a. measure). So let

$$(\pi_q)_{q \in Q} \colon Q \to B(E, \mathcal{L}) \tag{9.4.1}$$

be a given parametrized family of random variables. In addition, in accordance with (9.2.1), n random variables $S_1, \ldots, S_n$ are fixed. The object of the investigator's choice is the step-valued probability density with respect to the measure η. Throughout this section, we denote by $\mathfrak{F}$ the set of all these densities, so that

$$\mathfrak{F} \triangleq \left\{ f \in B_0^+(E, \mathcal{L}) \,\Big|\, \int_E f \, d\eta = 1 \right\}. \tag{9.4.2}$$

Then every density $f \in \mathfrak{F}$ is associated with two systems of expectations, one of which is connected with family (9.4.1) and the other with $S_1, \ldots, S_n$. We shall assume, within this section, that $\forall f \in \mathfrak{F}, g \in B(E, \mathcal{L})$:

$$\mathbf{M}_f(g) \triangleq \int_E g f \, d\eta, \tag{9.4.3}$$

getting (in the form of (9.4.3)) the expectation of the random variable g with respect to the probability $f * \eta$: see (3.4.11). Then, $\forall f \in \mathfrak{F}$

$$(\mathbf{M}_f(\pi_q))_{q \in Q} \in \mathbf{R}^Q \tag{9.4.4}$$

is, generally speaking, an infinite-dimensional vector of expectations of random variables (9.4.1) which, for brevity, will be called objective. We shall be concerned with minimizing (9.4.4) with respect to f in the sense of the order $\leq$ (see Sec. 2.6). At the same time, the vector with the components $\mathbf{M}_f(S_1), \ldots, \mathbf{M}_f(S_n)$, each of which is an expectation, will be subjected to inclusion-type constraints

$$(\mathbf{M}_f(S_i))_{i \in \overline{1,n}} \in \mathbf{Y}. \tag{9.4.5}$$

Combining (9.4.4) and (9.4.5), we get the extremal problem

$$(\mathbf{M}_f(\pi_q))_{q \in Q} \to [\leq - \mathrm{MIN}], \qquad f \in \mathfrak{F}, \quad (\mathbf{M}_f(S_i))_{i \in \overline{1,n}} \in \mathbf{Y}. \tag{9.4.6}$$

Let us observe some of its peculiarities. First of all, we have to bear in mind that (9.4.3) is the expectation of g with respect to the probability space $(E, \mathcal{L}, f * \eta)$, which changes depending on the choice of f. The following interpretation of the step choice of f is possible. Namely, the realization of the random mechanism can be identified with the choice of the m-tuple

$$m \in \mathcal{N}, \quad (L_i)_{i \in \overline{1,m}} \in \Delta_m(E, \mathcal{L}), \quad (p_i)_{i \in \overline{1,m}} \in \mathbf{R}_+^m,$$
$$\forall j \in \overline{1, m}: \ \eta(L_j) = 0 \Rightarrow p_j = 0, \tag{9.4.7}$$

for which the sum of all p_j, $j \in \overline{1, m}$, equals 1. This means a finite decentralization of the basic space to $L_1, \ldots, L_m$ with a subsequent provision of every (measurable) set $L_k \in \mathcal{L}$ with the probability p_k $(k \in \overline{1, m})$. Assume that our possibilities of organizing a "generator of random points" from E are exhausted by the choice of some m-tuple (9.4.7) of the indicated type, and the "induced" randomness is thus characterized by the vector $(p_1, \ldots, p_m)$. This vector cannot be used directly for averaging since the structure of the functionals π_q and S_i may turn out to be finer, and this will require dealing with subsets of L_k. To extend the coarse probability $(p_1, \ldots, p_m)$ to the subsets of L_k, and onto the entire σ-algebra $\mathcal{L}$, we can uniformly distribute every weight p_k on the set L_k by introducing on L_k a density as the constant $p_k(\eta(L_k))^{-1}$ for $L_k \notin \mathfrak{N}$ (see Sec. 4.3) and setting it to zero otherwise. Thus, we come to the procedure of selecting a step-valued probability density, which is postulated in problem (9.4.6). This procedure is nothing but a means of realizing the m-tuple (9.4.7) which is used for generating a randomness. Note that problem (9.4.6) itself can be reduced to the form (9.2.4) for $\varepsilon = 0$. Indeed, let us define $S_{n+1} \in B_0^+(E, \mathcal{L})$ as an identity function on E: $S_{n+1}(x) \equiv 1$. Then we introduce in $\mathbf{R}^{n+1}$ the subset

$$Y \triangleq \{(y_i)_{i \in \overline{1,n+1}} \in \mathbf{R}^{n+1} | \ ((y_i)_{i \in \overline{1,n}} \in \mathbf{Y}) \& (y_{n+1} = 1)\}; \tag{9.4.8}$$

throughout this section, the symbol Y is used only in the sense of (9.4.8). Then problem (9.4.6) reduces to the form

$$\left(\int_E \pi_q f \, d\eta \right)_{q \in Q} \to [\leqq - \mathrm{MIN}], \qquad f \in B_0^+(E, \mathcal{L}), \qquad \left(\int_E S_i f \, d\eta \right)_{i \in \overline{1,n+1}} \in Y.$$
$$\tag{9.4.9}$$

Problem (9.4.9) is an example of a problem of the type (9.4.9), where $\varepsilon = 0$. Furthermore, it should be noted that the function S_{n+1} participating in (9.4.9) is obviously step-valued. This allows us to use the construction of Sec. 5.2 as applied to family (5.2.11). We ought to bear in mind that in the case under consideration the construction of Sec. 5.2 is substantially simplified and has, in fact, the form of Secs. 9.2 and 9.3. Here, we deal with the perturbation of the Y-constraints to the respective neighborhoods and, in this case, it is natural to choose these neighborhoods so that at the level of perturbed problems the normalization condition defining the elements of set (9.4.2) not be violated. In this connection, the neighborhoods Y can be specified following the type (9.2.17), taking into account that the dimension is modified by unity. We shall not repeat now the corresponding constructions of Secs. 9.2 and 9.3 but shall only note at the substantive level the inferences that are typical of problem (9.4.6), (9.4.9) and can be established with the aid of these constructions. Namely, in problem (9.4.6), it is of interest to clarify the issues of stability with respect to perturbing the $\mathbf{Y}$-constraints toward weakening; if problem (9.4.6) does not possess this property, then the question arises of the sensitivity of $\mathbf{Y}$ with respect to the perturbation in various directions. In the first case, we use neighborhoods (9.2.2), obtaining the problems

$$(\mathbf{M}_f(\pi_q))_{q \in Q} \to [\leqq - \mathrm{MIN}], \qquad f \in \mathfrak{F}, \quad (\mathbf{M}_f(S_i))_{i \in \overline{1,n}} \in \mathbf{Y}_\varepsilon, \qquad (9.4.10)$$

where $\varepsilon \geqslant 0$. For $\varepsilon = 0$, we have as a specimen of (9.4.10) the unperturbed problem (9.4.6), or, alternatively, problem (9.4.9). To the family of all ε-problems (9.4.10), $\varepsilon > 0$, we can assign an asymptotic value similar to (9.2.24). The coincidence of this value with the conventional one, defined in terms of problem (9.4.6) by analogy with Sec. 9.3, can be qualified as computational stability. An answer to this question is actually contained in Sec. 9.3, since we are able to model problem (9.4.6) in the form (9.4.9), and this is in complete agreement with the setting of Sec. 9.2 up to the notation and dimension of the space in which the constraints on the vector integrand are introduced. The key role is played here by the condition of the $\mathcal{L}$-step-valuedness of $S_1, \ldots, S_n$, as is shown in Sec. 9.3. A simple example of problem (9.4.6) satisfying this condition has the form

$$(\mathbf{M}_f(\pi_q))_{q \in Q} \to [\leqq - \mathrm{MIN}], \qquad f \in \mathfrak{F}, \quad \left(\int_{L_i} f \, d\eta \right)_{i \in \overline{1,n}} \in \mathbf{Y},$$

where $L_1 \in \mathcal{L}, \ldots, L_n \in \mathcal{L}$. If this condition of step-valuedness is absent, so that there are functionals among $S_1, \ldots, S_n$ which are not elements of $B_0(E, \mathcal{L})$, then it is advisable to resort to replacing in (9.4.6) the set $\mathbf{Y}$ by (9.2.17), i.e., turn to the analysis of the problems

$$(\mathbf{M}_f(\pi_q))_{q \in Q} \to [\leqq - \mathrm{MIN}], \qquad f \in \mathfrak{F}, \quad (\mathbf{M}_f(S_i))_{i \in \overline{1,n}} \in \mathbf{Y}_\varepsilon^0, \qquad (9.4.11)$$

$\varepsilon \geqslant 0$. Now, two asymptotic values arise, one of which corresponds to the family of ε-problems (9.4.10) and the other of which corresponds to the family of ε-problems (9.4.11), and in both cases ε runs over $(0, \infty)$. The coincidence of these two values can be qualified as asymptotic nonsensitivity to perturbation of a part of the constraints, namely, perturbation with respect to the coordinates labelled $i \in \overline{1, n}$ for which $S_i \in B_0(E, \mathcal{L})$. The character of the inference obtained in this case has the meaning of relation (9.2.25) with due change of notation which takes into account, in particular, the change in the dimension from n to $n + 1$. Various "pathological" effects in problem (9.4.6) can be realized already in the simplest examples of the scalar optimization problem, similar to those considered in Sec. 1.2.

9.5. Extension in the Class of Vector Finitely Additive Measures

We shall now consider an extension of constructions of an abstract extremal problem generalizing the setting of Sec. 1.4 in the cone optimization part.

Furthermore, we shall consider some issues concerning the scalar optimization problem, in particular, some properties of stability type, not only for the values but also for the solutions of the corresponding extremal problem. Here we mean a problem in the class of conventional solutions, which, as we already know, do not make up a compact set. The peculiarity of this section lies in the fact that here we study the extremal problems in which the conventional solution f is a vector-function. Namely, we shall further assume that we are given a number $r \in \mathcal{N}$ (the dimension of conventional "controls") and consider as solutions the collections

$$f = (f_1, \ldots, f_r), \tag{9.5.1}$$

where $f_1 \in B_0^+(E, \mathcal{L}), \ldots, f_r \in B_0^+(E, \mathcal{L})$. In Sec. 1.4, we already came across the necessity of using precisely this notion of conventional solution. The settings of this section can be motivated by various optimal control problems for which the typical case is where the control systems are multichannel. Throughout this section, we shall assume that we are given a number $c \in [0, \infty)$ defining the "resource" constraint

$$\sum_{i=1}^{r} \int_E f_i \, d\eta \leqslant c \tag{9.5.2}$$

on the choice of f (9.5.1), which is conceptually similar to condition (1.4.2). As for the substantive meaning of condition (9.5.2), we can give a number of examples similar to the control problem from Sec. 1.4, in which it determines the total resource of a multichannel control scheme: there is a common "tank" from which the "fuel" can be fed during a given period of time into channels labelled $1, \ldots, r$. We are controlling the intensities $f_1, \ldots, f_r$ of the inflow of the "fuel" into the channels during the respective period of time, achieving some goals from the standpoint of criterion optimization. Precisely this meaning can be given to condition (1.4.2), which is a special case of (9.5.2). Other interpretations of constraint (9.5.2) are also possible, but we shall not discuss them now. We shall give the setting of this problem in general form, confining ourselves, however, to using conditions on a "finite" integrand, in contrast to the infinite systems of conditions of this kind considered in Chap. 5. Then we shall consider the solution and basic extension constructions for a cone optimization problem and at the end of the section, a problem of scalar optimization. We shall need some notation and a number of conventions concerning vector-functions and matrix-valued mappings; we think it necessary to specify them before setting the problem. As for the substantive setting of the problem, we suggest that the interested reader get acquainted with its special case in Sec. 1.4. In the sequel, we shall use a constraint similar to (1.4.3). We shall further fix two numbers

$$(n \in \mathcal{N}) \& (r \in \mathcal{N}),$$

the first of which plays, in fact, the same role as in Secs. 9.1–9.4; the number r defines the dimension of control vectors $f(x)$, $x \in E$. To shorten the main assertions, we shall introduce special symbols for the respective finite-dimensional spaces:

$$(\mathbf{N} \triangleq \mathbf{R}^n) \& (\mathcal{R} \triangleq \mathbf{R}^r) \& (\mathbf{M} \triangleq \mathbf{R}^{\overline{1,n} \times \overline{1,r}}).$$

Throughout this section, the symbols $\mathbf{M}$, $\mathbf{N}$, and $\mathcal{R}$ are used only in this sense. The elements of $\mathbf{M}$ are $(n \times r)$-matrices; let us fix a mapping

$$S \colon E \to \mathbf{M},$$

whose components will be denoted by $S_{i,j}$, $i \in \overline{1,n}$, $j \in \overline{1,r}$. In other words, the mapping

$$(i,j) \mapsto S_{i,j} \colon \overline{1,n} \times \overline{1,r} \to \mathbf{R}^E$$

has the following property. Namely, $\forall\, x \in E$: $S(x)$ is the matrix

$$(i,j) \mapsto S_{i,j}(x) \colon \overline{1,n} \times \overline{1,r} \to \mathbf{R}.$$

This agreement will be adhered to throughout this section. We shall postulate in the sequel that $\forall\, i \in \overline{1,n},\, j \in \overline{1,r}$:

$$S_{i,j} \in B(E,\mathcal{L}); \tag{9.5.3}$$

thus (9.5.3) defines the stratum property of S. We shall need to use vector-functions as controls, or, more precisely, to use the elements of the space of mappings of the set E into $\mathcal{R}$. If $f \in \mathcal{R}^E$ and $i \in \overline{1,r}$, then we shall denote by f_i the ith component of the function f: $f_i \in \mathbf{R}^E$; here f_i assigns to each element $x \in E$ the coordinate of the vector $f(x)$ with the subscript i. In this case, $\forall\, f \in \mathcal{R}^E$ the mapping

$$i \mapsto f_i\colon \overline{1,r} \to \mathbf{R}^E$$

obviously has the property that $\forall\, x \in E$: $f(x)$ coincides with the vector

$$i \mapsto f_i(x)\colon \overline{1,r} \to \mathbf{R}.$$

We further assume that

$$B_0^+[E,\mathcal{L},\mathcal{R}] \triangleq \{f \in \mathcal{R}^E \,|\, \forall\, i \in \overline{1,r}\colon f_i \in B_0^+(E,\mathcal{L})\}; \tag{9.5.4}$$

the space (9.5.4) will be used as the space of conventional solutions. If $f \in \mathbf{N}^E$ and $i \in \overline{1,n}$, then we shall again denote by f_i the ith component (coordinate function) of f; this definition is similar to the preceding one. Let

$$B[E,\mathcal{L},\mathbf{N}] \triangleq \{f \in \mathbf{N}^E \,|\, \forall\, i \in \overline{1,n}\colon f_i \in B(E,\mathcal{L})\}. \tag{9.5.5}$$

In (9.5.4) and (9.5.5), we introduced the sets of step-valued and stratum vector-functions on E assuming values in $\mathcal{R}$ and $\mathbf{N}$ respectively. Now we must introduce an appropriate set of vector measures with the aim of closing some subsets of (9.5.4) that are defined by the natural resource constraints. Moreover, if $\mu \in \mathcal{R}^{\mathcal{L}}$ and $i \in \overline{1,r}$, then we denote by μ_i the ith component of the vector-function of the sets μ: μ_i assigns to every set $L \in \mathcal{L}$ the ith coordinate of the vector $\mu(L) \in \mathcal{R}$. Then

$$(\mathcal{R} - \mathrm{add})^+[\mathcal{L},\eta] \triangleq \{\mu \in \mathcal{R}^{\mathcal{L}} \,|\, \forall\, i \in \overline{1,n}\colon \mu_i \in (\mathrm{add})^+[\mathcal{L},\eta]\}. \tag{9.5.6}$$

We note that we defined solutions (conventional and generalized) in (9.5.4) and (9.5.6) as vector-functions, namely, as functions mapping the sets E and $\mathcal{L}$, respectively, into the r-dimensional arithmetic space $\mathcal{R}$. In (9.5.1), however, we used a different form of notation: the control f was actually given by listing its components. Of course, this difference in notation is not essential. The notation used in (9.5.4)–(9.5.6) is, in our opinion, more convenient for writing down the linear algebra operations. For example, $\forall f \in \mathcal{R}^E, x \in E$,

$$S(x)f(x) \triangleq \left(\sum_{j=1}^{r} S_{i,j}(x)f_j(x) \right)_{i \in \overline{1,n}} \in \mathbf{N} \qquad (9.5.7)$$

is a typical product of a matrix and a vector or a linear transformation of the vector $f(x)$ by virtue of the matrix $S(x)$. If, in (9.5.7), f is an element of (9.5.4), then the vector-function

$$(S(x)f(x))_{x \in E} \in B[E, \mathcal{L}, \mathbf{N}] \qquad (9.5.8)$$

is realized so that we can define componentwise the η-integral of the vector-function (9.5.8)

$$\int_E S(x)f(x)\,\eta(dx) \in \mathbf{N}, \qquad (9.5.9)$$

identified with an n-dimensional vector of the following form: if $i \in \overline{1,n}$, then the ith coordinate of vector (9.5.9) is, by definition, the ordinary η-integral of the ith component of (9.5.8), defined according to the scheme of Chap. 3. Throughout this section, the notation of (9.5.9) has only this meaning. Furthermore, we define the integral of the matrix-valued function S (9.5.2) by vector measure-elements (9.5.6). Namely, $\forall \mu \in (\mathcal{R} - \mathrm{add})^{+}[\mathcal{L}, \eta]$

$$\int_E S(x)\,\mu(dx) \triangleq \left(\sum_{j=1}^{r} \int_E S_{i,j}\,d\mu_j \right)_{i \in \overline{1,n}} \in \mathbf{N}; \qquad (9.5.10)$$

for all subsequent purposes the simplest definition (9.5.10) is sufficient and no other more complicated constructions of vector integration will be required (see, for example, [9, 56], and others). This notation does not lead to ambiguity. In principle, we could even now introduce the main elements of the extremal problem under examination. However, for the purposes of greater generality, we shall consider, while defining the criterion, some questions of the topological structure of set (9.5.6). We shall not need here all the topologies of the triad $\mathfrak{M}$ of Chap. 4, and so we shall confine ourselves to just a short summary of necessary notions. Actually, we shall deal with the use of the product of r copies of the topology $\tau_\eta^*(\mathcal{L})$ from Sec. 4.2. If K is a nonempty set,

$$h\colon K \to (\mathcal{R} - \mathrm{add})^{+}[\mathcal{L}, \eta]$$

and $j \in \overline{1,r}$, then we denote by $(h(\cdot))_j$ the operator

$$t \mapsto (h(t))_j \colon K \to (\mathrm{add})^+ [\mathcal{L}, \eta].$$

In the sequel, we shall denote by $\Theta_\eta^*(\mathcal{L})$ the (unique) topology on set (9.5.6), for which

$$\mathop{\forall}_{T} \; S[T \neq \varnothing] \, \forall \; \ll \in (\mathrm{DIR})\,[T], \quad h \in (\mathcal{R} - \mathrm{add})^+ [\mathcal{L}, \eta]^T,$$

$$\mu \in (\mathcal{R} - \mathrm{add})^+ [\mathcal{L}, \eta]$$

the net $(T, \ll, h)$ converges to μ with respect to $\Theta_\eta^*(\mathcal{L})$ if and only if $\forall\, j \in \overline{1,r}$

$$(T, \ll, (h(\cdot))_j)$$

converges to μ_j with respect to the topology $\tau_*(\mathcal{L})$ as a net in $(\mathrm{add})^+ [\mathcal{L}, \eta]$. By its meaning, the space

$$((\mathcal{R} - \mathrm{add})^+ [\mathcal{L}, \eta], \Theta_\eta^*(\mathcal{L})) \tag{9.5.11}$$

is identifiable with the product of r copies of set (4.2.2) equipped with the topology $\tau_\eta^*(\mathcal{L})$. We shall not check this obvious assertion; let us only note that the representation in terms of the Tychonoff product corresponds to representing a vector measure in the form of a collection of its real-valued components, whereas we used, in this case, the notion of a vector-function on $\mathcal{L}$.

By analogy with (9.5.11), we introduce one more topology on set (9.5.6), namely, we denote by $\Theta_\eta^0(\mathcal{L})$ the (unique) topology on set (9.5.6) such that

$$\mathop{\forall}_{K} \; S[K \neq \varnothing] \, \forall \angle \in (\mathrm{DIR})\,[K], \quad l \in (\mathcal{R} - \mathrm{add})^+ [\mathcal{L}, \eta]^K,$$

$$\nu \in (\mathcal{R} - \mathrm{add})^+ [\mathcal{L}, \eta]$$

the net $(K, \angle, l)$ converges to ν in the sense of $\Theta_\eta^0(\mathcal{L})$ if and only if $\forall\, j \in \overline{1,r}$

$$(K, \angle, (l(\cdot))_j)$$

is a net in $(\mathrm{add})^+ [\mathcal{L}, \eta]$ converging to ν_j in the sense of $\tau_0(\mathcal{L})$. Again we should note that the construction of the space

$$((\mathcal{R} - \mathrm{add})^+ [\mathcal{L}, \eta], \Theta_\eta^0(\mathcal{L})) \tag{9.5.12}$$

actually repeats the operation of the Cartesian product of topological spaces, as in the case of (9.5.11).

In the extremal problems considered in the sequel, we use resource constraints similar to (9.5.2) and their generalized analogs. Let us start with the latter, setting $\forall\, b \in [0, \infty)$

$$\widehat{\Xi}^{+}_{\mathcal{R}}[b] \triangleq \left\{ \mu \in (\mathcal{R} - \mathrm{add})^{+}[\mathcal{L}, \eta]\,\Big|\, \sum_{i=1}^{r} \mu_i(E) \leqslant b \right\}. \tag{9.5.13}$$

This set is compact in the topological space (9.5.11). Indeed, (9.5.13) is, obviously, a closed set in the sense of (9.5.11) and is contained in the set

$$\Xi^{\times}_{\mathcal{R}}[b] \triangleq \{ \mu \in (\mathcal{R} - \mathrm{add})^{+}[\mathcal{L}, \eta]\,|\, \forall\, i \in \overline{1, r}\colon \mu_i \in \Xi_{+}[b] \}, \tag{9.5.14}$$

which is compact by Tychonoff's theorem (it is easy to see that (9.5.13) is also closed in the subspace of (9.5.11) corresponding to set (9.5.14)). Thus, $\forall\, b \in [0, \infty)$

$$\tilde{\vartheta}^{*}_{\eta}[\mathcal{L}, b] \triangleq \Theta^{*}_{\eta}(\mathcal{L})|_{\widehat{\Xi}^{+}_{\mathcal{R}}[b]}$$

is the compact topology on (9.5.13), so that

$$(\widehat{\Xi}^{+}_{\mathcal{R}}[b], \tilde{\vartheta}^{*}_{\eta}[\mathcal{L}, b]) \tag{9.5.15}$$

is a compact space. Note that similar subspaces could also be considered with respect to the topological space (9.5.12), but we shall not need this now. If $f \in B^{+}_{0}[E, \mathcal{L}, \mathcal{R}]$, then, by definition,

$$f * \eta \in (\mathcal{R} - \mathrm{add})^{+}[\mathcal{L}, \eta]$$

is a vector f.a. measure, such that $\forall\, L \in \mathcal{L}$

$$(f * \eta)(L) \triangleq ((f_i * \eta)(L))_{i \in \overline{1, r}};$$

here, if $j \in \overline{1, r}$, then the jth component of $f * \eta$ is $f_j * \eta$. We have thus introduced an indefinite integral of a step vector-function; we use the former notation, which, however, does not lead to ambiguity. We shall now introduce sets, similar to (9.5.13), in the space of conventional solutions, setting $\forall\, b \in [0, \infty)$

$$M^{+}_{\mathcal{R}}[b] \triangleq \left\{ f \in B^{+}_{0}[E, \mathcal{L}, \mathcal{R}]\,\Big|\, \sum_{i=1}^{r} \int_{E} f_i\, d\eta \leqslant b \right\}, \tag{9.5.16}$$

$$\widetilde{M}^{+}_{\mathcal{R}}[b] \triangleq \{ f * \eta\colon f \in M^{+}_{\mathcal{R}}[b] \}; \tag{9.5.17}$$

we can regard (9.5.17) as an inessential transformation of set (9.5.16), whose role in the sequel is very significant. Now we can introduce into consideration an objective operator for the cone optimization problem by duly modifying the definitions from Sec. 8.2. We fix again, if the contrary is not specified, the nonempty set Q from Sec. 2.6, so that (2.6.1) will again play the role of the space of estimates with duly coordinated structures. Let

$$w: Q \to \mathbf{C}((\mathcal{R} - \mathrm{add})^+ [\mathcal{L}, \eta], \Theta_\eta^*(\mathcal{L})); \tag{9.5.18}$$

the symbol w will be used only in the sense of (9.5.18) if the contrary is not assumed. With the aid of (9.5.18), we now introduce, by definition, the operator

$$W: Q \to \mathbf{R}^{B_0^+[E,\mathcal{L},\mathcal{R}]}, \tag{9.5.19}$$

by setting $\forall q \in Q,\ f \in B_0^+[E, \mathcal{L}, \mathcal{R}]$

$$W(q)(f) \triangleq w(q)(f * \eta). \tag{9.5.20}$$

Again, W will be used only in the sense of (9.5.19), (9.5.20) until the contrary is specified. Moreover, we adopt the conventions

$$\mathbf{W} \triangleq (\mathrm{term})\,[W], \qquad \mathcal{W} \triangleq (\mathrm{term})\,[w]. \tag{9.5.21}$$

Here, we certainly take into account the fact that (see (2.6.2), (9.5.19))

$$W \in \mathcal{Q}_{B_0^+[E,\mathcal{L},\mathcal{R}]}. \tag{9.5.22}$$

But, in this case, in view of (2.6.6) and (9.5.18), we have

$$w \in \mathcal{Q}^0_{(\mathcal{R}-\mathrm{add})^+[\mathcal{L},\eta]}(\Theta_\eta^*(\mathcal{L})). \tag{9.5.23}$$

In this case, it follows from (9.5.23) that

$$w \in \mathcal{Q}_{(\mathcal{R}-\mathrm{add})^+[\mathcal{L},\eta]}. \tag{9.5.24}$$

Combining (2.6.3), (2.6.4), and (9.5.21)–(9.5.24), we find that

$$\mathbf{W}: B_0^+[E, \mathcal{L}, \mathcal{R}] \to \mathbf{R}^Q,$$
$$\mathcal{W}: (\mathcal{R} - \mathrm{add})^+ [\mathcal{L}, \eta] \to \mathbf{R}^Q$$

are operators defined by the following conditions. Namely, $\forall f \in B_0^+[E, \mathcal{L}, \mathcal{R}]$, $q \in Q$,

$$\mathbf{W}(f)(q) = W(q)(f). \tag{9.5.25}$$

Similarly, $\forall \mu \in (\mathcal{R} - \mathrm{add})^+ [\mathcal{L}, \eta],\ q \in Q$

$$\mathcal{W}(\mu)(q) = w(q)(\mu). \tag{9.5.26}$$

We shall now consider problems corresponding to the objective operators
(9.5.25) and (9.5.26): conventional and generalized. Here we assume that
we are given a set $\mathbf{Y}$, $\mathbf{Y} \subset \mathbf{R}^n$, satisfying the conditions of Sec. 9.2; the
conventions of Sec. 9.2, concerning the notation and definitions relative to
$\mathbf{N} = \mathbf{R}^n$, the norm $\| \cdot \|_n$, and perturbations (9.2.2) are fulfilled in what
follows. Thus, the main extremal problem, which is also called conventional,
has the form

$$\mathbf{W}(f) \to [\leqq - \mathrm{MIN}\,], \qquad f \in M_{\mathcal{R}}^{+}[c],$$
$$\int_E S(x)f(x)\,\eta(dx) \in \mathbf{Y}. \tag{9.5.27}$$

The generalized problem is defined as follows:

$$\mathcal{W}(\mu) \to [\leqq - \mathrm{MIN}\,], \qquad \mu \in \hat{\Xi}_{\mathcal{R}}^{+}[c],$$
$$\int_E S(x)\,\mu(dx) \in \mathbf{Y}. \tag{9.5.28}$$

From the subsequent constructions it will follow that (9.5.28) can be regarded
as a regularization of problem (9.5.27). The exact meaning of this statement
will be clarified below, after introducing the corresponding constraints of
asymptotic character. We shall, however, first make some general remarks
concerning problem (9.5.27) and some generalizations of the assertions from
Chap. 4 to the case of vector measures. We observe, first of all, that the
representation of the objective operator (9.5.19) in terms of (9.5.18) is, at
first glance, too general. Nevertheless, this representation embraces many
extremal problems that are interesting for practical settings. In particular,
the example of Sec. 1.4 admits reduction to the form of (9.5.18)–(9.5.20).
Let us discuss, at the substantive level, one more situation of this kind.

Namely, let us assume that in an m-dimensional space ($m \in \mathcal{N}$) we have
a collection g_q, $q \in Q$, of continuous real functions of m variables and a
collection of matrix-valued functions G_q, $q \in Q$, each of which maps E into
the set of all $(m \times r)$-matrices and has elements of $B(E, \mathcal{L})$ as its components.
Let us suppose in this special case that the values of operator (9.5.19) have
the form

$$W(q)(f) \triangleq g_q\!\left(\int_E G_q(x)f(x)\,\eta(dx) \right),$$

where $q \in Q$, $f \in B_0^{+}[E, \mathcal{L}, \mathcal{R}]$. Then there exists an operator (9.5.18) for
which representation (9.5.20) is valid. Indeed, we can introduce $\forall q \in Q$ a

functional

$$w(q)\colon (\mathcal{R} - \mathrm{add})^+[\mathcal{L}, \eta] \to \mathbf{R},$$

assuming that $\forall \mu \in (\mathcal{R} - \mathrm{add})^+[\mathcal{L}, \eta]$

$$w(q)(\mu) \triangleq g_q\left(\int_E G_q(x)\,\mu(dx)\right); \tag{9.5.29}$$

here the integral of a matrix-valued mapping can be defined by analogy with (9.5.10). The continuity of $w(q)$ with respect to (9.5.11) is obvious, so that the operator w defined by (9.5.29) is in full agreement with (9.5.18). Equality (9.5.20) holds by virtue of the simplest properties of the integral from Chap. 3. We shall refrain from further detailing of this rather frequently encountered (and admitting various generalizations specified by quite concrete conditions) case. Let us only note that many detailed elaborations of this kind appear precisely in control problems; this circumstance was illustrated in Sec. 1.4 (observe that for reducing the control problem mentioned above to the form defined by (9.5.29), one has to use the arrow-space and Cauchy's formula). Returning to the general case, let us note, first of all, the basic density property of the conventional controls f in the space of generalized control-measures μ from set (9.5.6).

THEOREM 9.5.1. *Let $b \in [0, \infty)$. Then set (9.5.16) admits an everywhere dense imbedding into (9.5.13) in the sense of both topological spaces (9.5.11) and (9.5.12), i.e.,*

$$\widehat{\Xi}_{\mathcal{R}}^+[b] = \mathrm{cl}\,(\widetilde{M}_{\mathcal{R}}^+[b], \Theta_\eta^*(\mathcal{L})) = \mathrm{cl}\,(\widetilde{M}_{\mathcal{R}}^+[b], \Theta_\eta^0(\mathcal{L})).$$

SCHEME OF THE PROOF. We obviously have the inclusion

$$\widetilde{M}_{\mathcal{R}}^+[b] \subset \widehat{\Xi}_{\mathcal{R}}^+[b],$$

so that the equality

$$\mathrm{cl}\,(\widetilde{M}_{\mathcal{R}}^+[b], \Theta_\eta^*(\mathcal{L})) = \mathrm{cl}\,(\widetilde{M}_{\mathcal{R}}^+[b], \tilde{\vartheta}_\eta^*[\mathcal{L}, b])$$

holds true. Furthermore, let us observe that set (9.5.13) is also closed in the sense of (9.5.12). Indeed, if $(T, \ll, h)$ is a net in set (9.5.13), converging to the measure μ from set (9.5.6) in the sense of (9.5.12), then $\forall j \in \overline{1, r}$ the net $(T, \ll, (h(\cdot))_j)$ converges to μ_j in the sense of the topology $\tau_0(\mathcal{L})$. This means that $(h(t))_j(E) = \mu_j(E)$ from some moment (f.s.m.). But in this case, $h(t)(E) = \mu(E)$ f.s.m., so that the sum of $\mu_1(E), \ldots, \mu_r(E)$ does not exceed the number b, and this means that μ belongs to set (9.5.13), and hence, is closed. If now

$$\tilde{\vartheta}^0_\eta[\mathcal{L}, b] \triangleq \Theta^0_\eta(\mathcal{L})|_{\widehat{\Xi}^+_{\mathcal{R}}[b]}, \tag{9.5.30}$$

then, as one can easily verify,

$$\mathrm{cl}\,(\widetilde{M}^+_{\mathcal{R}}[b], \Theta^0_\eta(\mathcal{L})) = \mathrm{cl}\,(\widetilde{M}^+_{\mathcal{R}}[b], \tilde{\vartheta}^0_\eta[\mathcal{L}, b])$$

because (9.5.13) is a superset of $\widetilde{M}^+_{\mathcal{R}}[b]$ closed in space (9.5.12). Let us now compare the topologies

$$\tilde{\vartheta}^*_\eta[\mathcal{L}, b], \qquad \tilde{\vartheta}^0_\eta[\mathcal{L}, b].$$

Suppose that there is a net $(U, \angle, s)$ in set (9.5.13) and $\tilde{\mu}$ is its element, and let $(U, \angle, s)$ converge to $\tilde{\mu}$ in the sense of topology (9.5.30). Then $(U, \angle, s)$ also converges to $\tilde{\mu}$ as a net in space (9.5.12). In this case, $\forall\, i \in \overline{1, r}$ the coordinate net $(U, \angle, (s(\cdot))_i)$ converges to $\tilde{\mu}_i$ in the sense of $\tau_0(\mathcal{L})$. We should bear in mind that these coordinate nets assume values in cone (4.2.2), which follows directly from (9.5.6). But, in this case, every net $(U, \angle, (s(\cdot))_i)$, $i \in \overline{1, r}$, converges to $\tilde{\mu}_i \in (\mathrm{add})_+[\mathcal{L}]$ in the sense of the topology $\tau_0^+(\mathcal{L})$ from Sec. 4.2. Now, taking into account (4.2.12), we get the convergence of these coordinate nets to the measures $\tilde{\mu}_1, \ldots, \tilde{\mu}_r$, respectively, in the sense of $\tau_*^+(\mathcal{L})$ and, hence, of space (4.2.6). This means that $(U, \angle, s)$ converges to $\tilde{\mu}$ in space (9.5.11), and, hence, in space (9.5.15) since $s(u) \in \widehat{\Xi}^+_{\mathcal{R}}[b]$ for $u \in U$. We have established that the (generalized) convergence in the sense of $\tilde{\vartheta}^0_\eta[\mathcal{L}, b]$ leads to a similar convergence in space (9.5.15). This means, as it is easy to see, that

$$\tilde{\vartheta}^*_\eta[\mathcal{L}, b] \subset \tilde{\vartheta}^0_\eta[\mathcal{L}, b].$$

Taking into account the earlier representations in terms of relative closures, we now find that

$$\mathrm{cl}\,(\widetilde{M}^+_{\mathcal{R}}[b], \Theta^0_\eta(\mathcal{L})) \subset \mathrm{cl}\,(\widetilde{M}^+_{\mathcal{R}}[b], \Theta^*_\eta(\mathcal{L})) \subset \widehat{\Xi}^+_{\mathcal{R}}[b].$$

Suppose now that there is chosen an arbitrary vector measure $\nu \in \widehat{\Xi}^+_{\mathcal{R}}[b]$. Then, in particular,

$$\nu_1 \colon \mathcal{L} \to [0, \infty), \ldots, \nu_r \colon \mathcal{L} \to [0, \infty)$$

are elements of (4.2.2), and the sum of values $\nu_i(E)$, $i \in \overline{1, r}$, does not exceed b. Fixing the directed set $(\mathcal{D}, \prec)$ from Sec. 4.3, we construct the operators

$$\Theta^+_{\nu_1}[\cdot]\colon \mathcal{D} \to M^+_{\nu_1(E)}, \ldots, \Theta^+_{\nu_r}[\cdot]\colon \mathcal{D} \to M^+_{\nu_r(E)}. \tag{9.5.31}$$

The notation corresponds to that from Sec. 4.3, namely, we deal with operators each of which acts from $\mathcal{D}$ into the set $B_0^+(E, \mathcal{L})$. But, in this case, $\forall \mathcal{K} \in \mathcal{D}$

$$(\Theta_{\nu_i}^+[\mathcal{K}])_{i \in \overline{1,r}} \colon \overline{1,r} \to B_0^+(E, \mathcal{L}).$$

In other words, $\forall \mathcal{K} \in \mathcal{D}$, $x \in E$,

$$(\Theta_{\nu_i}^+[\mathcal{K}](x))_{i \in \overline{1,r}} \in \mathcal{R}; \tag{9.5.32}$$

the mapping associating (for the fixed $\mathcal{K} \in \mathcal{D}$) the element $x \in E$ with vector (9.5.32) has the functionals $\Theta_{\nu_i}^+[\mathcal{K}]$ as its components. Thus, this mapping is an element of (9.5.4), i.e., a step-valued vector-function on E. We have thus defined the operator acting from $\mathcal{D}$ into $B_0^+[E, \mathcal{L}, \mathcal{R}]$ according to the rule

$$\mathcal{K} \mapsto (\Theta_{\nu_i}^+[\mathcal{K}](\cdot))_{i \in \overline{1,r}} \colon \mathcal{D} \to B_0^+[E, \mathcal{L}, \mathcal{R}].$$

For brevity, let us denote this operator by θ, obtaining a mapping that transforms $\mathcal{K} \in \mathcal{D}$ into

$$x \mapsto (\Theta_{\nu_i}^+[\mathcal{K}](x))_{i \in \overline{1,r}} \colon E \to \mathcal{R};$$

if $j \in \overline{1,r}$, then $(\theta(\cdot))_j$ coincides with the operator $\Theta_{\nu_j}^+[\cdot]$. Here, as one can see from (4.3.8) and (9.5.31), the η-integral of the functional $\Theta_{\nu_j}^+[\mathcal{K}]$ does not exceed $\nu_j(E)$, $\mathcal{K} \in \mathcal{D}$. But then the sum of the above integrals over j, $j \in \overline{1,r}$, does not exceed the number b, as one can see from (9.5.13). This means that $\theta(\mathcal{K}) \in M_{\mathcal{R}}^+[b]$, $\mathcal{K} \in \mathcal{D}$. Therefore,

$$\mathcal{K} \mapsto \theta(\mathcal{K}) * \eta \colon \mathcal{D} \to \widetilde{M}_{\mathcal{R}}^+[b],$$

which follows from (9.5.17). This new operator, mapping $\mathcal{D}$ into $\widetilde{M}_{\mathcal{R}}^+[b]$, will be denoted, for brevity, by $\tilde{\theta}$, so that $\tilde{\theta}(\mathcal{K}) = \theta(\mathcal{K}) * \eta$ for $\mathcal{K} \in \mathcal{D}$. This obviously yields the net $(\mathcal{D}, \prec, \theta)$. Then $(\tilde{\theta}(\cdot))_j$ denotes, as before, the operator mapping $\mathcal{D}$ into $(\mathrm{add})^+[\mathcal{L}, \eta]$ and associating every element $\mathcal{K} \in \mathcal{D}$ with the measure $(\tilde{\theta}(\mathcal{K}))_j$, $j \in \overline{1,r}$. But, in this case, we have

$$(\tilde{\theta}(\mathcal{K}))_j = (\theta(\mathcal{K}))_j * \eta = \Theta_{\nu_j}^+[\mathcal{K}] * \eta$$

for $\mathcal{K} \in \mathcal{D}$. Now, taking account of Lemma 4.3.1, we get the assertion about the convergence of $(\mathcal{D}, \prec, (\tilde{\theta}(\cdot))_j)$ to ν_j in the sense of any topology from the triad $\mathfrak{M}$; in particular, we have the convergence of this net in the sense of $\tau_0(\mathcal{L})$. By the definition of the topological space (9.5.12), we now obviously have the convergence of the net $(\mathcal{D}, \prec, \tilde{\theta})$ in $\widetilde{M}_{\mathcal{R}}^+[b]$ to the vector measure

ν. By Birkhoff's theorem, we thus obtain, in the form of ν, an element of $\mathrm{cl}\,(\widetilde{M}_{\mathcal{R}}^{+}[b], \Theta_{\eta}^{0}(\mathcal{L}))$. The inclusion

$$\widehat{\Xi}_{\mathcal{R}}^{+}[b] \subset \mathrm{cl}\,(\widetilde{M}_{\mathcal{R}}^{+}[b], \Theta_{\eta}^{0}(\mathcal{L}))$$

is established, and this completes the proof.

Let us now observe some important facts connected with continuous dependence. First of all, in view of (9.5.10) and the definitions of Chap. 3, the operator

$$\mu \mapsto \int_{E} S(x)\,\mu(dx)\colon (\mathcal{R} - \mathrm{add})^{+}\,[\mathcal{L}, \eta] \to \mathbf{N} \qquad (9.5.33)$$

is continuous as the mapping of space (9.5.11) into $(\mathbf{N}, \|\cdot\|_{n})$; this fact follows directly from (9.5.10) and from the definition of the convergence of nets in the topological space (9.5.11). In the sequel, the property of continuity of vector-function (9.5.33) is used without additional explanations.

LEMMA 9.5.1. *Given* $\forall i \in \overline{1, n}$, $j \in \overline{1, r}$: $S_{i,j} \in B_{0}(E, \mathcal{L})$. *Let the net* $(T, \ll, h)$ *in* $(\mathcal{R} - \mathrm{add})^{+}\,[\mathcal{L}, \eta]$ *converge to* $\mu \in (\mathcal{R} - \mathrm{add})^{+}\,[\mathcal{L}, \eta]$ *in the topological space* (9.5.12). *Then f.s.m.* (*see Sec. 2.2*)

$$\int_{E} S(x)\,h(t)(dx) = \int_{E} S(x)\,\mu(dx).$$

THE SCHEME OF THE PROOF is rather obvious; nevertheless, we shall discuss its main points, taking into account that, by definition of the space (9.5.12), every coordinate net $(T, \ll, (h(\cdot))_{i})$ converges to μ_{i}, $i \in \overline{1, r}$, in the sense of the topology $\tau_{0}(\mathcal{L}) \in \mathfrak{M}$. This means, in view of the definitions from Sec. 4.2, that $\forall i \in \overline{1, r}$, $L \in \mathcal{L}$,

$$(h(t))_{i}(L) = \mu_{i}(L)$$

f.s.m. This implies, as a consequence, the property of coincidence

$$\int_{E} g\,d(h(t))_{i} = \int_{E} g\,d\mu_{i} \qquad (9.5.34)$$

f.s.m. if only $g \in B_{0}(E, \mathcal{L})$. The last property is obvious by virtue of the definitions from Sec. 3.4, since we actually use elementary integrals in (9.5.34). Now $\forall i \in \overline{1, n}$, $j \in \overline{1, r}$,

$$\int_{E} S_{i,j}\,d(h(t))_{j} = \int_{E} S_{i,j}\,d\mu_{j}$$

f.s.m.; the final assertion is obtained by applying (9.5.10).

REMARK. The property established in the lemma can also be interpreted as continuity, but only in the case where the real line is endowed with a discrete topology, which generates, after performing the operation of Tychonoff product, the respective topology in $\mathbf{N}$. We shall not use this interpretation in the sequel, and, therefore, we do not discuss it more extensively.

The following two cases (for more details, see [45, 10]) supplement Theorem 9.5.1 in a natural way by extending the density property of conventional solutions to the set of values of the vector integrand. In the sequel, we shall have to consider the following two copies of the compact space (9.5.15):

$$(\widehat{\Xi}_{\mathcal{R}}^{+}[c], \tilde{\vartheta}_{\eta}^{*}[\mathcal{L}, c]), \tag{9.5.35}$$

$$(\widehat{\Xi}_{\mathcal{R}}^{+}[c+1], \tilde{\vartheta}_{\eta}^{*}[\mathcal{L}, c+1]), \tag{9.5.36}$$

though the choice of the compact space (9.5.36) is subjective (we could use in (9.5.15) one version corresponding to the case $b = c + \varepsilon$, $\varepsilon > 0$). Let us now consider relaxations of problem (9.5.27). For this purpose, we imbed it into a parametrized family of extremal problems of the following form:

$$\mathbf{W}(f) \to [\leqq - \mathrm{MIN}], \qquad f \in M_{\mathcal{R}}^{+}[c+\varepsilon],$$
$$\int_{E} S(x)f(x)\,\eta(dx) \in \mathbf{Y}_{\varepsilon}, \tag{9.5.37}$$

where $\varepsilon \geqslant 0$. We use in (9.5.37) convention (9.2.2) and take into account (9.2.3). In addition, we consider the problems

$$\mathbf{W}(f) \to [\leqq - \mathrm{MIN}], \qquad f \in M_{\mathcal{R}}^{+}[c],$$
$$\int_{E} S(x)f(x)\,\eta(dx) \in \mathbf{Y}_{\varepsilon}, \tag{9.5.38}$$

where $\varepsilon \geqslant 0$. For $\varepsilon = 0$, each problem (9.5.37) and (9.5.38) degenerates into (9.5.27). Throughout this section, we set $\forall \varepsilon \in [0, \infty)$

$$\left(F^{+}(\varepsilon) \triangleq \left\{ f \in M_{\mathcal{R}}^{+}[c+\varepsilon] \,\middle|\, \int_{E} S(x)f(x)\,\eta(dx) \in \mathbf{Y}_{\varepsilon} \right\} \right)$$

$$\&(F_{*}^{+}(\varepsilon) \triangleq F^{+}(\varepsilon) \cap M_{\mathcal{R}}^{+}[c]),$$

obtaining the admissible sets of problems (9.5.37) and (9.5.38), respectively (we have to take into account (9.5.16)). In this case

$$F^{+}(0) = F_{*}^{+}(0) \in \mathcal{P}(M_{\mathcal{R}}^{+}[c])$$

is the admissible set of the initial problem (9.5.27). As one can easily verify, with due account of Theorem 9.5.1, the admissible set of the generalized problem is the common regularization of the families $\{F^+(\varepsilon)\colon \varepsilon \in (0,\infty)\}$ and $\{F_*^+(\varepsilon)\colon \varepsilon \in (0,\infty)\}$:

$$\widehat{\Omega}_{\mathcal{R}}^+ \triangleq \left\{\mu \in \widehat{\Xi}_{\mathcal{R}}^+[c]\Big|\ \int_E S(x)\,\mu(dx) \in \mathbf{Y}\right\}$$

$$= \bigcap_{\varepsilon\in(0,\infty)} \mathrm{cl}\left(\{f * \eta\colon f \in F^+(\varepsilon)\}, \Theta_\eta^*(\mathcal{L})\right)$$

$$= \bigcap_{\varepsilon\in(0,\infty)} \mathrm{cl}\left(\{f * \eta\colon f \in F_*^+(\varepsilon)\}, \tilde{\vartheta}_\eta^*[\mathcal{L}, c]\right). \tag{9.5.39}$$

Relation (9.5.39) will be used in the interests of compactification of an essential part of the space of conventional solutions according to the prescriptions of Sec. 2.6.

Let us note one more case that is conceptually similar to Lemma 9.5.1.

LEMMA 9.5.2. *Let* $\forall\, i \in \overline{1,n},\ j \in \overline{1,r}\colon S_{i,j} \in B_0(E, \mathcal{L})$. *Then*

$$\widehat{\Omega}_{\mathcal{R}}^+ = \mathrm{cl}\left(\{f * \eta\colon f \in F^+(0)\}, \tilde{\vartheta}_\eta^*[\mathcal{L}, c]\right) = \mathrm{cl}\left(\{f * \eta\colon f \in F^+(0)\}, \Theta_\eta^0(\mathcal{L})\right).$$

PROOF. It is easy to see that

$$\mathrm{cl}\left(\{f * \eta\colon f \in F^+(0)\}, \Theta_\eta^0(\mathcal{L})\right) \subset \mathrm{cl}\left(\{f * \eta\colon f \in F^+(0)\}, \vartheta_\eta^*[\mathcal{L}, c]\right).$$

The proof of this inclusion can be realized according to the scheme presented in the proof of Theorem 9.5.1. Let us also observe that, by virtue of (9.5.39), the set $\widehat{\Omega}_{\mathcal{R}}^+$ is closed in space (9.5.11), and

$$\{f * \eta\colon f \in F^+(0)\} \subset \widehat{\Omega}_{\mathcal{R}}^+.$$

Hence, we have the inclusion

$$\mathrm{cl}\left(\{f * \eta\colon f \in F^+(0)\}, \Theta_\eta^*(\mathcal{L})\right) = \mathrm{cl}\left(\{f * \eta\colon f \in F^+(0)\}, \tilde{\vartheta}_\eta^*[\mathcal{L}, c]\right) \subset \widehat{\Omega}_{\mathcal{R}}^+.$$

Therefore, it suffices to establish the inclusion

$$\widehat{\Omega}_{\mathcal{R}}^+ \subset \mathrm{cl}\left(\{f * \eta\colon f \in F^+(0)\}, \Theta_\eta^0(\mathcal{L})\right). \tag{9.5.40}$$

Take an arbitrary $\mu \in \widehat{\Omega}_{\mathcal{R}}^+$. Then it follows from (9.5.39) that

$$\mu \in \widehat{\Xi}_{\mathcal{R}}^+[c], \tag{9.5.41}$$

and then

$$\int_E S(x)\, \mu(dx) \in \mathbf{Y}. \tag{9.5.42}$$

Now it follows from (9.5.41) and Theorem 9.5.1 that there is a net $(D, \ll, h)$ in $M_{\mathcal{R}}^+[c]$ for which the net

$$(D, \ll, (h(\delta) * \eta)_{\delta \in D}) \tag{9.5.43}$$

converges to the measure μ in the sense of the topological space (9.5.12). Taking account of Lemma 9.5.1, we find that

$$\int_E S(x)h(\delta)(x)\, \eta(dx) = \int_E S(x)\,(h(\delta) * \eta)(dx) = \int_E S(x)\, \mu(dx)$$

f.s.m., whence we get, by virtue of (9.5.42), that

$$\int_E S(x)h(\delta)(x)\, \eta(dx) \in \mathbf{Y}$$

f.s.m. But with allowance for (9.2.3), this means that

$$h(\delta) \in F^+(0)$$

f.s.m. Hence, by virtue of the convergence of net (9.5.43) in the sense of (9.5.12), every neighborhood of the point μ in the topological space (9.5.12) meets the set $\{f * \eta:\ f \in F^+(0)\}$, i.e.,

$$\mu \in \mathrm{cl}\,(\{f * \eta:\ f \in F^+(0)\}, \Theta_\eta^0(\mathcal{L})).$$

Inclusion (9.5.40) is thus established, which completes the proof of the lemma.

Now, we shall consider asymptotic values and relations between them. We shall use the constructions from Sec. 2.6. For convenience of subsequent notation, we introduce the families

$$\mathcal{X}_1 \triangleq \{F^+(\varepsilon):\ \varepsilon \in (0, \infty)\}; \tag{9.5.44}$$

$$\mathcal{X}_2 \triangleq \{F_*^+(\varepsilon):\ \varepsilon \in (0, \infty)\}; \tag{9.5.45}$$

$$\widetilde{\mathcal{X}}_1 \triangleq \{F^+(\varepsilon):\ \varepsilon \in (0, 1]\}. \tag{9.5.46}$$

It is easy to see that each one of them is nonempty and contains, with every pair of its sets, a subset of their intersection. Family (9.5.46) is auxiliary with respect to (9.5.44); the comparison of asymptotic properties of families

(9.5.44), (9.5.45) and their relationship with the set $F^+(0)$ constitutes our main objective. Note that these asymptotic values, as are those from Sec. 2.6, will be defined in the form of respective sets of minimal elements on the sets of attainable estimates. Expressly, we mean the sets

$$\bigcap_{U \in \mathcal{X}_1} \mathrm{cl}\,(\mathbf{W}^1(U), \otimes^Q(\tau_{\mathbf{R}})) = \bigcap_{\varepsilon \in (0,\infty)} \mathrm{cl}\,(\mathbf{W}^1(F^+(\varepsilon)), \otimes^Q(\tau_{\mathbf{R}}))$$

$$= \bigcap_{\varepsilon \in (0,1]} \mathrm{cl}\,(\mathbf{W}^1(F^+(\varepsilon)), \otimes^Q(\tau_{\mathbf{R}}))$$

$$= \bigcap_{U \in \widetilde{\mathcal{X}}_1} \mathrm{cl}\,(\mathbf{W}^1(U), \otimes^Q(\tau_{\mathbf{R}})) \in \mathcal{P}(\mathbf{R}^Q); \tag{9.5.47}$$

$$\bigcap_{U \in \widetilde{\mathcal{X}}_2} \mathrm{cl}\,(\mathbf{W}^1(U), \otimes^Q(\tau_{\mathbf{R}}))$$

$$= \bigcap_{\varepsilon \in (0,\infty)} \mathrm{cl}\,(\mathbf{W}^1(F_*^+(\varepsilon)), \otimes^Q(\tau_{\mathbf{R}})) \in \mathcal{P}(\mathbf{R}^Q). \tag{9.5.48}$$

The elements of sets (9.5.47) and (9.5.48) are precisely the (asymptotically) attainable estimates. The asymptotic values are determined on the basis of minimizing the elements of (9.5.47) and (9.5.48) in the sense of $\leq$. For our purposes, an important role is also played by the generalized value corresponding to problem (9.5.28); the set of all generalized estimates has the form

$$\mathcal{W}^1(\widehat{\Omega}_{\mathcal{R}}^+) \in \mathcal{P}(\mathbf{R}^Q). \tag{9.5.49}$$

Finally, for determining the conventional value, we shall need the set

$$\mathrm{cl}\,(\mathbf{W}^1(F^+(0)), \otimes^Q(\tau_{\mathbf{R}})) = \mathrm{cl}\,(\mathbf{W}^1(F_*^+(0)), \otimes^Q(\tau_{\mathbf{R}})) \in \mathcal{P}(\mathbf{R}^Q). \tag{9.5.50}$$

The minimization on sets (9.5.47)–(9.5.50) will be identified with determination of the respective set of minimal elements. Note that from the definition of space (9.5.15) it easily follows that $\forall g \in \mathbf{C}((\mathcal{R} - \mathrm{add})^+[\mathcal{L}, \eta], \Theta_\eta^*(\mathcal{L}))$, $b \in [0, \infty)$,

$$(g \mid \widehat{\Xi}_{\mathcal{R}}^+[b]) \in \mathbf{C}(\widehat{\Xi}_{\mathcal{R}}^+[b], \widetilde{\vartheta}_\eta^*[\mathcal{L}, b]).$$

This inference is applicable, in particular, to functionals that are obtained as images by virtue of operator (9.5.18), so that $\forall b \in [0, \infty)$, $q \in Q$,

$$(w(q) \mid \widehat{\Xi}_{\mathcal{R}}^+[b]) \in \mathbf{C}(\widehat{\Xi}_{\mathcal{R}}^+[b], \widetilde{\vartheta}_\eta^*[\mathcal{L}, b]).$$

Taking account of this circumstance, we introduce a pair of auxiliary generalized abridged operators. Namely, let, by definition,

$$w_1 \colon Q \to \mathbf{C}(\widehat{\Xi}_{\mathcal{R}}^{+}[c+1], \tilde{\vartheta}_{\eta}^{*}[\mathcal{L}, c+1])$$

be an operator such that $\forall\, q \in Q$

$$w_1(q) \triangleq (w(q) \mid \widehat{\Xi}_{\mathcal{R}}^{+}[c+1]). \qquad (9.5.51)$$

Furthermore, let, by definition,

$$w_2 \colon Q \to \mathbf{C}(\widehat{\Xi}_{\mathcal{R}}^{+}[c], \tilde{\vartheta}_{\eta}^{*}[\mathcal{L}, c])$$

be an operator such that $\forall\, q \in Q$

$$w_2(q) \triangleq (w(q) \mid \widehat{\Xi}_{\mathcal{R}}^{+}[c]). \qquad (9.5.52)$$

In what follows, the symbols w_1 and w_2 are used in the sense of (9.5.51) and (9.5.52), respectively, if the contrary is not specified. Furthermore, let us introduce, as objective operators, the respective conventional abridged operators. Namely, we assume that

$$W_1 \colon Q \to \mathbf{R}^{M_{\mathcal{R}}^{+}[c+1]}$$

is, by definition, the operator for which $\forall\, q \in Q$

$$W_1(q) \triangleq (W(q) \mid M_{\mathcal{R}}^{+}[c+1]). \qquad (9.5.53)$$

Furthermore, let, by definition, the operator

$$W_2 \colon Q \to \mathbf{R}^{M_{\mathcal{R}}^{+}[c]}$$

be such that $\forall\, q \in Q$

$$W_2(q) \triangleq (W(q) \mid M_{\mathcal{R}}^{+}[c]). \qquad (9.5.54)$$

Unless otherwise specified, the symbols W_1 and W_2 will be used in the sequel only in the sense of (9.5.53) and (9.5.54), respectively. Observe that this kind of construction using a pair of abridged operators was already used, for instance, in (8.3). Note that

$$(W_1 \in \mathcal{Q}_{M_{\mathcal{R}}^{+}[c+1]}) \& (W_2 \in \mathcal{Q}_{M_{\mathcal{R}}^{+}[c]}),$$

and therefore, following Sec. 2.6, we can introduce their inessential transforms

$$(\mathbf{W}_1 \triangleq (\text{term})\,[W_1]) \& (\mathbf{W}_2 \triangleq (\text{term})\,[W_2]),$$

getting operators acting, respectively, from $M_{\mathcal{R}}^+[c+1]$ and $M_{\mathcal{R}}^+[c]$ into $\mathbf{R}^Q$. Here, we find, in view of (2.6.3) and (2.6.4), that $\forall f \in M_{\mathcal{R}}^+[c+1]$, $q \in Q$,

$$\mathbf{W}_1(f)(q) = W_1(q)(f) = W(q)(f) = \mathbf{W}(f)(q).$$

We have taken into account (9.5.25) and obtained the equality

$$\mathbf{W}_1 = (\mathbf{W} \mid M_{\mathcal{R}}^+[c+1]). \tag{9.5.55}$$

By analogy, we find that $\forall f \in M_{\mathcal{R}}^+[c]$, $q \in Q$,

$$\mathbf{W}_2(f)(q) = W_2(q)(f) = W(q)(f) = \mathbf{W}(f)(q).$$

In other words,

$$\mathbf{W}_2 = (\mathbf{W} \mid M_{\mathcal{R}}^+[c]). \tag{9.5.56}$$

Relations similar to (9.5.55) and (9.5.56) can also be obtained for generalized operators. Here we have to take into account that, according to (2.6.6), (9.5.51), and (9.5.52), we have

$$w_1 \in \mathcal{Q}^0_{\widehat{\Xi}_{\mathcal{R}}^+[c+1]}(\tilde{\vartheta}_\eta^*[\mathcal{L}, c+1]),$$
$$w_2 \in \mathcal{Q}^0_{\widehat{\Xi}_{\mathcal{R}}^+[c]}(\tilde{\vartheta}_\eta^*[\mathcal{L}, c]).$$

In this case, we have, according to (2.6.3), (2.6.4), and (2.6.7), in the form of

$$(\mathcal{W}_1 \triangleq (\text{term})\,[w_1]) \,\&\, (\mathcal{W}_2 \triangleq (\text{term})\,[w_2])$$

operators mapping the sets $\widehat{\Xi}_{\mathcal{R}}^+[c+1]$ and $\widehat{\Xi}_{\mathcal{R}}^+[c]$ into $\mathbf{R}^Q$:

$$\mathcal{W}_1 \in \mathbf{C}(\widehat{\Xi}_{\mathcal{R}}^+[c+1], \tilde{\vartheta}_\eta^*[\mathcal{L}, c+1], \mathbf{R}^Q, \otimes^Q(\tau_{\mathbf{R}})),$$
$$\mathcal{W}_2 \in \mathbf{C}(\widehat{\Xi}_{\mathcal{R}}^+[c], \tilde{\vartheta}_\eta^*[\mathcal{L}, c], \mathbf{R}^Q, \otimes^Q(\tau_{\mathbf{R}})).$$

Recall that $\forall \mu \in \widehat{\Xi}_{\mathcal{R}}^+[c+1]$, $q \in Q$,

$$\mathcal{W}_1(\mu)(q) = w_1(q)(\mu) = w(q)(\mu) = \mathcal{W}(\mu)(q).$$

This means that the equality

$$\mathcal{W}_1 = (\mathcal{W} \mid \widehat{\Xi}_{\mathcal{R}}^+[c+1])$$

holds true. Moreover, $\forall \mu \in \widehat{\Xi}_{\mathcal{R}}^+[c]$, $q \in Q$,

$$\mathcal{W}_2(\mu)(q) = w_2(q)(\mu) = w(q)(\mu) = \mathcal{W}(\mu)(q).$$

Therefore,

$$\mathcal{W}_2 = (\mathcal{W} \mid \widehat{\Xi}^+_{\mathcal{R}}[c]).$$

However, we recall that (see (9.5.25), (9.5.26)) $\forall f \in B^+_0[E, \mathcal{L}, \mathcal{R}]$, $q \in Q$,

$$\mathbf{W}(f)(q) = w(q)(f * \eta) = \mathcal{W}(f * \eta)(q).$$

This relation implies that $\forall f \in B^+_0[E, \mathcal{L}, \mathcal{R}]$

$$\mathbf{W}(f) = \mathcal{W}(f * \eta).$$

But in this case, taking into account (9.5.55), we find that $\forall f \in M^+_{\mathcal{R}}[c+1]$

$$\mathbf{W}_1(f) = \mathcal{W}_1(f * \eta).$$

In addition, by virtue of (9.5.56), we obviously find that $\forall f \in M^+_{\mathcal{R}}[c]$

$$\mathbf{W}_2(f) = \mathcal{W}_2(f * \eta).$$

The last two relations correspond to the consistency conditions from Sec. 2.5. The symbols $\mathbf{W}_1$, $\mathbf{W}_2$, $\mathcal{W}_1$, and $\mathcal{W}_2$ are used only in the sense of the last definitions. From the standpoint of the pointwise consistency conditions from Sec. 2.6, we find from (9.5.20) and (9.5.53) that $\forall q \in Q$

$$W_1(q) = (W(q) \mid M^+_{\mathcal{R}}[c+1]) = (w(q)(f * \eta))_{f \in M^+_{\mathcal{R}}[c+1]}$$

$$= (w_1(q)(f * \eta))_{f \in M^+_{\mathcal{R}}[c+1]}. \tag{9.5.57}$$

Similarly, we find that $\forall q \in Q$

$$W_2(q) = (w(q)(f * \eta))_{f \in M^+_{\mathcal{R}}[c]} = (w_2(q)(f * \eta))_{f \in M^+_{\mathcal{R}}[c]}. \tag{9.5.58}$$

Here we took account of (9.5.20) and (9.5.54). Relations (9.5.57) and (9.5.58) play an important role from the standpoint of verification of the pointwise consistency conditions in the compactification of an essential part of the space of solutions. Expressly, let us introduce the abridged integration operators

$$\mathbf{I}_1 \triangleq (f * \eta)_{f \in M^+_{\mathcal{R}}[c+1]}, \qquad \mathbf{I}_2 \triangleq (f * \eta)_{f \in M^+_{\mathcal{R}}[c]},$$

that assume values in the sets $\widehat{\Xi}^+_{\mathcal{R}}[c+1]$ and $\widehat{\Xi}^+_{\mathcal{R}}[c]$, respectively. Now we can introduce two extension models corresponding, in full measure, to Sec. 2.6 since (see (9.5.57), (9.5.58)) $\forall q \in Q$

$$(W_1(q) = w_1(q) \circ \mathbf{I}_1) \& (W_2(q) = w_2(q) \circ \mathbf{I}_2).$$

Indeed, the indicated models can be characterized by means of the following two specifications of the basic notions from Sec. 2.6:

(1) $X = M_{\mathcal{R}}^+[c+1]$, $\mathcal{X} = \tilde{\mathcal{X}}_1$, $s = W_1$, $Y = \hat{\Xi}_{\mathcal{R}}^+[c+1]$, $\tau = \tilde{\vartheta}_\eta^*[\mathcal{L}, c+1]$, $m = \mathbf{I}_1$, $g = w_1$,

(2) $X = M_{\mathcal{R}}^+[c]$, $\mathcal{X} = \mathcal{X}_2$, $s = W_2$, $Y = \hat{\Xi}_{\mathcal{R}}^+[c]$, $\tau = \tilde{\vartheta}_\eta^*[\mathcal{L}, c]$, $m = \mathbf{I}_2$, $g = w_2$.

Then, from Theorem 2.6.1 we obviously have

$$\bigcap_{\varepsilon \in (0,1]} \mathrm{cl}\,(\mathbf{W}_1^1(F^+(\varepsilon)), \otimes^Q(\tau_{\mathbf{R}})) = \bigcap_{U \in \tilde{\mathcal{X}}_1} \mathrm{cl}\,(\mathbf{W}_1^1(U), \otimes^Q(\tau_{\mathbf{R}}))$$

$$= W_1^1\left(\bigcap_{U \in \tilde{\mathcal{X}}_1} \mathrm{cl}\,(\mathbf{I}_1^1(U), \tilde{\vartheta}_\eta^*[\mathcal{L}, c+1]) \right), \tag{9.5.59}$$

$$\bigcap_{\varepsilon \in (0,\infty)} \mathrm{cl}\,(\mathbf{W}_2^1(F_*^+(\varepsilon)), \otimes^Q(\tau_{\mathbf{R}})) = \bigcap_{U \in \mathcal{X}_2} \mathrm{cl}\,(\mathbf{W}_2^1(U), \otimes^Q(\tau_{\mathbf{R}}))$$

$$= W_2^1\left(\bigcap_{U \in \mathcal{X}_2} \mathrm{cl}\,(\mathbf{I}_2^1(U), \tilde{\vartheta}_\eta^*[\mathcal{L}, c]) \right). \tag{9.5.60}$$

Now, taking account of (9.5.47), (9.5.48), and of the relationships mentioned above, concerning, in particular, representations in terms of restriction (see (9.5.55) and (9.5.56)), we can transform (9.5.59) and (9.5.60), obtaining

$$\bigcap_{\varepsilon \in (0,\infty)} \mathrm{cl}\,(\mathbf{W}^1(F^+(\varepsilon)), \otimes^Q(\tau_{\mathbf{R}}))$$

$$= W^1\left(\bigcap_{\varepsilon \in (0,1]} \mathrm{cl}\,(\{f * \eta \colon f \in F^+(\varepsilon)\}, \tilde{\vartheta}_\eta^*[\mathcal{L}, c+1]) \right),$$

$$\bigcap_{\varepsilon \in (0,\infty)} \mathrm{cl}\,(\mathbf{W}^1(F_*^+(\varepsilon)), \otimes^Q(\tau_{\mathbf{R}}))$$

$$= W^1\left(\bigcap_{\varepsilon \in (0,\infty)} \mathrm{cl}\,(\{f * \eta \colon f \in F_*^+(\varepsilon)\}, \tilde{\vartheta}_\eta^*[\mathcal{L}, c]) \right).$$

Let us now consider the right-hand sides of the last two relations. In the first of them, we should obviously take into account that (see (9.5.16)) $\forall \varepsilon \in (0,1]\colon F^+(\varepsilon) \subset M_{\mathcal{R}}^+[c+1]$, $\{f * \eta \colon f \in F^+(\varepsilon)\} \subset \hat{\Xi}_{\mathcal{R}}^+[c+1]$. This yields

$$\mathrm{cl}\,(\{f * \eta \colon f \in F^+(\varepsilon)\}, \tilde{\vartheta}_\eta^*[\mathcal{L}, c+1]) = \mathrm{cl}\,(\{f * \eta \colon f \in F^+(\varepsilon)\}, \Theta_\eta^*(\mathcal{L}))$$

$$\cap \hat{\Xi}_{\mathcal{R}}^+[c+1] = \mathrm{cl}\,(\{f * \eta \colon f \in F^+(\varepsilon)\}, \Theta_\eta^*(\mathcal{L}))$$

for $\varepsilon \in (0, 1]$. But in this case, by (9.5.39), we get the equality

$$\bigcap_{\varepsilon \in (0,1]} \mathrm{cl}\left(\{f * \eta \colon f \in F^+(\varepsilon)\}, \vartheta_\eta^*[\mathcal{L}, c+1]\right)$$

$$= \bigcap_{\varepsilon \in (0,1]} \mathrm{cl}\left(\{f * \eta \colon f \in F^+(\varepsilon)\}, \Theta_\eta^*(\mathcal{L})\right)$$

$$= \bigcap_{\varepsilon \in (0,\infty)} \mathrm{cl}\left(\{f * \eta \colon f \in F^+(\varepsilon)\}, \Theta_\eta^*(\mathcal{L})\right) = \widehat{\Omega}_{\mathcal{R}}^+.$$

Now, taking account of the last relation and of (9.5.39), we have for the following common integral representation the sets of attainable estimates:

$$\bigcap_{\varepsilon \in (0,\infty)} \mathrm{cl}\left(\mathbf{W}^1(F^+(\varepsilon)), \otimes^Q(\tau_{\mathbf{R}})\right)$$

$$= \bigcap_{\varepsilon \in (0,\infty)} \mathrm{cl}\left(\mathbf{W}^1(F_*^+(\varepsilon)), \otimes^Q(\tau_{\mathbf{R}})\right) = \mathcal{W}^1(\widehat{\Omega}_{\mathcal{R}}^+). \tag{9.5.61}$$

Relation (9.5.61) implies the now obvious inference about asymptotic equivalence with respect to the result of families (9.5.44) and (9.5.45).

THEOREM 9.5.2. *There is a coincidence of asymptotic values corresponding to the two types of relaxation defined by families (9.5.44) and (9.5.45), respectively, so that*

$$(\leqq - \mathrm{MIN})[\mathcal{W}^1(\widehat{\Omega}_{\mathcal{R}}^+)] = (\leqq - \mathrm{MIN})\left[\bigcap_{\varepsilon \in (0,\infty)} \mathrm{cl}\left(\mathbf{W}^1(F^+(\varepsilon)), \otimes^Q(\tau_{\mathbf{R}})\right)\right]$$

$$= (\leqq - \mathrm{MIN})\left[\bigcap_{\varepsilon \in (0,\infty)} \mathrm{cl}\left(\mathbf{W}^1(F_*^+(\varepsilon)), \otimes^Q(\tau_{\mathbf{R}})\right)\right].$$

From the practical standpoint, Theorem 9.5.2 has the meaning of inference about some structural stability of the problem with respect to the weakening of the resource constraint. We shall not dwell, in detail, on Theorem 9.5.2 and its diverse consequences concerning, in particular, the construction of asymptotically efficient approximate solutions since the reasoning is similar to the constructions from Chap. 8, realized in the coordinatewise version. We shall concentrate our attention on comparing values understood as sets in the estimates space.

THEOREM 9.5.3. *Let* $\forall i \in \overline{1, n}$, $j \in \overline{1, r}$: $S_{i,j} \in B_0(E, \mathcal{L})$. *Then*

$$\mathrm{cl}\left(\mathbf{W}^1(F^+(0)), \otimes^Q(\tau_{\mathbf{R}})\right) = \mathcal{W}^1(\widehat{\Omega}_{\mathcal{R}}^+).$$

The proof is actually a straightforward consequence of Lemma 9.5.2. Nevertheless, we shall point out some main aspects of the proof, returning for this purpose to some propositions from Secs. 2.5 and 2.6. First of all (see (9.5.17)), we have

$$\widetilde{F}^+(0) \triangleq \{f * \eta \colon f \in F^+(0)\} \in \mathcal{P}(\widetilde{M_{\mathcal{R}}^+}[c]).$$

This means that $\widetilde{F}^+(0)$ is a set in the compact space (9.5.35). We shall now make use of the property of continuity of $\mathcal{W}_2$ mentioned earlier in the sense of the topologies $\tilde{\vartheta}_\eta^*[\mathcal{L}, c]$ and $\otimes^Q(\tau_{\mathbf{R}})$. But in this case $\mathcal{W}_2$ is a closed mapping since it maps a compact space into a Hausdorff one. This means that $\mathcal{W}_2$ preserves the closure operation, so that, by Lemma 9.5.2, we have

$$\mathcal{W}^1(\widehat{\Omega}_{\mathcal{R}}^+) = \mathcal{W}_2^1(\widehat{\Omega}_{\mathcal{R}}^+) = \mathcal{W}_2^1(\mathrm{cl}\,(\widetilde{F}^+(0), \tilde{\vartheta}_\eta^*[\mathcal{L}, c]))$$

$$= \mathrm{cl}\,(\mathcal{W}_2^1(\widetilde{F}^+(0)), \otimes^Q(\tau_{\mathbf{R}})) = \mathrm{cl}\,(\mathcal{W}^1(\widetilde{F}^+(0)), \otimes^Q(\tau_{\mathbf{R}}))$$

$$= \mathrm{cl}\,(\{\mathcal{W}(f * \eta) \colon f \in F^+(0)\}, \otimes^Q(\tau_{\mathbf{R}})) = \mathrm{cl}\,(\{\mathbf{W}(f) \colon f \in F^+(0)\}, \otimes^Q(\tau_{\mathbf{R}}))$$

$$= \mathrm{cl}\,(\mathbf{W}^1(F^+(0)), \otimes^Q(\tau_{\mathbf{R}})).$$

The theorem is proved.

As an obvious consequence of Theorem 9.5.3, we have the implication

$$(\forall\, i \in \overline{1, n},\ j \in \overline{1, r}\colon S_{i,j} \in B_0(E, \mathcal{L}))$$

$$\Rightarrow ((\leqq - \mathrm{MIN}\,)[\mathrm{cl}\,(\mathbf{W}^1(F^+(0)), \otimes^Q(\tau_{\mathbf{R}}))] = (\leqq - \mathrm{MIN}\,)[\mathcal{W}^1(\widehat{\Omega}_{\mathcal{R}}^+)]).$$
$$(9.5.62)$$

Theorem 9.5.2 and relation (9.5.62) yield a natural sufficient condition of computational stability which reduces to the requirement of step-valuedness of the matrix-valued function S. Conceptually, this situation is equivalent to the conditions of Sec. 9.3. A natural concrete example of this situation was considered in Sec. 1.4, and a reduction to the general scheme under consideration can be easily achieved by specifying $(E, \mathcal{L})$ as an arrow-space; the measure η should be defined in this case as a function of length. Here we must use representations of the type (9.5.29).

SCALAR OPTIMIZATION PROBLEM. Throughout the remainder of this section, we shall consider a special case connected with the study of a scalar optimization problem. We shall use a very short presentation, specifying only those aspects that are typical precisely of this special case. We assume in the sequel that

$$w \in \mathbf{C}((\mathcal{R} - \mathrm{add})^+[\mathcal{L}, \eta], \Theta_\eta^*(\mathcal{L})),$$
$$W \colon B_0^+[E, \mathcal{L}, \mathcal{R}] \to \mathbf{R},$$
$$(9.5.63)$$

and $\forall\, f \in B_0^+[E, \mathcal{L}, \mathcal{R}]$

$$W(f) \triangleq w(f * \eta) \tag{9.5.64}$$

(actually, we consider in the sequel only those notions that were introduced in (9.5.18)–(9.5.20), but only for the case of a single-element set Q; for this reason, we resort to a new notation in (9.5.63) and (9.5.64)). The system of constraints of the problem considered below is assumed to be the same as in problem (9.5.27). Naturally, the definitions of asymptotic values will be simplified. Thus the conventional problem has, in our case, the form

$$W(f) \to \inf, \qquad f \in M_{\mathcal{R}}^+[c],$$
$$\int_E S(x) f(x)\, \eta(dx) \in \mathbf{Y}. \tag{9.5.65}$$

Respectively, the generalized problem acquires the form

$$w(\mu) \to \min, \quad \mu \in \widehat{\Xi}_{\mathcal{R}}^+[c], \quad \int_E S(x)\, \mu(dx) \in \mathbf{Y}. \tag{9.5.66}$$

The relationship between problems (9.5.65) and (9.5.66) is similar to that between problems (9.5.27) and (9.5.28). As before, we also consider, along with (9.5.65), perturbed problems of two types: in one of the versions, both the c-constraint and the $\mathbf{Y}$-constraint are perturbed, in the other version, only the $\mathbf{Y}$-constraint is perturbed. The corresponding admissible sets are defined by analogy with the general case; relation (9.5.39) and Lemma 9.5.2 provide a generalized representation for the asymptotics of perturbed constraints. The following assertion provides a supplement.

THEOREM 9.5.4. *The following three conditions are equivalent:* (1) $\forall\, \varepsilon \in (0, \infty)$ *we have* $F^+(\varepsilon) \neq \varnothing$; (2) $\forall\, \varepsilon \in (0, \infty)$ *we have* $F_*^+(\varepsilon) \neq \varnothing$; (3) $\widehat{\Omega}_{\mathcal{R}}^+ \neq \varnothing$.

The proof actually follows from (9.5.39). Nevertheless, we shall present a short scheme of it. The implications (3) $\Rightarrow$ (1) and (3) $\Rightarrow$ (2) easily follow from (9.5.39). Next, since $F_*^+(\varepsilon) \subset F^+(\varepsilon)$ for $\varepsilon > 0$, we have (2) $\Rightarrow$ (1). It remains to verify the validity of the implication (1) $\Rightarrow$ (3). Let $F^+(\varepsilon) \neq \varnothing$ for $\varepsilon > 0$. Then, in particular, $\widetilde{\mathcal{X}}_1$ (9.5.46) is a filter-base for $M_{\mathcal{R}}^+[c + 1]$. Therefore, the family $\widetilde{\mathcal{X}}_1^*$ of all sets

$$\mathrm{cl}\left(\{f * \eta\colon f \in F^+(\varepsilon)\}, \Theta_\eta^*(\mathcal{L})\right), \qquad \varepsilon \in (0, 1],$$

turns out, as one can easily check, to be a system of subsets of $\widehat{\Xi}_{\mathcal{R}}[c + 1]$ with the finite intersection property. This can be verified with allowance for Theorem 9.5.1. Since $\forall\, \varepsilon \in (0, 1]$

$$\mathrm{cl}\left(\{f * \eta\colon f \in F^+(\varepsilon)\}, \widetilde{\vartheta}_\eta^*[\mathcal{L}, c + 1]\right) = \mathrm{cl}\left(\{f * \eta\colon f \in F^+(\varepsilon)\}, \Theta_\eta^*(\mathcal{L})\right),$$

we have in the form $\widetilde{\mathcal{X}}_1^*$ a family of closed sets in the compact space (9.5.36). Therefore, $\widetilde{\mathcal{X}}_1^*$ is a family of closed sets with the finite intersection property in a compact space. This means (see Sec. 2.2) that the intersection of all sets from $\widetilde{\mathcal{X}}_1^*$ is nonempty, so that we have, as a consequence,

$$\bigcap_{\varepsilon\in(0,\infty)} \mathrm{cl}\,(\{f * \eta\colon f \in F^+(\varepsilon)\}, \Theta_\eta^*(\mathcal{L})) \neq \varnothing.$$

It now follows from (9.5.39) that $\widehat{\Omega}_{\mathcal{R}}^+ \neq \varnothing$. The implication $(1) \Rightarrow (3)$ is now established.

Throughout the remainder of this section, we assume the following condition to be fulfilled.

CONDITION 9.5.1. The generalized problem (9.5.66) is consistent: $\widehat{\Omega}_{\mathcal{R}}^+ \neq \varnothing$.

Taking this convention into account, we have $\forall\,\varepsilon \in (0,\infty)$

$$(F^+(\varepsilon) \neq \varnothing)\&(F_*^+(\varepsilon) \neq \varnothing).$$

From Lemma 9.5.2 it follows that

$$(\forall\, i \in \overline{1,n},\ j \in \overline{1,r}\colon S_{i,j} \in B_0(E,\mathcal{L})) \Rightarrow (F^+(0) \neq \varnothing).$$

We further observe that, as before, the objective functional (9.5.63) of the generalized problem admits continuous restrictions, $\forall\,b \in [0,\infty)$

$$(w \mid \widehat{\Xi}_{\mathcal{R}}^+[b]) \in \mathbf{C}(\widehat{\Xi}_{\mathcal{R}}^+[b], \tilde{\vartheta}_\eta^*[\mathcal{L}, b]).$$

Taking into account the compactness of (9.5.13), we find, in particular, from the last relation that the restrictions of the objective functional (9.5.64) are bounded, i.e., $\forall\,b \in [0,\infty)$

$$(W \mid M_{\mathcal{R}}^+[b]) \in \mathbf{B}(M_{\mathcal{R}}^+[b]).$$

In turn, this allows us to determine the finite values of W-minimization problems under the conditions of perturbed constraints. Namely, $\forall\,\varepsilon \in (0,\infty)$

$$\hat{v}_\varepsilon \triangleq \inf_{f\in F^+(\varepsilon)} W(f) \in \mathbf{R}, \tag{9.5.67}$$

$$\hat{v}_\varepsilon^* \triangleq \inf_{f\in F_*^+(\varepsilon)} W(f) \in \mathbf{R}, \tag{9.5.68}$$

with $\hat{v}_\varepsilon \leqslant \hat{v}_\varepsilon^*$. But the set $\{\hat{v}_\varepsilon^*\colon \varepsilon \in (0,\infty)\}$ is bounded from above since the functional $(W|M_{\mathcal{R}}^+[c])$ is bounded. This allows us to introduce the finite asymptotic value

$$\mathbf{V}^* \triangleq \sup(\{\hat{v}_\varepsilon^*\colon \varepsilon \in (0,\infty)\}) \in \mathbf{R}, \tag{9.5.69}$$

that majorizes the set $\{\hat{v}_\varepsilon \colon \varepsilon \in (0,\infty)\,\}$. Now,

$$\mathbf{V} \triangleq \sup(\{\hat{v}_\varepsilon \colon \varepsilon \in (0,\infty)\,\}) \in \mathbf{R}, \qquad (9.5.70)$$

$\mathbf{V} \leqslant \mathbf{V}^*$. Each of the values (9.5.69) and (9.5.70) can be interpreted as an asymptotic value. Furthermore,

$$\tilde{\mathbf{V}} \triangleq \min_{\mu \in \widehat{\Omega}^+_{\mathcal{R}}} w(\mu) \in \mathbf{R}; \qquad (9.5.71)$$

relation (9.5.71) defines the value of generalized problem (9.5.66); the minimum in (9.5.71) is certainly attained (see Sec. 2.2). The following theorem is valid.

THEOREM 9.5.5. *The values* (9.5.69)–(9.5.71) *coincide, i.e.,* $\mathbf{V} = \mathbf{V}^* = \tilde{\mathbf{V}}$

This assertion is a special case of Theorem 9.5.2, and therefore we shall not discuss the proof. Let us only note that the coincidence of values (9.5.69) and (9.5.70) can be interpreted as some property of structural stability of the problem with respect to the resource parameter or as the asymptotic nonsensitivity with respect to its perturbation; for small ε, $\varepsilon > 0$, the values $\hat{v}_\varepsilon$ and $\hat{v}_\varepsilon^*$ are close. We know from the examples of Chap. 1 that with respect to the $\mathbf{Y}$-constraint, this inference turns out, in general, to be invalid. We use the asymptotic nonsensitivity for regularizing the value function. Namely, consider extremal problems of the form

$$W(f) \to \inf, \qquad f \in M^+_{\mathcal{R}}[c + \delta],$$
$$\int_E S(x) f(x)\, \eta(dx) \in \mathbf{Y}, \qquad (9.5.72)$$

where $\delta \geqslant 0$. The examples of Sec. 1.2 show that the value of problem (9.5.72), as a function of δ, $\delta \geqslant 0$, may be a discontinuity at the point $\delta = 0$. A fairly detailed analysis of the setting was done in Sec. 7.3 (see, in particular, the system of constraints (7.3.23)); in the same section, we outlined the general scheme of regularization of the value function. Therefore, we shall not dwell on discussing the substantial side of the matter, referring the reader to Sec. 7.3, where a conceptually similar construction was used for investigating the value function of problem (7.3.15). Note that the situation considered in Sec. 7.3 was even more complicated from the conceptual point of view. We can obtain an analog of reduction (7.3.15), (7.3.16) if we replace problem (9.5.72) by the following problem with weakened constraints:

$$W(f) \to \inf, \qquad f \in M_{\mathcal{R}}^+[c+\delta],$$

$$\int_E S(x)f(x)\,\eta(dx) \in \mathbf{Y}_\varepsilon, \tag{9.5.73}$$

where $\varepsilon > 0$ and $\delta \geqslant 0$. In this case, δ specifies in (9.5.73) a real perturbation and ε is used as a "correction" parameter. Taking account of this circumstance, we shall tend ε to zero, obtaining a certain asymptotic property of problem (9.5.72). Thus, let $\forall \varepsilon \in [0, \infty)$, $\delta \in [0, \infty)$

$$F_+(\varepsilon, \delta) \triangleq \left\{ f \in M_{\mathcal{R}}^+[c+\delta] \,\middle|\, \int_E S(x)f(x)\,\eta(dx) \in \mathbf{Y}_\varepsilon \right\}. \tag{9.5.74}$$

As one can easily check, set (9.5.74) monotonically depends on its parameters, extending with an increase in any one of them. Here it is obvious that $\forall \varepsilon \in (0, \infty)$

$$(F^+(\varepsilon) = F_+(\varepsilon, \varepsilon)) \& (F_*^+(\varepsilon) = F_+(\varepsilon, 0)).$$

This allows us to identify $\mathcal{X}_2$ (9.5.45) with the family $\{ F_+(\varepsilon, 0) \colon \varepsilon \in (0, \infty) \}$. Altogether, we have to bear in mind that $\forall \varepsilon \in (0, \infty)$, $\delta \in [0, \infty)$

$$F_*^+(\varepsilon) \subset F_+(\varepsilon, \delta).$$

Therefore, every problem (9.5.73) is consistent, so that $\forall \varepsilon \in (0, \infty)$, $\delta \in [0, \infty)$

$$F_+(\varepsilon, \delta) \in 2^{M_{\mathcal{R}}^+[c+\delta]}.$$

Thus, every one of the problems (9.5.73) has a finite value: $\forall \varepsilon \in (0, \infty)$, $\delta \in [0, \infty)$

$$\hat{v}_0(\varepsilon, \delta) \triangleq \inf_{f \in F_+(\varepsilon, \delta)} W(f) \in \mathbf{R};$$

and the equality $\hat{v}_0(\varepsilon, 0) = \hat{v}_\varepsilon^*$ is valid. Recall that for a fixed δ, $\delta \geqslant 0$, the dependence

$$\varepsilon \mapsto \hat{v}_0(\varepsilon, \delta) \colon (0, \infty) \to \mathbf{R} \tag{9.5.75}$$

is monotonic; it should also be pointed out that $\hat{v}_0(\tilde{\varepsilon}, \delta) \leqslant \hat{v}_0(\tilde{\varepsilon}, 0), \tilde{\varepsilon} > 0$. But then, according to (9.5.69) and Theorem 9.5.5, we have that $\mathbf{V}$ is a majorant of every dependence (9.5.75), and this allows us to introduce finite upper bounds as distinctive quality estimates. To be more precise, $\forall \delta \in [0, \infty)$

$$\hat{\mathbf{V}}[\delta] \triangleq \sup(\{ \hat{v}_0(\varepsilon, \delta) \colon \varepsilon \in (0, \infty) \}) \in \mathbf{R}, \tag{9.5.76}$$

$\widehat{\mathbf{V}}[\delta] \leqslant \mathbf{V}$. We intend to replace the conventional value of problem (9.5.72) by the value (9.5.76). Then

$$\widehat{\mathbf{V}}[\cdot] \triangleq (\widehat{\mathbf{V}}[\delta])_{\delta \in [0,\infty)}$$

plays the role of a value function whose argument is the perturbing parameter in problem (9.5.72). Let us note, first of all, that (9.5.76) defines a monotonic dependence:

$$\forall \delta_1 \in [0,\infty), \ \ \delta_2 \in [\delta_1,\infty): \ \ \widehat{\mathbf{V}}[\delta_2] \leqslant \widehat{\mathbf{V}}[\delta_1]. \tag{9.5.77}$$

The second essential circumstance is that $\widehat{\mathbf{V}}[0] = \mathbf{V}$. This equality follows from (9.5.68), (9.5.69), and (9.5.76).

THEOREM 9.5.6. *The function* $\widehat{\mathbf{V}}[\cdot]$ *is continuous at the point* $\delta = 0$, *so that* $\forall \tilde{\varepsilon} \in (0,\infty) \ \exists \delta \in (0,\infty) \ \forall \varkappa \in [0,\tilde{\delta}]$

$$\widehat{\mathbf{V}}[0] - \tilde{\varepsilon} < \widehat{\mathbf{V}}[\varkappa] \leqslant \widehat{\mathbf{V}}[0].$$

PROOF. We fix an $\tilde{\varepsilon} \in (0,\infty)$, and then, with allowance for (9.5.70), select a $\theta \in (0,\infty)$ for which $\mathbf{V} - \tilde{\varepsilon} < \hat{v}_\theta$. Let us further choose an arbitrary number $\gamma \in (0,\theta]$. Then, as we already know (see (9.5.77)), $\widehat{\mathbf{V}}[\theta] \leqslant \widehat{\mathbf{V}}[\gamma]$. At the same time, as can be seen from (9.5.76), we have the inequality $\hat{v}_0(\theta,\theta) \leqslant \widehat{\mathbf{V}}[\theta]$. However, it is seen from the definitions that $\hat{v}_0(\theta,\theta) = \hat{v}_\theta$. We have thus obtained the inequality $\mathbf{V} - \tilde{\varepsilon} < \widehat{\mathbf{V}}[\gamma]$, and this completes the proof.

We have thus constructed a value function for passing to relaxed problems. The value for every problem (9.5.73) does not exceed the value for a similar problem (9.5.72). Moreover, the latter can, for example, for $\delta = 0$, turn out to be inconsistent. Theorem 9.5.5 provides for one more characteristic of the quantity $\mathbf{V} = \tilde{\mathbf{V}}$: this number is a regularization of the value by means of introducing small correcting perturbations.

Let us briefly consider some topological properties of the generalized problem. Here we shall use constructions that are conceptually similar to those in Sec. 9.2. Along with problem (9.5.66), we shall consider the relaxed generalized problems

$$w(\mu) \to \min, \quad \mu \in \widehat{\Xi}^{\pm}_{\mathcal{R}}[c+\varepsilon], \quad \int\limits_{E} S(x)\,\mu(dx) \in \mathbf{Y}_\varepsilon, \tag{9.5.78}$$

where $\varepsilon \geqslant 0$. Every one of the problems (9.5.78) is consistent; problem (9.5.66) is a special case of (9.5.78) corresponding to the situation where

$\varepsilon = 0$. Let $\forall \varepsilon \in [0, \infty)$

$$\tilde{\Omega}^{\pm}_{\mathcal{R}}[\varepsilon] \triangleq \left\{ \mu \in \hat{\Xi}^{\pm}_{\mathcal{R}}[c + \varepsilon] \middle| \int_E S(x)\,\mu(dx) \in \mathbf{Y}_\varepsilon \right\}; \tag{9.5.79}$$

set (9.5.79) is the admissible set for (9.5.78); it is nonempty and closed in the topological space (9.5.11). The last assertion easily follows from the continuity of operator (9.5.33) with due account of the fact that every one of the sets (9.2.2) is closed. In this case, set (9.5.79) is obviously also closed in the compact space (9.5.15) if we set $b = c + \varepsilon$, $\varepsilon \geqslant 0$. Taking account of (9.5.63), we find that the minimum is certainly attained in every problem (9.5.78), so that $\forall \varepsilon \in [0, \infty)$

$$\tilde{V}_*[\varepsilon] \triangleq \min_{\mu \in \tilde{\Omega}^{\pm}_{\mathcal{R}}[\varepsilon]} w(\mu) \in \mathbf{R},$$

$$\tilde{\Omega}^{(\text{opt})}_{\mathcal{R}}[\varepsilon] \triangleq \{ \mu \in \tilde{\Omega}^{\pm}_{\mathcal{R}}[\varepsilon] | \, w(\mu) = \tilde{V}_*[\varepsilon] \} \neq \varnothing.$$

Moreover, $\hat{\Omega}^{\pm}_{\mathcal{R}} = \tilde{\Omega}^{\pm}_{\mathcal{R}}[0]$, $\tilde{V} = \tilde{V}_*[0]$. Thus, in the case $\varepsilon = 0$, set (9.5.79) degenerates into an admissible set of the main generalized problem. The corresponding transformation occurs with the value of problem (9.5.78) under the condition $\varepsilon = 0$. Let us further observe that (9.5.79) defines a monotonic dependence, namely, an increase in ε, $\varepsilon \geqslant 0$, is associated with an extension of the admissible set. Here we have, as a consequence, $\forall \varepsilon \in [0, \infty)$, the inequality

$$\tilde{V}_*[\varepsilon] \leqslant \tilde{V} = \mathbf{V}.$$

Actually, we can claim more. Namely, the number $\mathbf{V} = \tilde{V}_*[0]$ satisfies the condition

$$\mathbf{V} = \sup(\{ \tilde{V}_*[\varepsilon] \colon \varepsilon \in (0, \infty) \}).$$

This relation is natural for extremal problems on compact spaces and we shall not discuss it here in greater detail (we should bear in mind that the intersection of all sets (9.5.79) with ε, $\varepsilon > 0$, running over its domain, coincides with $\hat{\Omega}^{\pm}_{\mathcal{R}}$). Thus, the family of relaxed generalized problems possesses computational stability. We note, in addition to this assertion, that the following easily verifiable proposition is true: if H_* is a neighborhood of the set $\tilde{\Omega}^{(\text{opt})}_{\mathcal{R}}[0]$ in the topological space (9.5.11), then

$$\exists \varepsilon_* \in (0, \infty) \, \forall \varepsilon \in (0, \varepsilon_*) \colon \tilde{\Omega}^{(\text{opt})}_{\mathcal{R}}[\varepsilon] \subset H_*.$$

Thus, the family of relaxed generalized problems has not only computational stability at the point $\varepsilon = 0$ but also stability in the sense of the corresponding optimal solutions.

Returning to set (9.5.39), we note that on the basis of approximation of its elements by the nets of conventional solutions (elements of $M_{\mathcal{R}}^+[c]$) in the sense of the topological space (9.5.11) (and this is always possible by virtue of Theorem 9.5.1), asymptotically optimal approximate solutions are realized. Moreover, on these solutions every constraint $f \in F_*^+(\varepsilon)$, $\varepsilon > 0$, is complied with f.s.m, and the quality criterion generates the corresponding real-valued net converging to $\mathbf{V}$. We shall not discuss this question more thoroughly because the relevant constructions were repeatedly used before for similar purposes. Let us now consider more comprehensively the case of a scalar optimization problem with the step-valued mapping S. Namely, we assume in the sequel that the following condition is fulfilled.

CONDITION 9.5.2. The mapping S is componentwise step-valued: $\forall\, i \in \overline{1,n}$, $j \in \overline{1,r}$,

$$S_{i,j} \in B_0(E,\mathcal{L}).$$

Let us note at once that now the conventional problem is also consistent, i.e., $F^+(0) \neq \varnothing$. Taking this into account, we have a finite value for problem (9.5.65), so that

$$\mathbf{V}^0 \triangleq \inf_{f \in F^+(0)} W(f) \in \mathbf{R}.$$

THEOREM 9.5.7. *The conventional, generalized, and asymptotic values coincide, i.e., $\mathbf{V} = \mathbf{V}^* = \tilde{\mathbf{V}} = \mathbf{V}^0$.*

The proof easily follows from Lemma 9.5.2 and relation (9.5.64). Let us observe, in addition to this assertion, that by approximating the elements of the (nonempty) set (9.5.39) by elements of $\widetilde{M}_{\mathcal{R}}^+[c]$ it is possible to construct, in the class of nets, asymptotically optimal approximate solutions that faithfully comply with the constraint $f \in F^+(0)$; here, it is required to realize an approximation in the sense of the topological space (9.5.12). This is practically obvious by virtue of Lemma 9.5.1. Note that in the problem under consideration, we can confine ourselves, without loss of optimality, to the class of sequential approximate solutions, using net-solutions as an intermediate link (see [49]). We shall not discuss this interesting possibility here, referring the reader for details to [49].

We shall concentrate our attention on examining an indirect characteristic of stability, defined in terms of limit elements of conventional solutions. For this purpose, we shall, first of all, define dependence (9.5.67) at the point

$\varepsilon = 0$, setting $\hat{v}_0 = \hat{v}_\varepsilon|_{\varepsilon=0} \triangleq \mathbf{V}^0$. Let $\forall \varepsilon \in [0,\infty)$, $\delta \in (0,\infty)$,

$$V_\varepsilon(\delta) \triangleq \{f \in F^+(\varepsilon)|\, W(f) \leqslant \hat{v}_\varepsilon + \delta\}. \qquad (9.5.80)$$

In (9.5.80), we defined the set of all δ-optimal solutions of the ε-perturbed problem. Let $\forall \varepsilon \in (0,\infty)$

$$V[\delta] \triangleq V_0(\delta) = V_\varepsilon(\delta)|_{\varepsilon=0}.$$

THEOREM 9.5.8. *Let $\varkappa \in (0,\infty)$ and $\gamma \in (\varkappa,\infty)$. Then*

$$\bigcap_{\varepsilon\in(0,\infty)} \mathrm{cl}\left(\left\{f * \eta\colon f \in \bigcup_{\theta\in(0,\varepsilon]} V_\theta(\varkappa)\right\}, \Theta_\eta^*(\mathcal{L})\right)$$

$$\subset \mathrm{cl}\left(\{f * \eta\colon f \in V[\gamma]\}, \Theta_\eta^0(\mathcal{L})\right). \qquad (9.5.81)$$

SCHEME OF THE PROOF. Let μ^* be an element of the set on the right-hand side of (9.5.81), $\alpha > 0$. Taking account of Birkhoff's theorem (see Chap. 2), we select a net $(D, \prec, \varphi)$ in the set

$$\Sigma \triangleq \left\{f * \eta\colon f \in \bigcup_{\theta\in(0,\alpha]} V_\theta(\varkappa)\right\},$$

converging to μ^* in the topological space (9.5.11). Furthermore, we select an operator

$$\psi\colon D \to \bigcup_{\theta\in(0,\alpha]} V_\theta(\varkappa)$$

from the condition $\varphi(d) = \psi(d) * \eta$, $d \in D$. It is easy to verify that ψ assumes values in $F^+(\alpha)$. In this case, we obviously always have the inequality

$$W(\psi(d)) = w(\varphi(d)) \leqslant \mathbf{V}^0 + \varkappa.$$

However, $(D, \prec, w \circ \varphi)$ converges to $w(\mu^*)$, so that $w(\mu^*) \leqslant \mathbf{V}^0 + \varkappa$. However, φ assumes values in $\hat{\Xi}_\mathcal{R}^+[c + \alpha]$, so that μ^* is an element of the same set, and

$$\int_E S(x)\,\mu^*(dx) \in \mathbf{Y}_\alpha.$$

This is a simple consequence of continuity of the operator (9.5.33). Since the (strictly) positive number α was arbitrary, we have

$$\mu^* \in \hat{\Omega}_\mathcal{R}^\pm\colon w(\mu^*) \leqslant \mathbf{V}^0 + \varkappa. \qquad (9.5.82)$$

Now we shall use Theorem 9.5.1 in the part concerning the approximation in the sense of space (9.5.12). We select a net $(T, \angle, \bar{\rho})$ in $\widetilde{M}_{\mathcal{R}}^{+}[c]$ that converges to μ^* in the sense of space (9.5.12). Furthermore, we select an operator

$$\rho\colon T \to M_{\mathcal{R}}^{+}[c]$$

with the property $\bar{\rho}(t) = \rho(t) * \eta$, $t \in T$. By virtue of (3.4.11) and Lemma 9.5.1, we have f.s.m that

$$\int\limits_{E} S(x)\rho(t)(x)\, \eta(dx) = \int\limits_{E} S(x)\, \mu^*(dx).$$

Then $\rho(t) \in F^{+}(0)$ f.s.m. Note that $(T, \angle, \bar{\rho})$ converges to μ^* in the sense of the topological space (9.5.11). We already used these arguments when proving Theorem 9.5.1, invoking relative topologies. In this reasoning, the property of integral boundedness of ρ, which makes it possible to use in proofs subspaces of topological spaces (9.5.11) and (9.5.12), is essential. Taking into account the continuity of (9.5.63), we now find that the net $(T, \angle, w \circ \bar{\rho})$ converges to $w(\mu^*)$. But $\forall\, t \in T$

$$(W \circ \rho)(t) = W(\rho(t)) = w(\rho(t) * \eta) = w(\bar{\rho}(t)) = (w \circ \bar{\rho})(t).$$

Hence, $w(\mu^*)$ is a generalized limit of the net $(T, \angle, W \circ \rho)$, and this means, in particular, that

$$|W(\rho(t)) - w(\mu^*)| < \gamma - \varkappa$$

f.s.m; taking account of the estimate in terms of $\mathbf{V}^0$ proved earlier, we find that

$$W(\rho(t)) \leqslant \mathbf{V}^0 + \gamma$$

f.s.m. Finally, we get from (9.5.80) that $\rho(t) \in \mathcal{V}[\gamma]$ f.s.m. This means that μ^* is an element of the set on the right-hand side of (9.5.81). This completes the proof.

This theorem has the meaning of a statement concerning some weak stability, which is, moreover, regularized by means of the introduction of the interlayer $\gamma - \varkappa > 0$. The fact is that we can roughen the right-hand side of (9.5.81) to the closure in the sense of (9.5.11), making use of the relationship for the relative topologies mentioned in the proof of Theorem 9.5.1. As for the condition $\varkappa < \gamma$, note that the statement of the theorem is valid only in the domain of small values of both indicated parameters. By analogy with Theorem 9.5.7, we can show the validity of the assertion $\forall\, \varepsilon \in [0, \infty)$

$$\widetilde{\mathbf{V}}_{*}[\varepsilon] = \hat{v}_{\varepsilon}. \tag{9.5.83}$$

Taking account of (9.5.83), we establish one important proposition with the following substantial meaning: optimal generalized solutions are limits of almost optimal conventional solutions and nothing else.

THEOREM 9.5.9. *The following equality holds:*

$$\widetilde{\Omega}_{\mathcal{R}}^{(\mathrm{opt})}[0] = \bigcap_{\delta \in (0,\infty)} \mathrm{cl}\left(\{f * \eta \colon f \in \mathcal{V}[\delta]\}, \Theta_{\eta}^{*}(\mathcal{L})\right)$$

$$= \bigcap_{\delta \in (0,\infty)} \mathrm{cl}\left(\{f * \eta \colon f \in \mathcal{V}[\delta]\}, \Theta_{\eta}^{0}(\mathcal{L})\right).$$

SCHEME OF THE PROOF. Taking account of the integral boundedness property for $\mathcal{V}[\tilde{\delta}]$, $\tilde{\delta} > 0$, we can use the repeatedly mentioned relation for the relative topologies, so that $\forall\, \delta \in (0, \infty)$

$$\mathrm{cl}\left(\{f * \eta \colon f \in \mathcal{V}[\delta]\}, \Theta_{\eta}^{0}(\mathcal{L})\right) \subset \mathrm{cl}\left(\{f * \eta \colon f \in \mathcal{V}[\delta]\}, \Theta_{\eta}^{*}(\mathcal{L})\right).$$

As a consequence, we have the inclusion

$$\bigcap_{\delta \in (0,\infty)} \mathrm{cl}\left(\{f * \eta \colon f \in \mathcal{V}[\delta]\}, \Theta_{\eta}^{0}(\mathcal{L})\right)$$

$$\subset \bigcap_{\delta \in (0,\infty)} \mathrm{cl}\left(\{f * \eta \colon f \in \mathcal{V}[\delta]\}, \Theta_{\eta}^{*}(\mathcal{L})\right). \tag{9.5.84}$$

We shall now select an arbitrary element μ^* of the set on the right-hand side of (9.5.84). If $\delta_0 \in [0, \infty)$, then, according to the assumption, we can indicate a net $(D, \prec, \varphi)$ in the set $\widetilde{\Gamma} \triangleq \{f * \eta \colon f \in \mathcal{V}[\delta_0]\}$, converging to μ^* in space (9.5.11). Let us select, in addition, an operator

$$\psi \colon D \to \mathcal{V}[\delta_0],$$

for which $\varphi(d) = \psi(d) * \eta$ if $d \in D$. Then, according to (9.5.80), for $d \in D$ we have $\psi(d) \in F^+(0)$:

$$W(\psi(d)) \leqslant \hat{v}_0 + \delta_0. \tag{9.5.85}$$

Note that $\mu^* \in \mathrm{cl}\left(\widetilde{\Gamma}, \Theta_{\eta}^{*}(\mathcal{L})\right)$, so that $\mu^* \in \widehat{\Xi}_{\mathcal{R}}^{+}[c]$. Moreover,

$$\int_{E} S(x)\,\varphi(\tilde{d})(dx) = \int_{E} S(x)\psi(\tilde{d})(x)\,\eta(dx) \in \mathbf{Y}$$

for $d \in D$. But, owing to the convergence to μ^*, we find, by virtue of the continuity of operator (9.5.33), that

$$\int_{E} S(x)\,\mu^*(dx) \in \mathbf{Y},$$

so that $\mu^* \in \widehat{\Omega}_{\mathcal{R}}^+$ in accordance with (9.5.39). From (9.5.83) and (9.5.85), we find that $\forall\, d \in D$

$$(w \circ \varphi)(d) = w(\varphi(d)) = w(\psi(d) * \eta) = W(\psi(d)) \leqslant \widetilde{\mathbf{V}}_*[0] + \delta_0.$$

Since $(D, \prec, w \circ \varphi)$ converges to $w(\mu^*)$, it follows that $w(\mu^*) \leqslant \widetilde{\mathbf{V}}_*[0] + \delta_0$. However, δ_0, $\delta_0 > 0$, was arbitrary, so that $w(\mu^*) \leqslant \widetilde{\mathbf{V}}_*[0]$. This means that

$$\mu^* \in \widetilde{\Omega}_{\mathcal{R}}^{(\mathrm{opt})}[0];$$

thus we have established the inclusion

$$\bigcap_{\delta \in (0,\infty)} \mathrm{cl}\left(\{f * \eta \colon f \in \mathcal{V}[\delta]\}, \Theta_\eta^*(\mathcal{L})\right) \subset \widetilde{\Omega}_{\mathcal{R}}^{(\mathrm{opt})}[0]. \tag{9.5.86}$$

We take an arbitrary element λ from the set on the right-hand side of (9.5.86), so that we have $\lambda \in \widehat{\Omega}_{\mathcal{R}}^+$, and with this $w(\lambda) = \mathbf{V}$. Taking account of Theorem 9.5.1, we select a net $(T, \angle, \alpha)$ in $\widetilde{M}_{\mathcal{R}}^+[c]$, converging to the measure λ in the topological space (9.5.12). Let us also select an operator

$$\beta \colon T \to M_{\mathcal{R}}^+[c],$$

for which we have the equality $\alpha(t) = \beta(t) * \eta$ for $t \in T$. Then, with due account of Lemma 9.5.2, $\beta(t) \in F^+(0)$ f.s.m.; here we use the assertion of the lemma concerning the net $(T, \angle, \alpha)$, and also relation (3.4.11). Note that $(T, \angle, \alpha)$ converges to λ in the sense of $\Theta_\eta^*(\mathcal{L})$; this is established on the basis of comparing relative topologies by analogy with Theorem 9.5.1. But in that case, the net $(T, \angle, w \circ \alpha)$ converges to $w(\lambda)$. However, here we have

$$(w \circ \alpha)(t) = w(\alpha(t)) = w(\beta(t) * \eta) = W(\beta(t)) = (W \circ \beta)(t),$$

so that actually $w(\lambda)$ is a generalized limit of the net $(T, \angle, W \circ \beta)$. As a result, we get the convergence of the indicated net to $\mathbf{V}^0 = \hat{v}_0$ (see Theorem 9.5.7). Let $\gamma_* \in (0,\infty)$. Then $|W(\beta(t)) - \mathbf{V}^0| < \gamma_*$ f.s.m., but this means that $\beta(t) \in \mathcal{V}[\gamma_*]$ f.s.m. Then $\alpha(t)$ is f.s.m. an element of the set $\{f * \eta \colon f \in \mathcal{V}[\gamma_*]\}$. By virtue of the convergence of $(T, \angle, \alpha)$ to λ (in the sense of $\Theta_\eta^0(\mathcal{L})$), this means that

$$\lambda \in \mathrm{cl}\left(\{f * \eta \colon f \in \mathcal{V}[\gamma_*]\}, \Theta_\eta^0(\mathcal{L})\right).$$

Since γ_*, $\gamma_* > 0$, was arbitrary, we have established that λ is an element of the set on the right-hand side of (9.5.84), and therefore established the inclusion

$$\widetilde{\Omega}_{\mathcal{R}}^{(\mathrm{opt})}[0] \subset \bigcap_{\delta \in (0,\infty)} \mathrm{cl}\left(\{f * \eta \colon f \in \mathcal{V}[\delta]\}, \Theta_\eta^0(\mathcal{L})\right),$$

which, by virtue of (9.5.84) and (9.5.86), completes the proof.

Conclusion

In Chap. 5 of this monograph, we considered a general construction of weakening the conditions on the vector integrand of a control. Making use of the techniques of f.a. measure theory, we managed to obtain some limit representations of admissible sets. These limit representations have the meaning of "convergence attractors" (see (2.5.1) and (5.2.1)) and are not directly connected with the compactification procedures. At the same time, the example from Sec. 7.2 provides another attractor, a sort of "hidden chaos." Both attractors correspond to the asymptotics of conventional solutions that observe, with a growing accuracy, the constraints of the original problem. However, in other aspects, the properties of these attractors show marked distinctions, and the example from Sec. 7.2 confirms this. In practical problems, it is also of interest for us to compare the indicated attractors, this time not in the space of conventional controls or their idealizations, but in the space of some states generated by the controls. The constructions of Chap. 6 should be viewed from this point of view. A specific example of this kind is provided by the problem of investigating relaxations of the attainability domain for a linear controllable system under integral constraints on the choice of the control program. The effects arising in extremal problems and admitting a rather efficient description in generalized constructions are also connected with the convergence attractor and fundamentally also admit an exhaustive explanation in terms of asymptotic attainability. It should be noted that the last notion can be applied to the investigation of diverse problems of practical interest, for which the conditions of integral boundedness that are important for the statements of Chaps. 7–9 do not hold, although they provide a somewhat new quality in this class of problems (see Theorems 6.3.3 and 6.4.4). Many constructions of this kind were not included in this monograph because the orientation toward exploring issues of asymptotic optimization usually presupposes a certain possibility of extension (compactification) of the solution space, and this is realized in accordance with the constructions of Chap. 2 and in the principal Chaps. 7–9. The study of the convergence attractor undertaken recently led to the establishment for unbounded, generally speaking, in the integral sense, problems of conditions of the type of stability and asymptotic nonsensitivity in terms of the coincidence of the attractors mentioned above under various perturbations. A key role in these constructions is played by generalized problems in the class of f.a. measures, which do not satisfy, generally speaking, any natural conditions of strong boundedness (see (5.2.14)). Nevertheless, a certain systematization and additional investigations are needed in these cases.

Let us note in conclusion some studies which are less directly connected with the subject matter of this monograph but which also deal with issues of extension theory and with some aspects of the f.a. measure theory. We have already noted that the control theory widely uses constructions that use the representation (in some form) of sliding modes in nonlinear control systems as a result of the action of generalized controls (in this connection, we point out [74, 78], and there are many other works). Some statements that admit certain conceptual analogies with the constructions of Chap. 7 and concern the extension constructions for nonlinear controllable processes in the class of (c.a.) Radon measures, are given in [69, 70]. In a number of cases, this method of extending control problems with geometric constraints, realized in the class of c.a. regular Borel measures (or the corresponding measure-valued functions [4, 20, 63]), made it possible to obtain fundamental results; see, e.g., [65]. We can mention [61, 71] as a f.a. analog of these constructions. Here, we should also mention the studies in the field of impulse control problems [66–68], in which elements of the theory of generalized functions were used; in particular, the purely impulse control problem and its extension were considered in [66, 67]. The general extension constructions for abstract extremal problems were discussed in [76, 77], and issues of the reduction of a conventional extremal problem to its asymptotic setting were considered there, too (there are some conceptual analogies with [34, 49]). As for using f.a. measures in extension constructions that are somewhat different from those studied in this monograph, we must note [62] (there a procedure was considered that has the meaning of extension of a Markov process to a Feller one with the use of a scheme that is conceptually similar to that of Sec. 3.6). A detailed exposition of various issues of the f.a. measure theory can be found in [75]. In connection with other applications of the f.a. measure theory (game theory, utility theory), we recommend [60, 72, 73]. In connection with ill-posed problems in topological spaces, see, in particular, [64].

References

1. M. I. Alekseychik and G. E. Naumov, "On the existence of efficient points in the infinite-dimensional vector optimization problem," *Avtomat. Telemekh.*, No. 5 (1981).

2. E. G. Belov and A. G. Chentsov, "Extension and stability of some multicriteria problems," *Kibernetika*, No. 5, 44–49 (1990).

3. N. Bourbaki, *General Topology. Basic Structures* [Russian translation], Nauka, Moscow (1968).

4. J. Warga, *Optimal Control of Differential and Functional Equations*, Academic Press, New York (1972).

5. D. A. Vladimirov, *Boolean Algebras* [in Russian], Nauka, Moscow (1969).

6. R. V. Gamkrelidze, *Fundamentals of Optimal Control Theory* [in Russian], Tbilisi Univ. Press, Tbilisi (1977).

7. E. G. Gol'stein, *Duality Theory in Mathematical Programming and Its Applications* [in Russian], Nauka, Moscow (1971).

8. V. V. Gorokhovik, *Convex and Nonsmooth Problems of Vector Optimization* [in Russian], Nauka i Tekhnika, Minsk (1990).

9. N. Dunford and J. T. Schwartz, *Linear Operators. General Theory*, Interscience Publishers, New York–London (1958).

10. R. J. Duffin, "Infinite programs," In: *Linear Inequalities and Related Topics*, Moscow (1959), pp. 263–267.

11. M. Davis, *Applied Nonstandard Analysis* Wiley, New York (1977).

12. I. I. Eremin, V. D. Mazurov, and N. N. Astaf'ev, *Improper Problems of Linear and Convex Programming* [in Russian], Nauka, Moscow (1983).

13. S. T. Zavalishin and A. N. Sesekin, *Impulse Processes. Models and Applications* [in Russian], Nauka, Moscow (1991).

14. L. V. Zudikhin and A. G. Chentsov, "On a problem of successive control with bounded resources. I," *Avtomatika*, No. 2 (1990).

15. L. V. Zudikhin and A. G. Chentsov, "Compactification of a control problem in the class of finitely additive measures. I," In: *Functional-Differential Equations*, Perm (1989), pp. 151–161.

16. L. V. Zudikhin and A. G. Chentsov, "Compactification of a control problem in the class of finitely additive measures. II," In: *Functional-Differential Equations*, Perm (1990).

17. A. D. Ioffe and V. M. Tikhomirov, *Theory of Extremal Problems* [in Russian], Nauka, Moscow (1974).

18. L. V. Kantorovich and G. P. Akilov, *Functional Analysis* [in Russian], Nauka, Moscow (1977).

19. J. L. Kelley, *General Topology* [Russian translation], Nauka, Moscow (1981).

20. N. N. Krasovskii and A. I. Subbotin, *Positional Differential Games* [in Russian], Nauka, Moscow (1974).

21. N. N. Krasovskii, *Dynamic System Control. Problem of the Minimum of Guaranteed Result* [in Russian], Nauka, Moscow (1985).

22. N. N. Krasovskii, *Game Problems on Motion Encounter* [in Russian], Nauka, Moscow (1970).

23. K. Kuratowski and A. Mostowski, *Set Theory*, North-Holland, Amsterdam (1968).

24. M. Minu, *Mathematical Programming. Theory and Algorithms* [in Russian], Nauka, Moscow (1990).

25. J. Neveu, *Basis Mathematiques de Calcul des Probabilités*, Paris (1964).

26. J. von Neumann and O. Morgenstern, *Theory of Games and Economic Behavior* [Russian translation], Nauka, Moscow (1970).

27. G. Owen, *Game Theory*, Saunders, Philadelphia (1968).

28. A. I. Panasyuk and V. I. Panasyuk, *Asymptotic Turnpike Optimization of Controllable Systems*, Nauka i Tekhnika, Minsk (1986).

29. L. S. Pontryagin, *Ordinary Differential Equations* [in Russian], Nauka, Moscow (1974).

30. V. P. Serov and A. G. Chentsov, "On an extension construction for a control problem with integral constraints," *Diff. Uravn.*, **26**, No. 4, 607–618 (1990).

31. R. Sikorski, *Boolean Algebras*, FGR (1964).

32. A. I. Subbotin and A. G. Chentsov, *Guarantee Optimization in Control Problems* [in Russian], Nauka, Moscow (1981).

33. S. P. Trofimov, "On duality break for a class of convex programming problems," In: *Studies in Improper Optimization Problems* [in Russian], Sverdlovsk (1988), pp. 57–62.

34. A. G. Chentsov, *Optimization under Fuzzy Constraints* [in Russian], UNTs AN SSSR, Sverdlovsk (1986).

35. A. G. Chentsov, "On the problem of universal integrability of bounded functions," *Mat. Sb.*, **131**, No. 1, 73–93 (1986).

36. A. G. Chentsov, *Applications of Measure Theory to Control Problems* [in Russian], Sverdlovsk, Sredne-Ural. Publishing House (1985).

37. A. G. Chentsov, "Order structure of scalar finitely additive measures," UNTs AN SSSR, Institute of Mathematics and Mechanics, Sverdlovsk (1983), Dep. VINITI, No. 6590-83.

38. A. G. Chentsov, "Finitely additive measures and integrals (theory and applications)," Ural. Polytekh. Inst. Sverdlovsk, 1985, Dep. VINITI, No. 1143-85.

39. A. G. Chentsov, "Two-valued measures on a semialgebra of sets and some of their applications to infinite-dimensional problems of mathematical programming," *Kibernetika*, No. 6, 72–76 (1988).

40. A. G. Chentsov, "Extension of extremal problems in the class of finitely additive measures. I," Ural. Politekh. Institut, Sverdlovsk (1989), Dep. VINITI, No. 4201-B89.

41. A. G. Chentsov, "Extension of extremal problems in the class of finitely additive measures. II," Ural. Politekh. Institut, Sverdlovsk (1990), Dep. VINITI, No. 3952-B90.

42. A. G. Chentsov, "On some representations of positive finitely additive measures approximable by indefinite integrals," Ural. Politekh. Institut, Sverdlovsk (1987), Dep. VINITI, No. 8511-B87.

43. A. G. Chentsov, "On the problem of extension and computational stability in a class of extremal problems," *Kibernetika*, No. 4, 122–123 (1990).

44. A. G. Chentsov, "Asymptotic efficiency and extensions in the class of finitely additive measures," *Dokl. Akad. Nauk SSSR*, **314**, No. 5, pp. 1–1085 (1990).

45. A. G. Chentsov, "Extension of some cone optimization problems under perturbed constraints," Ural. Politekh. Institut, Sverdlovsk (1989), Dep. VINITI, No. 2046-B89.

46. A. G. Chentsov, "Normalized finitely additive measures and extensions in a class of nonlinear extremal problems," Ural. Politekh. Institut, Sverdlovsk (1989), Dep. VINITI, No. 2045-B89.

47. A. G. Chentsov, "Some asymptotic optimization problems and duality relations," Ural. Politekh. Institut, Sverdlovsk (1987), Dep. VINITI, No. 2070-B87.

48. A. G. Chentsov, "Finitely additive measures and minimum problems," *Kibernetika*, No. 3, 67–70 (1988).

49. A. G. Chentsov, "Fuzzy constraints and extremal problems," Ural. Politekh. Institut, Sverdlovsk (1985), Dep. VINITI, No. 8931-B85.

50. A. G. Chentsov, "Vector finitely additive measures and extensions in a class of nonlinear extremal problems," UrO AN SSSR, Institute of Mathematics amd Mechanics, Sverdlovsk (1988), Dep. VINITI, No. 8191-B88.

51. A. G. Chentsov, "On a finitely additive extension in a class of extremal problems," Ural. Politekh. Institut, Sverdlovsk (1988), Dep. VINITI, No. 1882-B88.

52. H. Schaefer, *Topological Vector Spaces*, Macmillan, New York (1966).

53. I. Eceland and R. Temam, *Convex Analysis and Variational Problems*, North-Holland, Amsterdam (1976).

54. R. Engelking, *General Topology*, PWN, Warszawa (1983).

55. L. Young, *Lectures on the Calculus of Variations and Optimal Control Theory*, Saunders, Philadelphia (1969).

56. J. Distel and J. J. Uhl, *Vector Measures*: Amer. Math. Soc., Providence (1977).

57. K. Yosida and E. H. Hewitt, "Finitely additive measures," *Trans. Am. Math. Soc.*, **72**, 44–66 (1952).

58. S. Leader, "On universally integrable functions," *Proc. Am. Math. Soc.*, **6**, 232–234 (1955).

59. H. Mahnard, "A Radon–Nikodym theorem," *Pacific J. Math.*, **83**, No. 2, 401–413 (1979).

Supplementary References

60. R. Aumann and L. Shapley, *Values of Non-Atomic Games*, Princeton Univ. Press, Princeton (1974).

61. E. G. Belov and A. G. Chentsov, *On an Extremal Problem on the Space of Finitely Additive Measures* [in Russian], UrO AN SSSR, Sverdlovsk (1986).

62. A. I. Zhdanok, "Ergodic theorems for nonsmooth Markov processes," In: *Topological Spaces and Their Mappings*, Riga (1981).

63. A. G. Ivanov, "Uniform local controllability of a nonlinear system in the class of generalized controls," In: *Nonlinear Oscillations and Control Theory* [in Russian], Izhevsk (1989), pp. 5–18.

64. V. K. Ivanov, "Ill-posed problems in topological spaces," *Sib. Mat. Zh.*, **10**, No. 5, 1065–1074 (1969).

65. A. V. Kryazhimskii, "On the theory of positional differential games of approach-deviation," *Dokl. Akad. Nauk SSSR*, **239**, No. 4, 779–782 (1978).

66. B. M. Miller, "Problem of nonlinear impulse control of objects described by ordinary differential equations. II," *Avtomat. Telemekh.*, No. 3, 34–41 (1978).

67. B. M. Miller, "Optimality condition in the control problem for a system described by a differential equation with measure," *Avtomat. Telemekh.*, No. 6, 60–71 (1982).

68. Yu. V. Orlov, *Theory of Optimal Systems with Generalized Controls* [in Russian], Nauka, Moscow (1988).

69. V. E. Pak and A. G. Chentsov, "Some topological properties of generalized solutions of nonlinear controlled systems," *Diff. Uravn.*, No. 11, 2004–2005 (1989).

70. V. E. Pak and A. G. Chentsov, "On a regularization procedure for a value function of a nonlinear control problem," *Diff. Uravn.*, No. 3 (1992).

71. A. B. Pashayev and A. G. Chentsov, "Generalized control problem in the class of finitely additive measures," *Kibernetika*, No. 2, 110–112 (1986).

72. P. Fishburn, *Utility Theory for Making Decisions* [in Russian], Nauka, Moscow (1978).

73. L. E. Dubbins and L. J. Savage, *Inequalities for Stochastic Processes. How to Gamble If You Must*, Dovers, New York (1976).

74. E. Mascolo and L. Migliaccio, "Relaxation methods in control theory," *Appl. Math. Optim.*, **20**, 97–103 (1989).

75. K. P. S. B. Rao and M. B. Rao, *Theory of Charges. A Study of Finitely Additive Measures*, Academic Press, New York (1983).

76. T. Roubičec, "Stable extensions of constrained optimization problems," *J. Math. Anal. Appl.*, **141**, No. 1, 120–135 (1989).

77. T. Roubičec, "Constrained optimization: a general tolerance approach," *Appl. Mat.*, **35**, No. 2, 99–128 (1990).

78. T. Roubičec, "Convex compactifications and special extensions of optimization problems," *Nonlinear Anal. Theory. Methods*, **16**, No. 12 (1991).

Index